UNDERGROUND STORAGE TANKS

A Primer on the Federal Regulatory Program

Second Edition

RICHARD P. FAHEY

SONREEL
SECTION OF NATURAL RESOURCES, ENERGY, AND ENVIRONMENTAL LAW
AMERICAN BAR ASSOCIATION

Printed in the United States of America.

Library of Congress Catalog Card Number 94-73107
ISBN 1-57073-101-2

00 99 98 97 96 5 4 3 2 1

ABOUT THE AUTHOR

Richard Fahey is a partner with the Columbus, Ohio, office of the law firm of Arter & Hadden. He is chair of the firm's Environmental Practice Group and served as an Adjunct Professor of Environmental Law at the Capital University Law School and Ohio State University College of Law from 1976 through 1987. Mr. Fahey received his juris doctor degree from the Northwestern University Law School in 1971.

CONTENTS

APPENDICES

I. INTRODUCTION

What comes to mind at the mention of the regulation of hazardous substances is a site stocked with piles of leaking drums polluting the air and waters near it. As initially enacted in 1976, the Resource Conservation and Recovery Act (RCRA)[1] limited environmental regulation of hazardous substances to such sites.[2]

However, in 1984, in the Hazardous and Solid Waste Amendments (HSWA),[3] Congress expanded the scope of its regulatory reach beyond waste sites so as to prevent "regulated substances" in underground storage tanks (USTs) from leaking into groundwater. Virtually every facility with a UST became regulated. Regulated substances include petroleum and a broad range of products and materials which have commercial value but exhibit hazardous characteristics.[4]

A UST is defined in section 9001(1) of the RCRA Amendments to include any tank that is used to accumulate a regulated substance, and is 10 percent or more beneath the surface of the ground.[5] An underground tank system (UST system) has been defined as a UST plus connected underground piping, underground ancillary equipment, and the associated containment system, if any.[6]

The expanded program also envisions that the individual states will obtain approval of the United States Environmental Protection Agency (EPA), and thereafter act as the primary regulators of the USTs within their boundaries.[7] Amended subtitle I includes the minimum standards to be enacted nationally, and each state's ability to establish a program more stringent than the EPA's is preserved, regardless of whether a state seeks authorization.[8]

By further amending subtitle I in the Superfund Amendments and Reauthorization Act of 1986 (SARA), Congress imposed financial responsibility requirements on owners and operators of USTs by requiring them to maintain evidence of financial responsibility to clean up leaking underground tanks and compensate third parties.[9] The EPA may clean up releases of petroleum from USTs and require tank owners and operators to also perform such clean up.[10] Further, a trust fund now finances these activities.[11] In contrast to the 1984 amendments aimed at regulating tank standards and operations to prevent leaks, the 1986 provisions adopt a "mini superfund" approach to correcting leaking petroleum underground tanks.

When and how is this entire program implemented? What does it require of the owner of a UST? What new types of liability does it create for the owner or operator? How can an owner or operator comply with new requirements? This monograph addresses these questions. One point, however, cannot be overstressed: any facility with a UST must give compliance with this regulatory program the highest priority.

II. SCOPE OF REGULATION

A. Regulated Tanks and Substances

Under RCRA as originally enacted, only USTs for hazardous wastes were subject to regulation.[12] However, as concern over groundwater contamination grew into a major public issue, Congress concluded that many of the sources of such contamination involved the underground storage of materials, which under RCRA were neither "solid waste"[13] nor "hazardous waste,"[14] and therefore were outside the purview of the regulatory system. Of particular concern was the storage of petroleum products in underground tanks, which has been a common practice due to fire and safety codes.

Subtitle I was included in the RCRA re-authorization legislation because of an increasing awareness of the problem of leakage from USTs used to store regulated substances. The Conference Committee Report on the 1984 amendments cited petroleum industry estimates that as many as 75,000 to 100,000 underground tanks were leaking, and that an additional 360,000 underground tanks may develop leaks within the next five years. The committee's report also states that, although it is difficult to determine the precise source of groundwater contamination, USTs are considered the source or probable source of a substantial number of groundwater contaminations.[15]

The seriousness of the leaking UST problem was confirmed in February 1993, when it was established that 166,000 USTs were confirmed as leaking and only 44,000 had been cleaned up.[16] The total cost of responding to the nation's leaking USTs is now estimated to be $4.1 billion.[17] While the various states had budgeted approximately $1.6 billion per year in 1992 to implement clean ups, the volume of claims filed totalled almost $28 billion.[18]

1. Statutory and Regulatory Exclusions

Congress recognized a limited number of exceptions to the definition of UST: private consumption heating oil tanks, home and farm motor fuel storage tanks of 1,100 gallons or less capacity, septic tanks, regulated pipelines, surface impoundments, storm and waste water collection systems, flow- through process tanks, liquid traps and associated gathering lines related to oil and gas production, and storage tanks located upon the floor of an underground area.[19] Substances already regulated as a hazardous waste under Subtitle C of RCRA, similarly, are not governed by these provisions.

Additionally, the following units meet the statutory definition of a UST system but are excluded from regulation under this program: any UST system holding hazardous waste that is listed on Subtitle C of the Solid Waste Disposal Act, waste water treatment tanks subject to regulation under the Clean Water Act, equipment or machinery containing regulated substances for operational purposes, any UST system with a capacity of 110 gallons or less, any UST system containing a *de minimis* concentration of regulated substances, and any emergency spill or overflow containment system that is expeditiously emptied after use.[20]

On February 12, 1993,[21] the EPA proposed to exempt, from certain portions of the EPA's hazardous waste regulations, petroleum-contaminated media and debris such as soils and ground water that are generated from UST corrective actions subject to subtitle I of RCRA.[22] The exemption would be limited to the twenty-five newly listed organic chemicals under the Toxicity Characteristics (i.e., Hazardous Waste Codes D018 through D043 only). The EPA will publish its final determination on this proposal after its review of the public comments.

2. Deferrals

The EPA, in addition to the exclusions provided by rule and by Congress, has set forth deferrals for a limited group of USTs, including wastewater treatment tanks not regulated under the Clean Water Act, field constructed tanks, airport hydrant fuel distribution systems, and certain tanks associated with nuclear power stations and radioactive materials.[23]

However, even these deferred tanks must meet three interim standards.[24] First, the tanks must prevent releases due to corrosion or structural failings for the operational life of the tank system.[25] Second, the tanks must be either cathodically protected against corrosion, constructed of non-corrodible material, steel clad with non-corrodible material, or designed in a manner to prevent the release or threatened release of any stored substance.[26] Third, the tanks must be constructed or lined with material that is compatible with stored substances.[27]

The only exception to meeting the interim standards requires that a corrosion expert[28] show that the UST can be installed at a site where the soils will not cause the UST to corrode and leak during its operating life.[29] If such a showing can be made, the deferred system need not be cathodically protected. Deferral tanks must also meet the corrective action standards discussed in section VI, *infra*.

B. Owners and Operators

Congress distinguished several key participants involved with USTs, and assigned them different types of responsibility and liability. The first class, operators, is a relatively familiar one. Section 9001(4) of the RCRA Amendments defines an operator as any person in control of, or having responsibility for, the daily operation of the USTs. For example, a lessee of a service station who oversees the day-to-day activities of the facility, including the USTs, would be an "operator" even though the lessee is not an owner.

The second major class of participants is owners. However, section 9001(3) of the RCRA Amendments sets forth two very different definitions of owners.[30] The distinction turns on whether the USTs were being used on November 8, 1984, the date of the enactment of the RCRA Amendments. Section 9001(3)(A) defines an owner as any person who owns such a tank used for the storage, use, or dispensing of regulated substances, and whose UST was in use on November 8, 1984, or brought into use after that date. Section 9001(3)(B) creates a group of owners who may no longer hold title to the tank but who are nevertheless governed by RCRA and responsible for complying with the applicable provisions of the program. When a tank was in use before November 8, 1984, but not in use on that date, an owner is any person who owned the tank immediately before the discontinuation of its use. Thus, there is now created trailing liability for anyone who may have sold a site with a tank no longer in use but left in place.

The practical question faced by the second class of owners is whether the tank was no longer in use on November 8, 1984. A tank could be shown to have been inoperable when it was filled with inert material, such as sand, or otherwise rendered unusable. Examples of rendering a tank unusable are abandoning the tank, paving over intakes and vents, or selling the tank to a person who has no use for it, such as a residential real estate developer who builds homes on the site.

Section 9003(h)(9) creates a security interest exemption by providing that:

> . . . the term "owner" does not include any person who, without participating in the management of an underground storage tank and otherwise not engaged in petroleum production, refining, and marketing, holds indicia of ownership primarily to protect the owner's security interest in the tank.

This statutory exemption is limited to liability for corrective action at petroleum-contaminated sites. There is limited legislative history to provide guidance to the types of activities that Congress considered to be consistent with the security exemption, or about the types of activities that might be impermissible participation in a UST or UST system's management by a lienholder.[31]

III. NOTIFICATION REQUIREMENTS

A. Owner's Duty

1. Tanks in Service

In the RCRA Amendments, Congress required owners of USTs that were in use on November 8, 1984, or that were still in the ground but taken out of service after January 1, 1974, to provide notification to the implementing agency of the tank's existence.[32] In 1986, the EPA promulgated regulations requiring all regulated tanks to be registered as of May 8, 1986.[33] Notification of new tanks placed in service after that date must be given within thirty days of becoming operational.[34]

Notifications are to be made using the EPA form set forth in Appendix C, or on a similar form provided by the appropriate state agency. At a minimum, notification forms are required to include the age, size, type, location, and uses of the tank.[35] Appendix D lists the state agencies designated to receive notifications. While the regulations set forth a federal notification form, each of the states has adopted its own forms that should be used to register USTs.

For state notifications, the responsible agency of the state in which a tank is located should be contacted to obtain the appropriate state form. If relevant information previously was submitted to the state, the practitioner should determine whether an update of that information is required. While the EPA does not require updates of tank registrations, states do have the option to require them. Any USTs for which notice was given pursuant to section 103(C) of CERCLA do not require notification to the EPA under the UST program.

Owners of tanks requiring notification may provide notice of several tanks using one form, but owners who own tanks located at more than one place of operation must file a separate notification form for each separate place of operation.[36] Moreover, when a tank is used to store a mixture of materials or more than one regulated substance, the owner should indicate the substance of greatest quantity.[37] Trade names for substances may not be used because they may prevent correct identification of the constituents of the contents.[38]

2. Out-of-Operation Tanks

For each UST taken out of operation after January 1, 1974, the owner must have provided notification of the existence of the tank on or before May 8, 1986, unless the owner knows that the tank subsequently was removed from the ground. The owner of a tank taken out of operation on or before January 1, 1974, is not required to provide any notification.[39] It is important to recall that the "owner" of these regulated USTs may not be the current owner of the property if the previous owner discontinued use before November 8, 1984.

The contents of notifications for USTs out of operation must include, to the extent known to the owner, the date that the tank was taken out of operation; the age of the tank when it was taken out of operation; the size, type, and location of the tank; and the type and quantity of substances left stored in the tank on the date it was taken out of operation.[40] Use the EPA form in Appendix C or the appropriate form from the appropriate state for notification of out-of-operation tanks.[41]

B. Seller's Duty

In addition to owners, sellers of USTs also have notification responsibilities.[42] Beginning October 24, 1988, any person who sells a tank that is intended to be used as a UST must notify the purchaser of its obligation to register the tank.[43] This obligation can be met by modifying the sales form, providing a statement on the invoice describing the tank notification requirements, or leaving a copy of the appropriate notification form with the purchaser of the tank.[44] Appendix E sets forth the form of the statement that the EPA has approved as meeting the obligation of the seller to provide such notice.

C. Duty of Depositors of Regulated Substances

Beginning December 8, 1985, through June 8, 1987, any person who deposited regulated substances in a UST was responsible to "reasonably notify" the owner or operator of the tank of the owner's notification requirements.[45] The definition of "regulated substances" found at 40 C.F.R. section 280.12 controls. Persons who put substances in a tank include operators, distributors, and transporters. If a person who deposited regulated substances in a tank left a copy of the notification form with the owner or operator, that copy constitutes a reasonable attempt to inform the owner of its obligation.[46]

IV. TANK CONSTRUCTION AND OPERATING REQUIREMENTS

A. General Standards and Requirements for New USTs

Effective December 22, 1988, all owners and operators of new UST systems must meet a number of requirements designed to prevent releases due to structural failure, corrosion, or spills and overfills during the useful life of the system.[47] All the equipment comprising the UST system must be installed in accordance with a code of practice developed by a nationally recognized association or independent testing laboratory and in accordance with the manufacturer's instructions.[48] Moreover, any such installation must be certified, tested, and inspected in accordance with EPA regulations.[49]

1. Underground Tanks

The underground tanks themselves must be properly designed, constructed, and protected from corrosion.[50] Consistent with a code of practice developed by a nationally recognized association or independent testing laboratory, the regulations provide that the following types of underground tanks meet the EPA-mandated standard:

- Tanks constructed of fiberglass-reinforced plastic;
- Tanks constructed of steel and cathodically protected;
- Tanks constructed of a steel-fiberglass-reinforced-plastic composition;
- Tanks constructed of metal without additional corrosive protection measures, provided that:
  - — the tanks are installed at a location that a corrosion expert determines has soils that will not corrode the tanks and cause leaks, and

— the owner or operator maintains records to show that the tank remains in compliance with UST corrosion standards during the course of its useful life; and

• Tanks that the EPA or an authorized state agency has determined are both designed to prevent a release and are no less protective of human health and the environment than the other standards specified in the new UST construction and operating regulations.[51]

2. Underground Pipes

Piping that is in contact with the ground and routinely contains regulated substances must also be protected from corrosion. This is to be accomplished by using pipes constructed of either steel with cathodic protection or fiberglass-reinforced plastic.[52] Unprotected metal pipes can be used if a corrosion expert determines that they will not corrode in the soil's environment at the site.[53] The implementing agency[54] can approve piping it determines is designed to prevent a release or threatened release, provided that the piping is no less protective of human health and the environment than the other piping specified in the UST regulation.[55]

3. Spill and Overfill Prevention Equipment

To prevent spills associated with regulated product transfers, owners and operators are now required to install and use equipment that (1) prevents releases to the environment when the transfer hose is detached from the fill pipe (e.g., a catchment basin); (2) automatically shuts off when the tank is 95 percent full or triggers an alarm when the tank is 90 percent full; or (3) restricts flow thirty minutes prior to overfilling, alerts the operator with a high level alarm one minute before overfilling, or automatically shuts off flow into the tank so that none of the fittings located on top of the tank are exposed to product due to overfilling.[56] These requirements do not apply if the tank is filled by transfers of no more than twenty-five gallons at one time or the implementing agency approves an alternative.[57]

Owners and operators are required to ensure that releases due to spilling or overfilling do not occur from tanks.[58] Should any spill or overfill occur, it must be immediately reported, investigated, and cleaned up.[59] Additionally, regulated materials stored in a UST system must be compatible with the materials from which the system is constructed.[60]

Tanks installed after December 22, 1988, must be fitted with systems for leak detection, corrosion protection, and spill and overfill prevention, while existing tanks must be retrofitted within ten years for corrosion protection and spill and overflow prevention. Corrosion protection systems are required to be continuously operated and maintained to protect metal parts of the UST.[61] A cathodic-type corrosion protection system must be tested within six months of installation and at least every three years thereafter.[62] The frequency of inspection for a system with an impressed current cathodic-type corrosion protection system is every sixty days to ensure that the equipment is running properly.[63]

All testing must be done by a qualified cathodic protection tester.[64] Additionally, records must be kept for at least the last two inspections, except for impressed current cathodic systems for which the records of the past three inspections must be maintained by the owner or operator.[65]

4. Repairs

Any repairs made to tanks must meet a rigid set of standards to ensure that the tanks will maintain the structural integrity of the system for their operating life. All such repairs must be made in accordance with nationally recognized codes of practice.[66] Repaired tanks and piping must be tightness tested within thirty days of completion,[67] while any cathodically protected system must have its corrosion protection system tested within six months of the repairs.[68] Records of any repairs are to be maintained by owners and operators for the operating life of the system.[69]

B. Leak Detection Requirements

1. All UST Systems

Owners or operators of a UST system are required to provide a method, or combination of methods, of leak detection that can detect a release from any portion of the system that routinely contains product.[70] The release detection system must be installed, calibrated, operated, and maintained as directed by the manufacturer's instructions.[71] Performance standards for release detection systems are specified, and any detection systems, except automatic line leak detectors, used after December 22, 1990, that were not previously permanently installed must be capable of detecting a leak with a probability of detection of 0.95 and a probability of false alarm of 0.05.[72] Automatic line leak detectors were required to meet the possibilities of detection and false alarm as of September 22, 1991.[73]

All new tanks installed after December 22, 1988, are required to have a leak detection system. Existing tanks must phase in the release detection in accordance with the time schedule set forth in Appendix F. Any existing UST system that cannot comply with the requirements must be closed by the date on which it would have been required to implement the release detection system.[74] All tanks installed after December 22, 1993, must comply with the release detection requirements immediately upon installation.[75]

2. Petroleum UST Systems

A. TANKS

Petroleum UST system owners and operators must monitor underground tanks at least once every thirty days.[76] New tanks may use automatic tank gauging, monitoring for vapors in the soil, interstitial monitoring, ground water monitoring, or other approved methods for the monthly monitoring.[77]

For existing tanks equipped with devices that prevent spills and overfills and that either meet new tank standards for corrosion protection or have internal lining, a monthly inventory control[78] combined with a tank tightness test every five years may be used.[79] This method, however, can only be used until December 22, 1998, or for a maximum of ten years from the date of installation, or upgrading of the tank, whichever date is later.[80]

Petroleum USTs that do not have corrosion protection, linings, or spill and overfill prevention devices may use a monthly inventory control in combination with an annual tank tightness test until December 22, 1998.[81] By that date, the tank must either be upgraded or permanently closed.[82] Tanks with a capacity of 550 gallons or less may use weekly tank gauging to comply with the monitoring requirements.[83]

B. PIPING

All UST systems with pressurized piping must have devices to automatically shut off or restrict flow or an alarm that signals a leak.[84] Additionally, either an annual tightness test must be conducted on the piping or monthly monitoring conducted using vapor monitoring, ground water monitoring, interstitial monitoring, or other approved methods.[85]

UST systems equipped with suction piping do not require leak detection if below-grade piping is sloped to force the flow to drain back into the tank, and there is only one check valve for each suction line that is located directly below the suction pump.[86] All other suction piping requires either monthly monitoring using any of the methods applicable to pressure piping, or tightness testing of the piping every three years.[87]

3. Hazardous Substance UST Systems

All new hazardous substance USTs must have a secondary containment system capable of retaining any release from the tank until the substance can be removed, and preventing a release from the UST into the environment at any time during the operational life of the tank.[88] Underground piping must be equipped with secondary containment, and if it is pressurized an automatic line leak detector is required.[89] The secondary containment system must be checked for evidence of a release at least every thirty days.[90] Existing systems must now meet the requirements for petroleum UST systems, and by December 22, 1998, they must meet the requirements for new hazardous substance USTs.[91]

C. Upgrading Existing Petroleum UST Systems

Within ten years of the effective date of the UST regulations, but not later than December 22, 1998, all existing petroleum UST systems must either meet the new UST system standards, be closed, or meet the upgrading requirements of those regulations.[92] The upgrading standards call for tanks to be fitted with either internal linings, cathodic protection, or a combination of internal lining and cathodic protection.[93] All existing metal underground pipes must be cathodically protected, and spill and overfill prevention equipment must be installed that meets new system standards.[94]

V. REPORTING AND RECORDKEEPING REQUIREMENTS

A. Agency Inspections and Monitoring

Representatives of the EPA or of an approved state program are granted broad authority to collect information needed from any owner or operator, not only for enforcement, but also for developing regulations or approved studies. Section 9005(a) of the RCRA Amendments provides agency representatives with the power to enter at reasonable times any establishment or other place where a UST is located; inspect and obtain samples from any person of any regulated substances contained in the tank; conduct monitoring or testing of the tank, associated equipment contents, or surrounding soils, air, surface water, or groundwater; and take corrective action.[95]

Information collected from these efforts is not assumed to be confidential, and is subject to disclosure. If a party wishes to maintain the confidentiality of such information, then it carries the burden of showing that it is entitled to protection under section 1905 of title 18 of the United States Code.[96] The data sought to be kept confidential must be designated at the time that it is submitted.[97]

B. Reporting Requirements for Owners and Operators

In addition to the EPA's general authority to make inspections and conduct monitoring, section 9003(c)(3) of the RCRA Amendments calls for owners and operators to maintain records and make reports to the agency. Requirements for reporting releases and spills are detailed at subpart E of the regulations. A complete listing of the report and recordkeeping requirements for owners and operators is contained in Appendix G of this monograph.

1. *Suspected Releases*

Suspected releases must be reported by owners and operators within twenty-four hours of discovery, or on observation of unusual operating conditions, or on observation of release detection monitoring results.[98] Any suspected release must be investigated and confirmed within seven days by the owner or operator, unless either party has been directed to do otherwise by the implementing agency.[99]

2. *Spills*

Any spill or overfill is to be immediately contained, cleaned up, and reported within twenty-four hours.[100] Corrective action is required to be commenced by the owner or operator if either a reportable quantity of a hazardous substance or more than twenty-five gallons of petroleum is released.[101] Spills below those corrective action limits must be cleaned up within twenty-four hours, and if that is not possible, the implementing agency must be notified.[102]

VI. RESPONSE AND CORRECTIVE ACTIONS TO CONFIRMED RELEASES

In addition to setting performance standards designed to prevent releases from UST systems, the EPA's regulations set forth in considerable detail the measures that owners and operators must take in response to a confirmed release.

A. Initial Response

A confirmed release must be reported to the implementing agency within twenty-four hours. Owners and operators must: (1) undertake immediate action to prevent any further release; and (2) identify and mitigate any fire, explosion, or vapor hazard.[103] Unless directed to do otherwise, the owner or operator must initiate abatement measures and a site check consisting of the following measures:

1. Remove any regulated substance from the UST system;
2. Visually inspect any assessable releases and prevent further migration;
3. Continue to monitor and mitigate any releases;
4. Remedy hazards posed by contaminated soils, including appropriate treatment and disposal;

5. Measure for the presence of a release where contamination is most likely to be present; and
6. Investigate to determine the presence of any free product, and remove it if present.[104]

Additionally, within twenty days of the release, a report must be submitted to the implementing agency summarizing the abatement steps taken.[105] Within forty-five days of the release, detailed information characterizing the site and release must be collected and reported by the owner or operator.[106] A report on any free product removal at the site must also be submitted to the agency within forty-five days of the release.[107] An investigation for soil and groundwater clean up can also be required and is to be submitted as soon as practicable or in accordance with a schedule established by the implementing agency.[108]

B. Corrective Action Plan

Based on its review of the information submitted as part of the initial response to a release, the implementing agency may call for a corrective action plan.[109] The agency can approve the plan submitted to it only when the agency has determined that the plan will adequately protect human health, safety, and the environment.[110] The regulations provide for public notice and an opportunity for public participation in the approval of any corrective action plan.[111] Following approval, owners and operators are required to implement the plan and report to the implementing agency on their efforts.

VII. CLOSURE OF OUT-OF-SERVICE UST SYSTEMS

A. Temporary Closure

The regulations recognize several stages of closure and specify measures to be taken to secure the UST system for each of those levels. What constitutes "temporary closure" is not specified in the regulations. However, in comments preceding the promulgation, the EPA discussed several factors it considers relevant, including whether normal operating and maintenance procedures are being followed and the nature and volume of material being stored in the system.[112]

When a UST system is temporarily closed, owners and operators must continue operation and maintenance of corrosion protection in accordance with 40 C.F.R. section 280.31[113] and any release detection in accordance with 40 C.F.R sections 280.40-.45.[114] If a release is suspected or detected, owners and operators must comply with the reporting and recordkeeping requirements[115] of 40 C.F.R. sections 280.50-.53 and the response and corrective action requirements[116] of 40 C.F.R. sections 280.60-.67. However, the release detection requirements are not required as long as the UST system is empty. The UST system is empty when all materials have been removed using commonly employed practices so that no more than 2.5 centimeters (one inch) of residue, or 0.3 percent by weight of the total capacity of the UST system, remain in the system.[117]

When an underground storage tank system has been temporarily closed for three months or more, vent lines must be left open and functioning and all lines, pumps, manways, and ancillary equipment must be capped and secured.[118] Unless an extension is obtained after twelve months, the system must be permanently closed.[119]

B. Permanent Closure and Changes-in-Service

At least thirty days prior to beginning a permanent closure or a change-in-service, owners and operators must give the implementing agency notice unless closure is part of a corrective action.[120] No requirements are specified in the EPA regulations as to the form of the notice, but prudence dictates that it be in writing and, if mailed, a return receipt obtained.

To be permanently closed, the UST system must be empty and cleaned by removing all liquids and accumulated sludges.[121] Unless the system is empty, the corrosion protection and release detection systems must be maintained and operating.[122] The system is empty "when all materials have been removed using commonly employed practices so that no more than 2.5 centimeters (one inch) of residue, or 0.3 percent by weight of the total capacity of the UST system, remain in the system."[123] All closed tanks must also be either removed or filled with an inert solid material and a site assessment done.[124]

The continued use of a system to store a nonregulated substance is considered a change in service.[125] To effect a change in service, the tank must be cleaned and a site assessment conducted by the owner or operator.[126]

Closure and change-in-service site assessments are required to measure for the presence of a release at a point in the system where contamination is most likely to be present.[127] If contamination is discovered, a corrective action must be undertaken.[128] The implementing agency is empowered to require assessments for systems closed before December 22, 1988, if there is reason to believe that the site poses a current or potential threat to human health and the environment.[129]

The results of site assessments are to be kept for at least three years and must be maintained by either the owners and operators who closed the system, or the current owners and operators of the site, or by mailing the results to the implementing agency if the results cannot be retained at the closed facility.[130]

VIII. FINANCIAL RESPONSIBILITY

A. Compliance Dates

The EPA Administrator is empowered to require evidence of financial responsibility of the UST owner or operator. The amount is to be sufficient for taking corrective action and for compensating third parties for bodily injury and property damage caused by sudden and nonsudden accidental releases arising from operating an underground storage tank.[131]

On October 26, 1988, the EPA issued financial assurance rules effective January 24, 1989, for USTs containing petroleum.[132] However, that compliance date was subsequently extended to December 31, 1993, because adequate financial assurance mechanisms had not developed to allow all petroleum marketing firms owning one to twelve USTs at more than one facility or fewer than one hundred USTs at a single facility and nonmarketers with net worth less than $20 million to comply with the requirements.[133] Local governments are required to comply with these requirements one year from the date of publication of final regulations that they may use to comply with the financial responsibility requirements for USTs containing petroleum in 40 C.F.R. part 280.[134] These final regulations were published on February 18, 1993 (copy attached as Appendix I).[135]

State and federal government entities (not local government entities) are exempt from the financial responsibility requirements, as are owners and operators of exempt and deferred UST systems.[136] If the owner and operator of a petroleum UST are different persons, only one of them needs to meet the financial responsibility requirements. However, both parties are liable for any noncompliance.[137]

Compliance with the financial responsibility requirements is phased in over a two-year period based on the size and financial resources of the owner or operator. Appendix I shows the compliance schedule for different classes of owners.

B. Amounts and Mechanisms

Section 205 of SARA amended the original RCRA UST financial assurance provisions[138] to allow the EPA Administrator to retain some discretion in promulgating financial responsibility regulations. However, section 205 specifically mandates that underground tanks containing petroleum cannot have less than $1 million of coverage for each "occurrence."[139] An occurrence is an accident "including continuous or repeated exposures to conditions, which results in a release from an underground tank."[140]

The Administrator can establish lower levels of coverage for tanks that are not located at facilities engaged in petroleum production, refining, or marketing and that are not used to handle substantial quantities of petroleum. Additionally, the Administrator can establish different classes of underground tanks for financial responsibility purposes, and suspend enforcement of the requirement when the various methods of satisfying financial responsibility are available.[141]

Owners or operators (including local governments) are generally required to have available at least $1 million in financial responsibility to cover corrective actions and third-party bodily injury and property damages[142] (see Appendix J). Whenever any additional tanks are acquired or installed, the amount of aggregate assurance provided must be reviewed by the owner or operator.[143]

The mandatory financial assurance can be shown by a variety of instruments, or a combination of instruments, including a financial test of self-insurance;[144] guarantees;[145] insurance and risk retention group coverages;[146] surety bonds;[147] letters of credit;[148] state-required mechanisms;[149] state funds or other state assurances;[150] trust funds;[151] or standby trust funds.[152] The only combination of mechanisms that is not allowed is the financial test of self-insurance and a guarantee where the financial statements of the owner or operator and the guarantor are not consolidated.[153]

The final rules regarding financial responsibility requirements, which were published on February 18, 1993, provide four additional mechanisms that allow local governments to comply with the financial responsibility requirements.[154] The mechanisms promulgated by the EPA are (1) a local government bond rating test;[155] (2) a local government financial test;[156] (3) a local government guarantee;[157] and (4) a local government fund.[158] Those mechanisms do not replace the existing mechanisms; rather, they supplement them. Local governments eligible to use the mechanisms may use them alone or in combination with other mechanisms.[159]

To qualify, each mechanism must comply with the specific provisions set forth in the regulations, including the exact wording of some of the instruments.[160] The various forms of financial assurance mechanisms may be substituted for one another as long as they continue to afford the required coverage.[161]

Owners or operators are required to maintain evidence of all financial assurance mechanisms used to comply with the regulations until the UST is closed, or if a corrective action is required, until the action has been completed as part of a closure.[162] Appropriate evidence of financial assurance mechanisms must be submitted to the implementing agency within thirty days of either a release, the commencement of any bankruptcy proceedings, or the inability to obtain or maintain the necessary assurance.[163]

While financial responsibility requirements are designed to be the primary means of cleaning up leaks and spills, section 205 of SARA establishes a "mini superfund" and authorizes the EPA Administrator to undertake corrective action or to order the owner or operator to take such action when there is a release of petroleum from a leaking UST.[164] The procedural rules governing the issuance of administrative corrective action orders issues under section 9003(h) of RCRA are set forth in 40 C.F.R. part 24.[165]

IX. ENFORCEMENT AND PENALTIES

Any owner who either knowingly submits false information or fails to notify the implementing agency of the existence or installation of an underground tank is subject to a civil penalty not to exceed $10,000 for each tank involved. Operators do not face this financial penalty. Failure-to-notify constitutes a single offense rather than separate offenses for each day of no notification.[166]

If the EPA Administrator determines that any person is in violation of any requirement of the UST provisions, the Administrator may either issue an administrative order requiring compliance within a reasonable specified time, or commence a civil action in the United States District Court in the jurisdiction in which the violation occurred, for appropriate relief including a temporary or permanent injunction. The Administrator in so acting may rely on information from any source.[167]

Administrative compliance orders are required to state with reasonable specificity the nature of the violation and the reasonable time for compliance. The Administrator may use a compliance order to assess a civil penalty in an amount to be determined after considering two factors, the seriousness of the violation, and any good faith efforts to comply with the applicable requirements.[168]

If the violation occurs in a state with an approved program, the Administrator must give notice to the state before issuing an order or commencing a civil action. A violator who fails to comply with an administrative compliance order is liable for a civil penalty of not more than $25,000 for each day of continued noncompliance with the order.[169]

The administrative compliance order becomes final unless the party named requests a public hearing within thirty days of service. Subpoenas may be issued for persons or for the production of documents and other matters necessary to such a hearing. Rules for discovery procedures can also be promulgated by the Administrator for such hearings.[170] The EPA has developed an enforcement guidance document that it uses to calculate civil penalties for violations of UST technical standards and financial responsibility regulations (see Appendix L).

There are no criminal penalties for noncompliance or violation of the various requirements for the regulation of USTs under subtitle I of RCRA.[171] Federal facilities, unless they are granted a presidential exemption, are subject to the requirements of the RCRA Amendments, but are not subject to civil penalties.[172]

X. CONCLUSION

The EPA UST regulations put in place one of the largest environmental regulatory programs ever undertaken in the United States. Congress's primary concern in mandating this program was to protect underground waters from contamination associated with UST facilities. Although underground hazardous waste tanks have been regulated for some time, the vast majority of USTs, estimated to exceed 2 million units at over 700,000 facilities, were not regulated prior to the promulgation of the UST regulations.[173]

A wide variety of facilities are affected by the UST program, from chemical plants to neighborhood service stations, and the cost of compliance is substantial. In particular, adequate financial responsibility assurance is likely to prove difficult and costly.

Alternatives to the underground storage of petroleum and regulated substances should be carefully considered in light of the growing costs and exposure to liability from such tanks. If no reasonable alternatives exist, the owner and operator must either implement a program to bring USTs into compliance or risk enforcement actions.

Finally, any due diligence by the buyer in real estate transactions involving facilities that contain regulated tanks must now include close scrutiny of the seller's compliance with operating, technical, and financial responsibility requirements of the EPA's UST program. While the tanks may be out of sight, they no longer can be out of mind.

Endnotes

1. Resource Conservation Recovery Act of 1976, Pub. L. No. 94-580, 90 Stat. 2795, 42 U.S.C.A. § 6901 *et seq.* (RCRA), *amended by* Pub. L. No. 95-609, Nov. 8, 1978; Pub. L. No. 96-463, Oct. 15, 1980; Pub. L. No. 96-482, Oct. 21, 1980; Pub. L. No. 96-510, Dec. 11, 1980; Pub. L. No. 97-272, Sept. 30, 1982; Pub. L. No. 97-375, Dec. 21, 1982; Pub. L. No. 98-45, July 12, 1983; Pub. L. No. 98-371, July 18, 1984; Hazardous and Solid Waste Amendments of 1984, Pub. L. No. 98-616, Nov. 8, 1984; Pub. L. No. 99-339, June 19, 1986; Superfund Amendments and Reauthorization Act of 1986 (SARA), Pub. L. No. 99-499, Oct. 17, 1986; Pub. L. No. 100-582, Nov. 1, 1988; Pub. L. No. 102-386, Oct. 6, 1992; Pub. L. No. 102-508, Oct. 24, 1992, (amendments collectively cited as RCRA Amendments). RCRA Amendments that treat USTs are included in Appendix A of this monograph. Appendix B includes codified regulations.

2. For a discussion of hazardous waste subject to RCRA regulation, see K. Woods, Identification of RCRA-Regulated Substances: How to Determine if Your Company Is Handling a Hazardous Waste and Possible Ways to Soften the Blow (Mono. No. 8, ABA Section of Natural Resources, Energy, and Environmental Law 1989).

3. RCRA Amendments, *supra* note 1, subch. IX §§ 6991–6991i (West Supp. 1993).

4. *Id.* § 6991(2). This provision also defines a "regulated substance" as either petroleum or any substance defined in § 9601(14) of the Comprehensive Environmental Response Compensation and Liability Act (Superfund or CERCLA), 42 U.S.C.A. §§ 9601–9675 (West 1983 & Supp. 1993). Petroleum is defined in § 6991 of RCRA to include crude oil or any fraction thereof which is liquid at standard conditions of temperature and pressure.

5. RCRA Amendments, *supra* note 1, § 9001(1), 42 U.S.C.A. § 6991(1) (West Supp. 1993). For purposes of § 9001(1), the volume of the tank below the surface of the ground does not include the volume of underground pipes attached to it.

6. 40 C.F.R. § 280.12 (1992).

7. *Id.* § 9004, 42 U.S.C.A. § 6991c. EPA approvals of state UST programs appear at 40 C.F.R. part 282 subpart B. As of this writing, EPA granted approval to three states: New Hampshire,58 Fed. Reg. 58, 624 (1993); Rhode Island, 58 Fed. Reg. 6894 (1993); and Washington, 58 Fed. Reg. 47, 217 (1993).

8. *Id.* § 9008, 42 U.S.C.A. § 6991g.

9. SARA, *supra* note 1, 42 U.S.C.A. 6991b(c)(4)–(6), 6991b(d)(1)–(5), 6691b(h) (West Supp. 1993).

10. RCRA Amendments, *supra* note 1, § 9003, 42 U.S.C.A. § 6991b(h) (West Supp. 1993). Other regulated substances can be cleaned up under Superfund authority; however, petroleum is not covered by Superfund.

11. The Leaking Underground Storage Tank Fund was established by the Tax Reform Act of 1986, 26 U.S.C.A. § 4081 (West 1989 & Supp. 1993). The Fund is financed by a 0.1 cent per gallon tax on gasoline and other motor fuels. The tax expires on December 31, 1995.

12. RCRA, *supra* note 1, 42 U.S.C.A. §§ 6921–6934 (West 1983). A hazardous waste is defined as any solid waste that, "because of its quantity, concentration, or physical, chemical, or infectious characteristics may (A) cause, or significantly contribute to an increase in mortality or an increase in serious irreversible, or incapacitating reversible, illness; or (B) pose a substantial present or potential hazard to human health or the environment when improperly treated, stored, transported, or disposed of, or otherwise managed." *Id.* § 6903(5).

13. *See id.* § 6903(27).

14. *See id.* § 6903(5).

15. *See* Notification Requirements for Owners of Underground Storage Tanks, 50 Fed. Reg. 21,772 (1985)(proposed rule).

16. *Remediation Costs*, 1 DRG Toxics Newsletter Feb. 1993, at 1 (reporting on a report issued by Environmental Information Ltd.).

17. *Id.*

18. *Id.*

19. *Id.*

20. 40 C.F.R. § 280.10 (1992).

21. 58 Fed. Reg. 8504 (Feb. 12, 1993).

22. 42 U.S.C.A. § 9001 *et seq.* (West 1983 & Supp. 1993).

23. 40 C.F.R. § 280.10(c) (1992).

24. *Id.* § 280.11 (1992).

25. *Id.* § 280.11(a)(1) (1992).

26. *Id.* § 280.11(a)(2) (1992). A "release" is virtually any migration of regulated substances from a UST into the groundwater, surface water, or subsurface soils. RCRA Amendments § 9001(5), 42 U.S.C.A. § 6991(5) (West Supp. 1993).

27. 40 C.F.R. § 280.11(a)(3) (1992).

28. A corrosion expert is "a person who, by reason of thorough knowledge of the physical sciences and the principles of engineering and mathematics acquired by a professional education and related practical experience, is qualified to engage in the practice of corrosion control on buried or submerged metal piping systems and metal tanks. Such a person must be accredited or certified as being qualified by the National Association of Corrosion Engineers or be a registered professional engineer who has certification or licensing that includes eduction metal piping systems and metal tanks." 40 C.F.R. § 280.12 (1992).

29. 40 C.F.R. § 280.11(b) (1992).

30. 42 U.S.C.A. § 6991(3) (West Supp. 1993).

31. The EPA has proposed a rule to implement the security interest exemption, which would provide that a holder who meets the criteria specified in the proposed rule (i.e., those whose only connection with the tank is as the bona fide holder of a security interest in the UST or UST system) is not subject to the UST technical standards and financial responsibility requirements otherwise applicabhle to a tank owner. 59 Fed. Reg. 30,448 (1994) (proposed to be codified at 40 C.F.R. §§ 280.200–280.250 and 281.39)

32. RCRA Amendments, *supra* note 1, § 9002(a), 42 U.S.C.A. § 6991a(a) (West. Supp. 1993).

33. 53 Fed. Reg. 37,082, 37,199 (1988) (codified at 40 C.F.R. § 280.22 (1992)). The information collected by the states from the UST system registrations must be kept in two separation inventories: one for USTs containing petroleum, and one for other regulated substances. This information was to have been submitted to the EPA within 270 days after Oct. 17, 1986, the enactment date of SARA, by each of the states. SARA, *supra* note 1, § 205(b), 42 U.S.C.A. § 6991a(c) (West Supp. 1993).

34. 40 C.F.R. § 280.22(a) (1992).

35. 40 C.F.R. § 280.22 (1992); 42 U.S.C.A. § 6991a(a)(2)(B) (West Supp. 1993); RCRA § 9002(a)(2)(B).

36. 40 C.F.R. § 280.22(c) (1992).

37. 50 Fed. Reg. 21,775 (1985).

38. *Id.*

39. RCRA Amendments, *supra* note 1, § 9002(a)(2)(A), 42 U.S.C.A. § 6991a(a)(2)(A) (West Supp. 1993).

40. *Id.* § 9002(a)(2)(B), 42 U.S.C.A. § 6991a(a)(2)(B) (West Supp. 1993).

41. 40 C.F.R. § 280.22(g) (1992).

42. RCRA Amendments, *supra* note 1, § 9002(a)(6), 42 U.S.C.A. § 6991a(a)(6) (West Supp. 1993); 40 C.F.R. § 280.22(g) (1992).

43. RCRA Amendments, *supra* note 1, § 9002(a)(6), 42 U.S.C.A. § 6991a(a)(6) (West Supp. 1993); 40 C.F.R. § 280.22(g) (1992).

44. 50 Fed. Reg. 21,774 (1985).

45. RCRA Amendments, *supra* note 1, § 9002(a)(5), 42 U.S.C.A. § 6991a(a)(5) (West Supp. 1993).

46. 40 Fed. Reg. 21,774 (1985).

47. In promulgating final UST technical standard regulations, the EPA Administrator is authorized to distinguish between types, classes, and ages of USTs. The Administrator can consider a number of factors in making these distinctions, including tank location; soil and climate conditions; use, size, age and maintenance history of the tank; industry standards and practices; hydrology; water table; technical capacity of owners and operators; and compatibility of the tank and material stored in it. RCRA Amendments, *supra* note 1, § 9003, 42 U.S.C.A. § 6991b (West Supp. 1993). In promulgating final regulations, Congress mandated that the EPA Administrator must include requirements for leak detection or an inventory control system and tank testing, recordkeeping and reporting, corrective action, financial responsibility for corrective action and third-party liability, and closure. On September 23, 1988, the EPA issued the long-awaited technical standards for underground storage tanks implementing these provisions that have been codified at 40 C.F.R. §§ 280.40–280.45 (1992).

48. 40 C.F.R. § 280.20(d) (1992).

49. *Id.* § 280.20(e) (1992).

50. *Id.* § 280.20(a) (1992).

51. 4*Id.* § 280.20(a)(1)–(5) (1992).

52. *Id.* § 280.20(b)(1),(2) (1992).

53. 4*Id.* § 280.20(b)(3) (1992).

54. An implementing agency may be either the EPA, or if a state has an approved program, the designated state or local agency responsible for carrying out an approved UST program. 40 C.F.R. § 280.12 (1992).

55. 40 C.F.R. § 280.20(b)(4) (1992).

56. *Id.* § 280.20(c)(1) (1992); 56 Fed. Reg. 38342, 38343 (Aug. 13, 1991).

57. *Id.* § 280.20(c)(2) (1992).

58. *Id.* § 280.30(a) (1992).

59. *Id.* § 280.30(b) (1992).

60. *Id.* § 280.32 (1992).

61. *Id.* § 280.31(a) (1992).

62. *Id.* § 280.31(b)(1) (1992). The EPA is currently compiling and reviewing public comments on a request to alter the current requirements for cathodic protection monitoring for steel USTs. *See* 58 Fed. Reg. 55066 (Oct. 25, 1993).

63. 40 C.F.R. § 280.31(c) (1992).

64. A cathodic protection tester is "a person who can demonstrate an understanding of the principles and measurements of all common types of cathodic protection systems as applied to buried or submerged metal piping and tank systems. At a minimum, such persons must have education and experience in soil resistivity, stray current, structure-to-soil potential, and component electrical isolation measurements of buried metal piping and tank systems." *Id.* § 280.12 (1992).

65. *Id.* § 280.31(d)(1) (1992).

66. *Id.* § 280.33(a) (1992).

67. *Id.* § 280.33(d) (1992).

68. *Id.* § 280.33(e) (1992).

69. *Id.* § 280.33(f) (1992).

70. *Id.* § 280.40(a)(1) (1992).

71. *Id.* § 280.40(a)(2) (1992).

72. *Id.* § 280.40(a)(3) (1992).

73. *Id.*, as amended at 56 Fed. Reg. 26 (Jan. 2, 1991).

74. 40 C.F.R. § 280.40(d).

75. *Id.* § 280.40(c).

76. *Id.* § 280.41(a) (1992).

77. *Id.* §§ 280.41(a)(1), 280.43(d)–(h) (1992).

78. Product inventory control must be conducted monthly to detect a release of at least 1.0% of flow-through plus 130 gallons on a monthly basis using any one of six prescribed methods. *Id.* § 280.43(a) (1992).

79. *Id.* § 280.41(a)(1) (1992). Interstitial is a technical term which refers to the space between the first and second wall of a secondary containment system.

80. *Id.*

81. *Id.* § 280.41(a)(2) (1992).

82. *Id.*

83. *Id.* § 280.41(a)(3) (1992).

84. *Id.* § 280.41(b)(1) (1992).

85. *Id.*

86. *Id.* § 280.41(b)(2) (1992).

87. *Id.*

88. *Id.* § 280.42(b)(1)(i),(ii) (1992).

89. *Id.* § 280.42(b)(4) (1992).

90. *Id.* § 280.42(b)(1)(iii) (1992).

91. *Id.* § 280.42(a) (1992).

92. *Id.*. § 280.21(a) (1992).

93. *Id.*. § 280.21(b) (1992).

94. *Id.*. § 280.21(c),(d) (1992).

95. RCRA Amendments, *supra* note 1, § 9005(a), 42 U.S.C.A. § 6991d(a) (West Supp. 1993).

96. *Id.* § 9005(b)(1), 42 U.S.C.A. § 6991d(b)(1) (West Supp. 1993).

97. *Id.* § 9005(b)(3), 42 U.S.C.A. § 6991d(b)(3) (West Supp. 1993).

98. 40 C.F.R. § 280.50(a) (1992).

99. *Id.* § 280.52 (1992).

100. *Id.* § 280.53(a) (1992).

101. *Id.* § 280.53(a)(1), (2) (1992).

102. *Id.* § 280.53(b) (1992).

103. *Id.* § 280.61 (1992).

104. *Id.* § 280.62(a) (1992).

105. *Id.* § 280.62(b) (1992).

106. *Id.* § 280.63(b) (1992).

107. *Id.* § 280.64(d) (1992).

108. *Id.* § 280.65 (1992).

109. *Id.* § 280.66(a) (1992).

110. *Id.* § 280.66(b) (1992).

111. *Id.* § 280.67 (1992).

112. *Id.* § 280.70 (1992).

113. *See* discussion *supra* section IV.A.3).

114. 40 C.F.R. § 280.70(a) (1992). *See* discussion *supra* section IV.B. for discussion of leak detection requirements.

115. *See* discussion *supra* section V.

116. *See* discussion *supra* of section VI.

117. 40 C.F.R. § 280.70(a) (1992).

118. *Id.* § 280.70(b) (1992).

119. *Id.* § 280.70(c) (1992).

120. *Id.* § 280.71(a) (1992).

121. *Id.* § 280.71(b) (1992).

122. *Id.* § 280.70(a) (1992).

123. *Id.*

124. *Id.* § 280.71(b) (1992).

125. *Id.* § 280.71(c) (1992).

126. *Id.*

127. *Id.* § 280.72(a) (1992).

128. *Id.* § 280.72(b) (1992).

129. *Id.* § 280.73 (1992).

130. *Id.* § 280.74 (1992).

131. RCRA Amendments, *supra* note 1, § 9003(d)(1), 42 U.S.C.A. § 6991b(d)(1) (West Supp. 1993).

132. 53 Fed. Reg. 43,321-83 (1988), as amended at 58 Fed. Reg. 9026 (Feb. 18, 1993).

133. 40 C.F.R. § 280.91(d) (1992), as amended at 55 Fed. Reg. 46022 (Oct. 31, 1990); 56 Fed. Reg. 66373 (Dec. 23, 1991).

134. 40 C.F.R. § 280.91(e) (1992), as amended at 55 Fed. Reg. 46022, Oct. 31, 1990.

135. 58 Fed. Reg. 9026 (Feb. 18, 1993).

136. 40 C.F.R. § 280.90(c),(d) (1992).

137. *Id.* § 280.90(e) (1992).

138. RCRA Amendments, *supra* note 1, § 9003(d)(1), 42 U.S.C.A. § 6991b(d)(1) (West Supp. 1993).

139. SARA, *supra* note 1, § 9003(d)(5)(A), 42 U.S.C.A. § 6991b(d)(5)(A) (West Supp. 1993).

140. 40 C.F.R. § 280.92(g) (1992). This section makes it clear that the definition of accident is not meant to limit or conflict with the standard insurance usage of the term. Although not defined, "aggregate" has its common insurance meaning.

141. RCRA Amendments, *supra* note 1, § 9003(d)(5)(D), 42 U.S.C.A. § 6991b(d)(5)(D) (West Supp. 1993).

142. 40 C.F.R. § 280.93 (1992).

143. *Id.* § 280.93(f) (1992).

144. *Id.* § 280.95 (1992).

145. *Id.* § 280.96 (1992).

146. *Id.* § 280.97 (1992).

147. *Id.* § 280.98 (1992).

148. *Id.* § 280.99 (1992).

149. *Id.* § 280.100 (1992).

150. *Id.* § 280.101 (1992). *See* Appendix K for a list of state financial assurance funds.

151. 40 C.F.R. § 280.102 (1992).

152. *Id.* § 280.103 (1992).

153. *Id.* § 280.94 (1992), as amended by 58 Fed. Reg. 9026 at 9051 (Feb. 18, 1993).

154. 58 Fed. Reg. 9026-9060 (Feb. 18, 1993). A copy is attached as Appendix I. Local governments were to have had financial insurance mechanisms in place by Feb. 18, 1994.

155. 58 Fed. Reg. at 9053 (to be codified at 40 C.F.R. § 280.104).

156. *Id.* at 9054 (to be codified at 40 C.F.R. § 280.105).

157. *Id.* at 9056 (to be codified at 40 C.F.R. § 280.106).

158. *Id.* at 9059 (to be codified at 40 C.F.R. § 280.107).

159. *Id.* 9026 at 9051 (to be codified at 40 C.F.R. § 280.94(a)(2)).

160. 40 C.F.R. § 280.94 to .103 (1992), as amended by 58 Fed. Reg. 9026 (Feb. 18, 1993).

161. *Id.* § 280.104 (1992), (renumbered § 280.108, as amended by 58 Fed. Reg. 9026).

162. *Id.* § 280.107 (1992), (renumbered § 280.111 as amended by 58 Fed. Reg. 9026).

163. *Id.* § 280.106 (1992), (renumbered § 280.110, as amended by 58 Fed. Reg. 9026).

164. RCRA Amendments, supra note 1, § 9003(h), 42 U.S.C.A. § 6991b(h) (West Supp. 1993).

165. 56 Fed. Reg. 49376 (Sept. 27, 1991).

166. *Id.* § 9006(d)(1), 42 U.S.C.A. § 6991e(d)(1) (West Supp. 1993).

167. *Id.* § 9006(a)(1), 42 U.S.C.A. § 6991e(a)(1) (West Supp. 1993).

168. *Id.* § 9006(c), 42 U.S.C.A. § 6991e(c) (West Supp. 1993).

169. *Id.* § 9006(a)(3), 42 U.S.C.A. § 6991e(a)(3) (West Supp. 1993).

170. *Id.* § 9006(b), 42 U.S.C.A. § 6991e(b) (West Supp. 1993).

171. *Id.* § 9006(d), 42 U.S.C.A. § 6991e(d) (West Supp. 1993).

172. *Id.* § 9007, 42 U.S.C.A. § 6991f (West Supp. 1993).

173. 53 Fed. Reg. 37,082, 37,083 (1988).

APPENDIX A

RCRA Amendments on Underground Storage Tanks

(42 U.S.C.A. §§ 6991–6991i (West Supp. 1994))

SUBCHAPTER IX—REGULATION OF UNDERGROUND STORAGE TANKS

§ 6991. Definitions and exemptions

For the purposes of this subchapter—

(1) The term "underground storage tank" means any one or combination of tanks (including underground pipes connected thereto) which is used to contain an accumulation of regulated substances, and the volume of which (including the volume of the underground pipes connected thereto) is 10 per centum or more beneath the surface of the ground. Such term does not include any—

(A) farm or residential tank of 1,100 gallons or less capacity used for storing motor fuel for noncommercial purposes,

(B) tank used for storing heating oil for consumptive use on the premises where stored,

(C) septic tank,

(D) pipeline facility (including gathering lines)—

(i) which is regulated under the Natural Gas Pipeline Safety Act of 1968 (49 U.S.C. App. 1671 et seq.),

(ii) which is regulated under the Hazardous Liquid Pipeline Safety Act of 1979 (49 U.S.C. App. 2001 et seq.), or

(iii) which is an intrastate pipeline facility regulated under State laws as provided in the provisions of law referred to in clause (i) or (ii) of this subparagraph,

and which is determined by the Secretary to be connected to a pipeline or to be operated or intended to be capable of operating at pipeline pressure or as an integral part of a pipeline,

(E) surface impoundment, pit, pond, or lagoon,

(F) storm water or waste water collection system,

(G) flow-through process tank,

(H) liquid trap or associated gathering lines directly related to oil or gas production and gathering operations, or

(I) storage tank situated in an underground area (such as a basement, cellar, mineworking, drift, shaft, or tunnel) if the storage tank is situated upon or above the surface of the floor.

The term "underground storage tank" shall not include any pipes connected to any tank which is described in subparagraphs (A) through (I).

(2) The term "regulated substance" means—

(A) any substance defined in section 9601(14) of this title (but not including any substance regulated as a hazardous waste under subchapter III of this chapter), and

(B) petroleum.

(3) The term "owner" means—

(A) in the case of an underground storage tank in use on November 8, 1984, or brought into use after that date, any person who owns an underground storage tank used for the storage, use, or dispensing of regulated substances, and

(B) in the case of any underground storage tank in use before November 8, 1984, but no longer in use on November 8, 1984, any person who owned such tank immediately before the discontinuation of its use.

(4) The term "operator" means any person in control of, or having responsibility for, the daily operation of the underground storage tank.

(5) The term "release" means any spilling, leaking, emitting, discharging, escaping, leaching, or disposing from an underground storage tank into ground water, surface water or subsurface soils.

(6) The term "person" has the same meaning as provided in section 6903(15) of this title, except that such term includes a consortium, a joint venture, and a commercial entity, and the United States Government.

(7) The term "nonoperational storage tank" means any underground storage tank in which regulated substances will not be deposited or from which regulated substances will not be dispensed after November 8, 1984.

(8) The term "petroleum" means petroleum, including crude oil or any fraction thereof which is liquid at standard conditions of temperature and pressure (60 degrees Fahrenheit and 14.7 pounds per square inch absolute).

(Pub.L. 89–272, Title II, § 9001, as added Pub.L. 98–616, Title VI, § 601(a), Nov. 8, 1984, 98 Stat. 3277, and amended Pub.L. 99–499, Title II, § 205(a), Oct. 17, 1986, 100 Stat. 1696; Pub.L. 102–508, Title III, § 302, Oct. 24, 1992, 106 Stat. 3307.)

§ 6991a. Notification

(a) Underground storage tanks

(1) Within 18 months after November 8, 1984, each owner of an underground storage tank shall notify the State or local agency or department designated pursuant to subsection (b)(1) of this section of the existence of such tank, specifying the age, size, type, location, and uses of such tank.

(2)(A) For each underground storage tank taken out of operation after January 1, 1974, the owner of such tank shall, within eighteen months after November 8, 1984, notify the State or local agency, or department designated pursuant to subsection (b)(1) of this section of the existence of such tanks (unless the owner knows the tank subsequently was removed from the ground). The owner of a tank taken out of operation on or before January 1, 1974, shall not be required to notify the State or local agency under this subsection.

(B) Notice under subparagraph (A) shall specify, to the extent known to the owner—

(i) the date the tank was taken out of operation,

(ii) the age of the tank on the date taken out of operation,

(iii) the size, type and location of the tank, and

(iv) the type and quantity of substances left stored in such tank on the date taken out of operation.

(3) Any owner which brings into use an underground storage tank after the initial notification period specified under paragraph (1), shall notify the designated State or local agency or department within thirty days of the existence of such tank, specifying the age, size, type, location and uses of such tank.

(4) Paragraphs (1) through (3) of this subsection shall not apply to tanks for which notice was given pursuant to section 9603(c) of this title.

(5) Beginning thirty days after the Administrator prescribes the form of notice pursuant to subsection (b)(2) of this section and for eighteen months thereafter, any person who deposits regulated substances in an underground storage tank shall reasonably notify the owner or operator of such tank of the owner's notification requirements pursuant to this subsection.

(6) Beginning thirty days after the Administrator issues new tank performance standards pursuant to section 6991b(c) of this title, any person who sells a tank intended to be used as an underground storage tank shall notify the purchaser of such tank of the owner's notification requirements pursuant to this subsection.

(b) Agency designation

(1) Within one hundred and eighty days after November 8, 1984, the Governors of each State shall designate the appropriate State agency or department or local agencies or departments to receive the notifications under subsection (a)(1), (2), or (3) of this section.

(2) Within twelve months after November 8, 1984, the Administrator, in consultation with State and local officials designated pursuant to subsection (b)(1) of this section, and after notice and opportunity for public comment, shall prescribe the form of the notice and the information to be included in the notifications under subsection (a)(1), (2), or (3) of this section. In prescribing the form of such notice, the Administrator shall take into account the effect on small businesses and other owners and operators.

(c) State inventories

Each State shall make 2 separate inventories of all underground storage tanks in such State containing regulated substances. One inventory shall be made with respect to petroleum and one with respect to other regulated substances. In making such inventories, the State shall utilize and aggregate the data in the notification forms submitted pursuant to subsections (a) and (b) of this section. Each State shall submit such aggregated data to the Administrator not later than 270 days after October 17, 1986.

(Pub.L. 89–272, Title II, § 9002, as added Pub.L. 98–616, Title VI, § 601(a), Nov. 8, 1984, 98 Stat. 3278, and amended Pub.L. 99–499, Title II, § 205(b), Oct. 17, 1986, 100 Stat. 1696.)

§ 6991b. Release, detection, prevention, and correction regulations

(a) Regulations

The Administrator, after notice and opportunity for public comment, and at least three months before the effective dates specified in subsection (f) of this section, shall promulgate release detection, prevention, and correction regulations applicable to all owners and operators of underground storage tanks, as may be necessary to protect human health and the environment.

(b) Distinction in regulations

In promulgating regulations under this section, the Administrator may distinguish between types, classes, and ages of underground storage tanks. In making such distinctions, the Administrator may take into consideration factors, including, but not limited to: location of the tanks, soil and climate conditions, uses of the tanks, history of maintenance, age of the tanks, current industry recommended practices, national consensus codes, hydrogeology, water table, size of the tanks, quantity of regulated substances periodically deposited in or dispensed from the tank, the technical capability of the owners and operators, and the compatibility of the regulated substance and the materials of which the tank is fabricated.

(c) Requirements

The regulations promulgated pursuant to this section shall include, but need not be limited to, the following requirements respecting all underground storage tanks—

(1) requirements for maintaining a leak detection system, an inventory control system together with tank testing, or a comparable system or method designed to identify releases in a manner consistent with the protection of human health and the environment;

(2) requirements for maintaining records of any monitoring or leak detection system or inventory control system or tank testing or comparable system;

(3) requirements for reporting of releases and corrective action taken in response to a release from an underground storage tank;

(4) requirements for taking corrective action in response to a release from an underground storage tank;

(5) requirements for the closure of tanks to prevent future releases of regulated substances into the environment; and

(6) requirements for maintaining evidence of financial responsibility for taking corrective action and compensating third parties for bodily injury and property damage caused by sudden and nonsudden accidental releases arising from operating an underground storage tank.

(d) Financial responsibility

(1) Financial responsibility required by this subsection may be established in accordance with regulations promulgated by the Administrator by any one, or any combination, of the following: insurance, guarantee, surety bond, letter of credit, qualification as a self-insurer or any other method satisfactory to the Administrator. In promulgating requirements under this subsection, the Administrator is authorized to specify policy or other contractual terms, conditions, or defenses which are necessary or are unacceptable in establishing such evidence of financial responsibility in order to effectuate the purposes of this chapter.

(2) In any case where the owner or operator is in bankruptcy, reorganization, or arrangement pursuant to the Federal Bankruptcy Code or where with reasonable diligence jurisdiction in any State court of the Federal Courts cannot be obtained over an owner or operator likely to be solvent at the time of judgment, any claim arising from conduct for which evidence of financial responsibility must be provided under this subsection may be asserted directly against the guarantor providing such evidence of financial responsibility. In the case of any action pursuant to this paragraph such guarantor shall be entitled to invoke all rights and defenses which would have been available to the owner or operator if any action had been brought against the owner or operator by the claimant and which would have been available to the guarantor if an action had been brought against the guarantor by the owner or operator.

(3) The total liability of any guarantor shall be limited to the aggregate amount which the guarantor has provided as evidence of financial responsibility to the owner or operator under this section. Nothing in this subsection shall be construed to limit any other State or Federal statutory, contractual or common law liability of a guarantor to its owner or operator including, but not limited to, the liability of such guarantor for bad faith either in negotiating or in failing to negotiate the settlement of any claim. Nothing in this subsection shall be construed to diminish the liability of any person under section 9607 or 9611 of this title or other applicable law.

(4) For the purpose of this subsection, the term "guarantor" means any person, other than the owner or operator, who provides evidence of financial responsibility for an owner or operator under this subsection.

(5)(A) The Administrator, in promulgating financial responsibility regulations under this section, may establish an amount of coverage for particular classes or categories of underground storage tanks containing petroleum which shall satisfy such regulations and which shall not be less than $1,000,000 for each occurrence with an appropriate aggregate requirement.

(B) The Administrator may set amounts lower than the amounts required by subparagraph (A) of this paragraph for underground storage tanks containing petroleum which are at facilities not engaged in petroleum production, refining, or marketing and which are not used to handle substantial quantities of petroleum.

(C) In establishing classes and categories for purposes of this paragraph, the Administrator may consider the following factors:

(i) The size, type, location, storage, and handling capacity of underground storage tanks in the class or category and the volume of petroleum handled by such tanks.

(ii) The likelihood of release and the potential extent of damage from any release from underground storage tanks in the class or category.

(iii) The economic impact of the limits on the owners and operators of each such class or category, particularly relating to the small business segment of the petroleum marketing industry.

(iv) The availability of methods of financial responsibility in amounts greater than the amount established by this paragraph.

(v) Such other factors as the Administrator deems pertinent.

(D) The Administrator may suspend enforcement of the financial responsibility requirements for a particular class or category of underground storage tanks or in a particular State, if the Administrator makes a determination that methods of financial responsibility satisfying the requirements of this subsection are not generally available for underground storage tanks in that class or category, and—

(i) steps are being taken to form a risk retention group for such class of tanks; or

(ii) such State is taking steps to establish a fund pursuant to section 6991c of this title to be submitted as evidence of financial responsibility.

A suspension by the Administrator pursuant to this paragraph shall extend for a period not to exceed 180 days. A determination to suspend may be made with respect to the same class or category or for the same State at the end of such period, but only if substantial progress has been made in establishing a risk retention group, or the owners or operators in the class or category demonstrate, and the Administrator finds, that the formation of such a group is not possible and that the State is unable or unwilling to establish such a fund pursuant to clause (ii).

(e) New tank performance standards

The Administrator shall, not later than three months prior to the effective date specified in subsection (f) of this section, issue performance standards for underground storage tanks brought into use on or after the effective date of such standards. The performance standards for new underground storage tanks shall include, but need not be limited to, design, construction, installation, release detection, and compatibility standards.

(f) Effective dates

(1) Regulations issued pursuant to subsection (c) and (d) of this section, and standards issued pursuant to subsection (e) of this section, for underground storage tanks containing regulated substances defined in section 6991(2)(B) of this title (petroleum, including crude oil or any fraction thereof which is liquid at standard conditions of temperature and pressure) shall be effective not later than thirty months after November 8, 1984.

(2) Standards issued pursuant to subsection (e) of this section (entitled "New Tank Performance Standards") for underground storage tanks containing regulated substances defined in section 6991(2)(A) of this title shall be effective not later than thirty-six months after November 8, 1984.

(3) Regulations issued pursuant to subsection (c) of this section (entitled "Requirements") and standards issued pursuant to subsection (d) of this section (entitled "Financial Responsibility") for underground storage tanks containing regulated substances defined in section 6991(2)(A) of this title shall be effective not later than forty-eight months after November 8, 1984.

(g) Interim prohibition

(1) Until the effective date of the standards promulgated by the Administrator under subsection (e) of this section and after one hundred and eighty days after November 8, 1984, no person may install an underground storage tank for the purpose of storing regulated substances unless such tank (whether of single or double wall construction)—

(A) will prevent releases due to corrosion or structural failure for the operational life of the tank;

(B) is cathodically protected against corrosion, constructed of noncorrosive material, steel clad with a noncorrosive material, or designed in a manner to prevent the release or threatened release of any stored substance; and

(C) the material used in the construction or lining of the tank is compatible with the substance to be stored.

(2) Notwithstanding paragraph (1), if soil tests conducted in accordance with ASTM Standard G57–78, or another standard approved by the Administrator, show that soil resistivity in an installation location is 12,000 ohm/cm or more (unless a more stringent standard is prescribed by the Administrator by rule), a storage tank without corrosion protection may be installed in that location during the period referred to in paragraph (1).

(h) EPA Response Program for Petroleum

(1) Before regulations

Before the effective date of regulations under subsection (c) of this section, the Administrator (or a State pursuant to paragraph (7)) is authorized to—

(A) require the owner or operator of an underground storage tank to undertake corrective action with respect to any release of petroleum when the Administrator (or the State) determines that such corrective action will be done properly and promptly by the owner or operator of the underground storage tank from which the release occurs; or

(B) undertake corrective action with respect to any release of petroleum into the environment from an underground storage tank if such action is necessary, in the judgment of the Administrator (or the State), to protect human health and the environment.

The corrective action undertaken or required under this paragraph shall be such as may be necessary to protect human health and the environment. The Administrator shall use funds in the Leaking Underground Storage Tank Trust Fund for payment of costs incurred for corrective action under subparagraph (B), enforcement action under subparagraph (A), and cost recovery under paragraph (6) of this subsection. Subject to the priority requirements of paragraph (3), the Administrator (or the State) shall give priority in undertaking such actions under subparagraph (B) to cases where the Administrator (or the State) cannot identify a solvent owner or operator of the tank who will undertake action properly.

(2) After regulations

Following the effective date of regulations under subsection (c) of this section, all actions or orders of the Administrator (or a State pursuant to paragraph (7)) described in paragraph (1) of this subsection shall be in conformity with such regulations. Following such effective date, the Administrator (or the State) may undertake corrective action with respect to any release of petroleum into the environment from an underground storage tank only if such action is necessary, in the judgment of the Administrator (or the State), to protect human health and the environment and one or more of the following situations exists:

(A) No person can be found, within 90 days or such shorter period as may be necessary to protect human health and the environment, who is—

(i) an owner or operator of the tank concerned,

(ii) subject to such corrective action regulations, and

(iii) capable of carrying out such corrective action properly.

(B) A situation exists which requires prompt action by the Administrator (or the State) under this paragraph to protect human health and the environment.

(C) Corrective action costs at a facility exceed the amount of coverage required by the Administrator pursuant to the provisions of subsections (c) and (d)(5) of this section and, considering the class or category of underground storage tank from which the release occurred, expenditures from the Leaking Underground Storage Tank Trust Fund are necessary to assure an effective corrective action.

(D) The owner or operator of the tank has failed or refused to comply with an order of the Administrator under this subsection or section 6991e of this title or with the order of a State under this subsection to comply with the corrective action regulations.

(3) Priority of corrective actions

The Administrator (or a State pursuant to paragraph (7)) shall give priority in undertaking corrective actions under this subsection, and in issuing orders requiring owners or operators to undertake such actions, to releases of petroleum from underground storage tanks which pose the greatest threat to human health and the environment.

(4) Corrective action orders

The Administrator is authorized to issue orders to the owner or operator of an underground storage tank to carry out subparagraph (A) of paragraph (1) or to carry out regulations issued under subsection (c)(4) of this section. A State acting pursuant to paragraph (7) of this subsection is authorized to carry out subparagraph (A) of paragraph (1) only until the State's program is approved by the Administrator under section 6991c of this title. Such orders shall be issued and enforced in the same manner and subject to the same requirements as orders under section 6991e of this title.

(5) Allowable corrective actions

The corrective actions undertaken by the Administrator (or a State pursuant to paragraph (7)) under paragraph (1) or (2) may include temporary or permanent relocation of residents and alternative household water supplies. In connection with the performance of any corrective action under paragraph (1) or (2), the Administrator may undertake an exposure assessment as defined in paragraph (10) of this subsection or provide for such an assessment in a cooperative agreement with a State pursuant to paragraph (7) of this subsection. The costs of any such assessment may be treated as corrective action for purposes of paragraph (6), relating to cost recovery.

(6) Recovery of costs

(A) In general

Whenever costs have been incurred by the Administrator, or by a State pursuant to paragraph (7), for undertaking corrective action or enforcement action with respect to the release of petroleum from an underground storage tank, the owner or operator of such tank shall be liable to the Administrator or the State for such costs. The liability under this paragraph shall be construed to be the standard of liability which obtains under section 311 of the Federal Water Pollution Control Act.

(B) Recovery

In determining the equities for seeking the recovery of costs under subparagraph (A), the Administrator (or a State pursuant to paragraph (7) of this subsection) may consider the amount of financial responsibility required to be maintained under subsections (c) and (d)(5) of this section and the factors considered in establishing such amount under subsection (d)(5) of this section.

(C) Effect on liability

(i) No transfers of liability

No indemnification, hold harmless, or similar agreement or conveyance shall be effective to transfer from the owner or operator of any underground storage tank or from any person who may be liable for a release or threat of release under this subsection, to any other person the liability imposed under this subsection. Nothing in this subsection shall bar any agreement to insure, hold harmless, or indemnify a party to such agreement for any liability under this section.

(ii) No bar to cause of action

Nothing in this subsection, including the provisions of clause (i) of this subparagraph, shall bar a cause of action that an owner or operator or any other person subject to liability under this section, or a guarantor, has or would have, by reason of subrogation or otherwise against any person.

→ Agricultural Excess or Surplus Insurance vs. ABD 878 F. Supp 1091 (1995)

(D) Facility

For purposes of this paragraph, the term "facility" means, with respect to any owner or operator, all underground storage tanks used for the storage of petroleum which are owned or operated by such owner or operator and located on a single parcel of property (or on any contiguous or adjacent property).

(7) State authorities

(A) General

A State may exercise the authorities in paragraphs (1) and (2) of this subsection, subject to the terms and conditions of paragraphs (3), (5), (9), (10), and (11), and including the authorities of paragraphs (4), (6), and (8) of this subsection if—

(i) the Administrator determines that the State has the capabilities to carry out effective corrective actions and enforcement activities; and

(ii) the Administrator enters into a cooperative agreement with the State setting out the actions to be undertaken by the State.

The Administrator may provide funds from the Leaking Underground Storage Tank Trust Fund for the reasonable costs of the State's actions under the cooperative agreement.

(B) Cost share

Following the effective date of the regulations under subsection (c) of this section, the State shall pay 10 per centum of the cost of corrective actions undertaken either by the Administrator or by the State under a cooperative agreement, except that the Administrator may take corrective action at a facility where immediate action is necessary to respond to an imminent and substantial endangerment to human health or the environment if the State fails to pay the cost share.

(8) Emergency procurement powers

Notwithstanding any other provision of law, the Administrator may authorize the use of such emergency procurement powers as he deems necessary.

(9) Definition of owner

As used in this subsection, the term "owner" does not include any person who, without participating in the management of an underground storage tank and otherwise not engaged in petroleum production, refining, and marketing, holds indicia of ownership primarily to protect the owner's security interest in the tank.

(10) Definition of exposure assessment

As used in this subsection, the term "exposure assessment" means an assessment to determine the extent of exposure of, or potential for exposure of, individuals to petroleum from a release from an underground storage tank based on such factors as the nature and extent of contamination and the existence of or potential for pathways of human exposure (including ground or surface water contamination, air emissions, and food chain contamination), the size of the community within the likely pathways of exposure, and the comparison of expected human exposure levels to the short-term and long-term health effects associated with identified contaminants and any available recommended exposure or tolerance limits for such contaminants. Such assessment shall not delay corrective action to abate immediate hazards or reduce exposure.

(11) Facilities without financial responsibility

At any facility where the owner or operator has failed to maintain evidence of financial responsibility in amounts at least equal to the amounts established by subsection (d)(5)(A) of this section (or a lesser amount if such amount is applicable to such facility as a result of subsection (d)(5)(B) of this section) for whatever reason the Administrator shall expend no monies from the Leaking Underground Storage Tank Trust Fund to clean up releases at such facility pursuant to the provisions of paragraph (1) or (2) of this subsection. At such facilities the Administrator shall use the authorities provided in subparagraph (A) of paragraph (1) and paragraph (4) of this subsection and section 6991e of this title to order corrective action to clean up such releases. States acting pursuant to paragraph (7) of this subsection shall use the authorities provided in subparagraph (A) of paragraph (1) and paragraph (4 of this subsection to order corrective action to clean up such releases. Notwithstanding the provisions of this paragraph, the Administrator may use monies from the fund to take the corrective actions authorized by paragraph (5) of this subsection to protect human health at such facilities and shall seek full recovery of the costs of all such actions pursuant to the provisions of paragraph (6)(A) of this subsection and without consideration of the factors in paragraph (6)(B) of this subsection. Nothing in this paragraph shall prevent the Administrator (or a State pursuant to paragraph (7) of this subsection) from taking corrective action at a facility where there is no solvent owner or operator or where immediate action is necessary to respond to an imminent and substantial endangerment of human health or the environment.

(Pub.L. 89–272, Title II, § 9003, as added Pub.L. 98–616, Title VI, § 601(a), Nov. 8, 1984, 98 Stat. 3279, and amended Pub.L. 99–499, Title II, § 205(c), (d), Oct. 17, 1986, 100 Stat. 1697, 1698.)

§ 6991c. Approval of State programs

(a) Elements of State programs

Beginning 30 months after November 8, 1984, any State may,[1] submit an underground storage tank release detection, prevention, and correction program for review and approval by the Administrator. The program may cover tanks used to store regulated substances referred to in 6991(2)(A) or (B) of this title or both. A State program may be approved by the Administrator under this section only if the State demonstrates that the State program includes the following requirements and standards and provides for adequate enforcement of compliance with such requirements and standards—

(1) requirements for maintaining a leak detection system, an inventory control system together with tank testing, or a comparable system or method designed to identify releases in a manner consistent with the protection of human health and the environment;

(2) requirements for maintaining records of any monitoring or leak detection system or inventory control system or tank testing system;

(3) requirements for reporting of any releases and corrective action taken in response to a release from an underground storage tank;

(4) requirements for taking corrective action in response to a release from an underground storage tank;

(5) requirements for the closure of tanks to prevent future releases of regulated substances into the environment;

(6) requirements for maintaining evidence of financial responsibility for taking corrective action and compensating third parties for bodily injury and property damage caused by sudden and nonsudden accidental releases arising from operating an underground storage tank;

(7) standards of performance for new underground storage tanks; and

(8) requirements—

(A) for notifying the appropriate State agency or department (or local agency or department) designated according to section 6991a(b)(1) of this title of the existence of any operational or non-operational underground storage tank; and

(B) for providing the information required on the form issued pursuant to section 6991a(b)(2) of this title.

(b) Federal standards

(1) A State program submitted under this section may be approved only if the requirements under paragraphs (1) through (7) of subsection (a) of this section are no less stringent than the corresponding requirements standards promulgated by the Administrator pursuant to section 6991b(a) of this title.

(2)(A) A State program may be approved without regard to whether or not the requirements referred to in paragraphs (1), (2), (3), and (5) of subsection (a) of this section are less stringent than the corresponding standards under section 6991b(a) of this title during the one-year period commencing on the date of promulgation of regulations under section 6991b(a) of this title if State regulatory action but no State legislative action is required in order to adopt a State program.

(B) If such State legislative action is required, the State program may be approved without regard to whether or not the requirements referred to in paragraphs (1), (2), (3), and (5) of subsection (a) of this section are less stringent than the corresponding standards under section 6991b(a) of this title during the two-year period commencing on the date of promulgation of regulations under section 6991b(a) of this title (and during an additional one-year period after such legislative action if regulations are required to be promulgated by the State pursuant to such legislative action).

(c) Financial responsibility

(1) Corrective action and compensation programs administered by State or local agencies or departments may be submitted for approval under subsection (a)(6) of this section as evidence of financial responsibility.

(2) Financial responsibility required by this subsection may be established in accordance with regulations promulgated by the Administrator by any one, or any combination, of the following: insurance, guarantee, surety bond, letter of credit, qualification as a self-insurer or any other method satisfactory to the Administrator. In promulgating requirements under this subsection, the Administrator is authorized to specify policy or other contractual terms including the amount of coverage required for various classes and categories of underground storage tanks pursuant to section 6991b(d)(5) of this title, conditions, or defenses which are necessary or are unacceptable in establishing such evidence of financial responsibility in order to effectuate the purposes of this subchapter.

(3) In any case where the owner or operator is in bankruptcy, reorganization, or arrangement pursuant to the Federal Bankruptcy Code or where with reasonable diligence jurisdiction in any State court of the Federal courts cannot be obtained over an owner or operator likely to be solvent at the time of judgment, any claim arising from conduct for which evidence of financial responsibility must be provided under this subsection may be asserted directly against the guarantor providing such evidence of financial responsibility. In the case of any action pursuant to this paragraph such guarantor shall be entitled to invoke all rights and defenses which would have been available to the owner or operator if any action had been brought against the owner or operator by the claimant and which would have been available to the guarantor if an action had been brought against the guarantor by the owner or operator.

(4) The total liability of any guarantor shall be limited to the aggregate amount which the guarantor has provided as evidence of financial responsibility to the owner or operator under this section. Nothing in this subsection shall be construed to limit any other State or Federal statutory, contractual or common law liability of a guarantor to its owner or operator including, but not limited to, the liability of such guarantor for bad faith either in negotiating or in failing to negotiate the settlement of any claim. Nothing in this subsection shall be construed to diminish the liability of any person under section 9607 or 9611 of this title or other applicable law.

(5) For the purpose of this subsection, the term "guarantor" means any person, other than the owner or operator, who provides evidence of financial responsibility for an owner or operator under this subsection.

(d) EPA determination

(1) Within one hundred and eighty days of the date of receipt of a proposed State program, the Administrator shall, after notice and opportunity for public comment, make a determination whether the State's program complies with the provisions of this section and provides for adequate enforcement of compliance with the requirements and standards adopted pursuant to this section.

(2) If the Administrator determines that a State program complies with the provisions of this section and provides for adequate enforcement of compliance with the requirements and standards adopted pursuant to this section, he shall approve the State program in lieu of the Federal program and the State shall have primary enforcement responsibility with respect to requirements of its program.

(e) Withdrawal of authorization

Whenever the Administrator determines after public hearing that a State is not administering and enforcing a program authorized under this subchapter in accordance with the provisions of this section, he shall so notify the State. If appropriate action is not taken within a reasonable time, not to exceed one hundred and twenty days after such notification, the Administrator shall withdraw approval of such program and reestablish the Federal program pursuant to this subchapter.

(Pub.L. 89–272, Title II, § 9004, as added Pub.L. 98–616, Title VI, § 601(a), Nov. 8, 1984, 98 Stat. 3282, and amended Pub.L. 99–499, Title II, § 205(e), Oct. 17, 1986, 100 Stat. 1702.)

[1] So in original.

§ 6991d. Inspections, monitoring, testing and corrective action

(a) Furnishing information

For the purposes of developing or assisting in the development of any regulation, conducting any study taking any corrective action, or enforcing the provisions of this subchapter, any owner or operator of an underground storage tank (or any tank subject to study under section 6991h of this title that is used for storing regulated substances) shall, upon request of any officer, employee or representative of the Environmental Protection Agency, duly designated by the Administrator, or upon request of any duly designated officer, employee, or representative of a State acting pursuant to subsection (h)(7) of section 6991b of this title or with an approved program, furnish information relating to such tanks, their associated equipment, their contents, conduct monitoring or testing, permit such officer at all reasonable times to have access to, and to copy all records relating to such tanks and permit such officer to have access for corrective action. For the purposes of developing or assisting in the development of any regulation, conducting any study, taking corrective action, or enforcing the provisions of this subchapter, such officers, employees, or representatives are authorized—

(1) to enter at reasonable times any establishment or other place where an underground storage tank is located;

(2) to inspect and obtain samples from any person of any regulated substances contained in such tank;

(3) to conduct monitoring or testing of the tanks, associated equipment, contents, or surrounding soils, air, surface water or ground water; and

(4) to take corrective action.

Each such inspection shall be commenced and completed with reasonable promptness.

(b) Confidentiality

(1) Any records, reports, or information obtained from any persons under this section shall be available to the public, except that upon a showing satisfactory to the Administrator (or the State, as the case may be) by any person that records, reports, or information, or a particular part thereof, to which the Administrator (or the State, as the case may be) or any officer, employee, or representative thereof has access under this section if made public, would divulge information entitled to protection under section 1905 of Title 18, such information or particular portion thereof shall be considered confidential in accordance with the purposes of that section, except that such record, report, document, or information may be disclosed to other officers, employees, or authorized representatives of the United States concerned with carrying out this chapter, or when relevent [1] in any proceeding under this chapter.

(2) Any person not subject to the provisions of section 1905 of Title 18 who knowingly and willfully divulges or discloses any information entitled to protection under this subsection shall, upon conviction, be subject to a fine of not more than $5,000 or to imprisonment not to exceed one year, or both.

(3) In submitting data under this subchapter, a person required to provide such data may—

(A) designate the data which such person believes is entitled to protection under this subsection, and

(B) submit such designated data separately from other data submitted under this subchapter.

A designation under this paragraph shall be made in writing and in such manner as the Administrator may prescribe.

(4) Notwithstanding any limitation contained in this section or any other provision of law, all information reported to, or otherwise obtained, by the Administrator (or any representative of the Administrator) under this chapter shall be made available, upon

written request of any duly authorized committee of the Congress, to such committee (including records, reports, or information obtained by representatives of the Environmental Protection Agency).

(Pub.L. 89–272, Title II, § 9005, as added Pub.L. 98–616, Title VI, § 601(a), Nov. 8, 1984, 98 Stat. 3284, and amended Pub.L. 99–499, Title II, § 205(f), Oct. 17, 1986, 100 Stat. 1702.)

1 So in original. Probably should be relevant.

§ 6991e. Federal enforcement

(a) Compliance orders

(1) Except as provided in paragraph (2), whenever on the basis of any information, the Administrator determines that any person is in violation of any requirement of this subchapter, the Administrator may issue an order requiring compliance within a reasonable specified time period or the Administrator may commence a civil action in the United States district court in which the violation occurred for appropriate relief, including a temporary or permanent injunction.

(2) In the case of a violation of any requirement of this subchapter where such violation occurs in a State with a program approved under section 6991c of this title the Administrator shall give notice to the State in which such violation has occurred prior to issuing an order or commencing a civil action under this section.

(3) If a violator fails to comply with an order under this subsection within the time specified in the order, he shall be liable for a civil penalty of not more than $25,000 for each day of continued noncompliance.

(b) Procedure

Any order issued under this section shall become final unless, no later than thirty days after the order is served, the person or persons named therein request a public hearing. Upon such request the Administrator shall promptly conduct a public hearing. In connection with any proceeding under this section the Administrator may issue subpoenas for the attendance and testimony of witnesses and the production of relevant papers, books, and documents, and may promulgate rules for discovery procedures.

(c) Contents of order

Any order issued under this section shall state with reasonable specificity the nature of the violation, specify a reasonable time for compliance, and assess a penalty, if any, which the Administrator determines is reasonable taking into account the seriousness of the violation and any good faith efforts to comply with the applicable requirements.

(d) Civil penalties

(1) Any owner who knowingly fails to notify or submits false information pursuant to section 6991a(a) of this title shall be subject to a civil penalty not to exceed $10,000 for each tank for which notification is not given or false information is submitted.

(2) Any owner or operator of an underground storage tank who fails to comply with—

(A) any requirement or standard promulgated by the Administrator under section 6991b of this title;

(B) any requirement or standard of a State program approved pursuant to section 6991c of this title; or

(C) the provisions of section 6991b(g) of this title (entitled "Interim Prohibition")

shall be subject to a civil penalty not to exceed $10,000 for each tank for each day of violation.

(Pub.L. 89-272, Title II, § 9006, as added Pub.L. 98-616, Title VI, § 601(a), Nov. 8, 1984, 98 Stat. 3286.)

Environmental Management, N.D.Ind.1990, 742 F.Supp. 1442.

§ 6991g. State authority

Nothing in this subchapter shall preclude or deny any right of any State or political subdivision thereof to adopt or enforce any regulation, requirement, or standard of performance respecting underground storage tanks that is more stringent than a regulation, requirement, or standard of performance in effect under this subchapter or to impose any additional liability with respect to the release of regulated substances within such State or political subdivision.

(Pub.L. 89-272, Title II, § 9008, as added Pub.L. 98-616, Title VI, § 601(a), Nov. 8, 1984, 98 Stat. 3286, and amended Pub.L. 99-499, Title II, § 205(g), Oct. 17, 1986, 100 Stat. 1702.)

§ 6991h. Study of underground storage tanks

(a) Petroleum tanks

Not later than twelve months after November 8, 1984, the Administrator shall complete a study of underground storage tanks used for the storage of regulated substances defined in section 6991(2)(B) of this title.

(b) Other tanks

Not later than thirty-six months after November 8, 1984, the Administrator shall complete a study of all other underground storage tanks.

(c) Elements of studies

The studies under subsections (a) and (b) of this section shall include an assessment of the ages, types (including methods of manufacture, coatings, protection systems, the compatibility of the construction materials and the installation methods) and locations (including the climate of the locations) of such tanks; soil conditions, water tables, and the hydrogeology of tank locations; the relationship between the foregoing factors and the likelihood of releases from underground storage tanks; the effectiveness and costs of inventory systems, tank testing, and leak detection systems; and such other factors as the Administrator deems appropriate.

(d) Farm and heating oil tanks

Not later than thirty-six months after November 8, 1984, the Administrator shall conduct a study regarding the tanks referred to in section 6991(1)(A) and (B) of this title. Such study shall include estimates of the number and location of such tanks and an analysis of the extent to which there may be releases or threatened releases from such tanks into the environment.

(e) Reports

Upon completion of the studies authorized by this section, the Administrator shall submit reports to the President and to the Congress containing the results of the studies and recommendations respecting whether or not such tanks should be subject to the preceding provisions of this subchapter.

(f) Reimbursement

(1) If any owner or operator (excepting an agency, department, or instrumentality of the United States Government, a State or a political subdivision thereof) shall incur costs, including the loss of business opportunity, due to the closure or interruption of operation of an underground storage tank solely for the purpose of conducting studies authorized by this section, the Administrator shall provide such person fair and equitable reimbursement for such costs.

(2) All claims for reimbursement shall be filed with the Administrator not later than ninety days after the closure or interruption which gives rise to the claim.

(3) Reimbursements made under this section shall be from funds appropriated by the Congress pursuant to the authorization contained in section 6916(g) of this title.

(4) For purposes of judicial review, a determination by the Administrator under this subsection shall be considered final agency action.

(Pub.L. 89–272, Title II, § 9009, as added Pub.L. 98–616, Title VI, § 601(a), Nov. 8, 1984, 98 Stat. 3287.)

§ 6991i. Authorization of appropriations

For authorization of appropriations to carry out this subchapter, see section 6916(g) of this title.

(Pub.L. 89–272, Title II, § 9010, as added Pub.L. 98–616, Title VI, § 601(a), Nov. 8, 1984, 98 Stat. 3287.)

APPENDIX B

Codified Regulations on Underground Storage Tanks

(40 C.F.R. Part 280 (1993))

PART 280—TECHNICAL STANDARDS AND CORRECTIVE ACTION REQUIREMENTS FOR OWNERS AND OPERATORS OF UNDERGROUND STORAGE TANKS (UST)

Subpart A—Program Scope and Interim Prohibition

Subpart B—UST Systems: Design, Construction, Installation and Notification

Subpart C—General Operating Requirements

Subpart D—Release Detection

Subpart E—Release Reporting, Investigation, and Confirmation

Subpart F—Release Response and Corrective Action for UST Systems Containing Petroleum or Hazardous Substances

Subpart G—Out-of-Service UST Systems and Closure

Subpart H—Financial Responsibility

APPENDICES TO PART 280

AUTHORITY: 42 U.S.C. 6912, 6991, 6991a, 6991b, 6991c, 6991d, 6991e, 6991f, and 6991h.

SOURCE: 53 FR 37194, Sept. 23, 1988, unless otherwise noted.

EDITORIAL NOTE: For a document removing the OMB control number wherever it appeared in part 280, see 58 FR 34370, June 25, 1993.

Subpart A—Program Scope and Interim Prohibition

§ 280.10 Applicability.

(a) The requirements of this part apply to all owners and operators of an UST system as defined in § 280.12 except as otherwise provided in paragraphs (b), (c), and (d) of this section. Any UST system listed in paragraph (c) of this section must meet the requirements of § 280.11.

(b) The following UST systems are excluded from the requirements of this part:

(1) Any UST system holding hazardous wastes listed or identified under Subtitle C of the Solid Waste Disposal Act, or a mixture of such hazardous waste and other regulated substances.

(2) Any wastewater treatment tank system that is part of a wastewater treatment facility regulated under section 402 or 307(b) of the Clean Water Act.

(3) Equipment or machinery that contains regulated substances for operational purposes such as hydraulic lift tanks and electrical equipment tanks.

(4) Any UST system whose capacity is 110 gallons or less.

(5) Any UST system that contains a *de minimis* concentration of regulated substances.

(6) Any emergency spill or overflow containment UST system that is expeditiously emptied after use.

(c) *Deferrals.* Subparts B, C, D, E, and G do not apply to any of the following types of UST systems:

(1) Wastewater treatment tank systems;

(2) Any UST systems containing radioactive material that are regulated under the Atomic Energy Act of 1954 (42 U.S.C. 2011 and following);

(3) Any UST system that is part of an emergency generator system at nuclear power generation facilities regulated by the Nuclear Regulatory Commission under 10 CFR part 50, appendix A;

(4) Airport hydrant fuel distribution systems; and

(5) UST systems with field-constructed tanks.

(d) *Deferrals.* Subpart D does not apply to any UST system that stores fuel solely for use by emergency power generators.

§ 280.11 Interim prohibition for deferred UST systems.

(a) No person may install an UST system listed in § 280.10(c) for the purpose of storing regulated substances unless the UST system (whether of single- or double-wall construction):

(1) Will prevent releases due to corrosion or structural failure for the operational life of the UST system;

(2) Is cathodically protected against corrosion, constructed of noncorrodible material, steel clad with a noncorrodible material, or designed in a manner to prevent the release or threatened release of any stored substance; and

(3) Is constructed or lined with material that is compatible with the stored substance.

(b) Notwithstanding paragraph (a) of this section, an UST system without corrosion protection may be installed at a site that is determined by a corrosion expert not to be corrosive enough to cause it to have a release due to corrosion during its operating life. Owners and operators must maintain records that demonstrate compliance with the requirements of this paragraph for the remaining life of the tank.

NOTE: The National Association of Corrosion Engineers Standard RP-02-85, "Control of External Corrosion on Metallic Buried, Partially Buried, or Submerged Liquid Storage Systems," may be used as guidance for complying with paragraph (b) of this section.

§ 280.12 Definitions.

Aboveground release means any release to the surface of the land or to surface water. This includes, but is not limited to, releases from the aboveground portion of an UST system and aboveground releases associated with overfills and transfer operations as the regulated substance moves to or from an UST system.

Ancillary equipment means any devices including, but not limited

API → (202) 682-8000
www.API.org

such devices as piping, fittings, flanges, valves, and pumps used to distribute, meter, or control the flow of regulated substances to and from an UST.

Belowground release means any release to the subsurface of the land and to ground water. This includes, but is not limited to, releases from the belowground portions of an underground storage tank system and belowground releases associated with overfills and transfer operations as the regulated substance moves to or from an underground storage tank.

Beneath the surface of the ground means beneath the ground surface or otherwise covered with earthen materials.

Cathodic protection is a technique to prevent corrosion of a metal surface by making that surface the cathode of an electrochemical cell. For example, a tank system can be cathodically protected through the application of either galvanic anodes or impressed current.

Cathodic protection tester means a person who can demonstrate an understanding of the principles and measurements of all common types of cathodic protection systems as applied to buried or submerged metal piping and tank systems. At a minimum, such persons must have education and experience in soil resistivity, stray current, structure-to-soil potential, and component electrical isolation measurements of buried metal piping and tank systems.

CERCLA means the Comprehensive Environmental Response, Compensation, and Liability Act of 1980, as amended.

Compatible means the ability of two or more substances to maintain their respective physical and chemical properties upon contact with one another for the design life of the tank system under conditions likely to be encountered in the UST.

Connected piping means all underground piping including valves, elbows, joints, flanges, and flexible connectors attached to a tank system through which regulated substances flow. For the purpose of determining how much piping is connected to any individual UST system, the piping that joins two UST systems should be allocated equally between them.

Consumptive use with respect to heating oil means consumed on the premises.

Corrosion expert means a person who, by reason of thorough knowledge of the physical sciences and the principles of engineering and mathematics acquired by a professional education and related practical experience, is qualified to engage in the practice of corrosion control on buried or submerged metal piping systems and metal tanks. Such a person must be accredited or certified as being qualified by the National Association of Corrosion Engineers or be a registered professional engineer who has certification or licensing that includes education and experience in corrosion control of buried or submerged metal piping systems and metal tanks.

Dielectric material means a material that does not conduct direct electrical current. Dielectric coatings are used to electrically isolate UST systems from the surrounding soils. Dielectric bushings are used to electrically isolate portions of the UST system (e.g., tank from piping).

Electrical equipment means underground equipment that contains dielectric fluid that is necessary for the operation of equipment such as transformers and buried electrical cable.

Excavation zone means the volume containing the tank system and backfill material bounded by the ground surface, walls, and floor of the pit and trenches into which the UST system is placed at the time of installation.

Existing tank system means a tank system used to contain an accumulation of regulated substances or for which installation has commenced on or before December 22, 1988. Installation is considered to have commenced if:

(a) The owner or operator has obtained all federal, state, and local approvals or permits necessary to begin physical construction of the site or installation of the tank system; and if,

(b)(1) Either a continuous on-site physical construction or installation program has begun; or,

(2) The owner or operator has entered into contractual obligations—which cannot be cancelled or modified without substantial loss—for physical construction at the site or installation of

the tank system to be completed within a reasonable time.

Farm tank is a tank located on a tract of land devoted to the production of crops or raising animals, including fish, and associated residences and improvements. A farm tank must be located on the farm property. "Farm" includes fish hatcheries, rangeland and nurseries with growing operations.

Flow-through process tank is a tank that forms an integral part of a production process through which there is a steady, variable, recurring, or intermittent flow of materials during the operation of the process. Flow-through process tanks do not include tanks used for the storage of materials prior to their introduction into the production process or for the storage of finished products or by-products from the production process.

Free product refers to a regulated substance that is present as a non-aqueous phase liquid (e.g., liquid not dissolved in water.)

Gathering lines means any pipeline, equipment, facility, or building used in the transportation of oil or gas during oil or gas production or gathering operations.

Hazardous substance UST system means an underground storage tank system that contains a hazardous substance defined in section 101(14) of the Comprehensive Environmental Response, Compensation and Liability Act of 1980 (but not including any substance regulated as a hazardous waste under subtitle C) or any mixture of such substances and petroleum, and which is not a petroleum UST system.

Heating oil means petroleum that is No. 1, No. 2, No. 4—light, No. 4—heavy, No. 5—light, No. 5—heavy, and No. 6 technical grades of fuel oil; other residual fuel oils (including Navy Special Fuel Oil and Bunker C); and other fuels when used as substitutes for one of these fuel oils. Heating oil is typically used in the operation of heating equipment, boilers, or furnaces.

Hydraulic lift tank means a tank holding hydraulic fluid for a closed-loop mechanical system that uses compressed air or hydraulic fluid to operate lifts, elevators, and other similar devices.

Implementing agency means EPA, or, in the case of a state with a program approved under section 9004 (or pursuant to a memorandum of agreement with EPA), the designated state or local agency responsible for carrying out an approved UST program.

Liquid trap means sumps, well cellars, and other traps used in association with oil and gas production, gathering, and extraction operations (including gas production plants), for the purpose of collecting oil, water, and other liquids. These liquid traps may temporarily collect liquids for subsequent disposition or reinjection into a production or pipeline stream, or may collect and separate liquids from a gas stream.

Maintenance means the normal operational upkeep to prevent an underground storage tank system from releasing product.

Motor fuel means petroleum or a petroleum-based substance that is motor gasoline, aviation gasoline, No. 1 or No. 2 diesel fuel, or any grade of gasohol, and is typically used in the operation of a motor engine.

New tank system means a tank system that will be used to contain an accumulation of regulated substances and for which installation has commenced after December 22, 1988. (See also "Existing Tank System.")

Noncommercial purposes with respect to motor fuel means not for resale.

On the premises where stored with respect to heating oil means UST systems located on the same property where the stored heating oil is used.

Operational life refers to the period beginning when installation of the tank system has commenced until the time the tank system is properly closed under Subpart G.

Operator means any person in control of, or having responsibility for, the daily operation of the UST system.

Overfill release is a release that occurs when a tank is filled beyond its capacity, resulting in a discharge of the regulated substance to the environment.

Owner means:

(a) In the case of an UST system in use on November 8, 1984, or brought into use after that date, any person who owns an UST system used for storage, use, or dispensing of regulated substances; and

(b) In the case of any UST system in use before November 8, 1984, but no longer in use on that date, any person who owned such UST immediately before the discontinuation of its use.

Person means an individual, trust, firm, joint stock company, Federal agency, corporation, state, municipality, commission, political subdivision of a state, or any interstate body. "Person" also includes a consortium, a joint venture, a commercial entity, and the United States Government.

Petroleum UST system means an underground storage tank system that contains petroleum or a mixture of petroleum with *de minimis* quantities of other regulated substances. Such systems include those containing motor fuels, jet fuels, distillate fuel oils, residual fuel oils, lubricants, petroleum solvents, and used oils.

Pipe or *Piping* means a hollow cylinder or tubular conduit that is constructed of non-earthen materials.

Pipeline facilities (including gathering lines) are new and existing pipe rights-of-way and any associated equipment, facilities, or buildings.

Regulated substance means:

(a) Any substance defined in section 101(14) of the Comprehensive Environmental Response, Compensation and Liability Act (CERCLA) of 1980 (but not including any substance regulated as a hazardous waste under subtitle C), and

(b) Petroleum, including crude oil or any fraction thereof that is liquid at standard conditions of temperature and pressure (60 degrees Fahrenheit and 14.7 pounds per square inch absolute).

The term "regulated substance" includes but is not limited to petroleum and petroleum-based substances comprised of a complex blend of hydrocarbons derived from crude oil though processes of separation, conversion, upgrading, and finishing, such as motor fuels, jet fuels, distillate fuel oils, residual fuel oils, lubricants, petroleum solvents, and used oils.

Release means any spilling, leaking, emitting, discharging, escaping, leaching or disposing from an UST into ground water, surface water or subsurface soils.

Release detection means determining whether a release of a regulated substance has occurred from the UST system into the environment or into the interstitial space between the UST system and its secondary barrier or secondary containment around it.

Repair means to restore a tank or UST system component that has caused a release of product from the UST system.

Residential tank is a tank located on property used primarily for dwelling purposes.

SARA means the Superfund Amendments and Reauthorization Act of 1986.

Septic tank is a water-tight covered receptacle designed to receive or process, through liquid separation or biological digestion, the sewage discharged from a building sewer. The effluent from such receptacle is distributed for disposal through the soil and settled solids and scum from the tank are pumped out periodically and hauled to a treatment facility.

Storm-water or wastewater collection system means piping, pumps, conduits, and any other equipment necessary to collect and transport the flow of surface water run-off resulting from precipitation, or domestic, commercial, or industrial wastewater to and from retention areas or any areas where treatment is designated to occur. The collection of storm water and wastewater does not include treatment except where incidental to conveyance.

Surface impoundment is a natural topographic depression, man-made excavation, or diked area formed primarily of earthen materials (although it may be lined with man-made materials) that is not an injection well.

Tank is a stationary device designed to contain an accumulation of regulated substances and constructed of non-earthen materials (e.g., concrete, steel, plastic) that provide structural support.

Underground area means an underground room, such as a basement, cellar, shaft or vault, providing enough space for physical inspection of the exterior of the tank situated on or above the surface of the floor.

Underground release means any belowground release.

341-150 O—93——33

Underground storage tank or *UST* means any one or combination of tanks (including underground pipes connected thereto) that is used to contain an accumulation of regulated substances, and the volume of which (including the volume of underground pipes connected thereto) is 10 percent or more beneath the surface of the ground. This term does not include any:

(a) Farm or residential tank of 1,100 gallons or less capacity used for storing motor fuel for noncommercial purposes;

(b) Tank used for storing heating oil for consumptive use on the premises where stored;

(c) Septic tank;

(d) Pipeline facility (including gathering lines) regulated under:

(1) The Natural Gas Pipeline Safety Act of 1968 (49 U.S.C. App. 1671, *et seq.*), or

(2) The Hazardous Liquid Pipeline Safety Act of 1979 (49 U.S.C. App. 2001, *et seq.*), or

(3) Which is an intrastate pipeline facility regulated under state laws comparable to the provisions of the law referred to in paragraph (d)(1) or (d)(2) of this definition;

(e) Surface impoundment, pit, pond, or lagoon;

(f) Storm-water or wastewater collection system;

(g) Flow-through process tank;

(h) Liquid trap or associated gathering lines directly related to oil or gas production and gathering operations; or

(i) Storage tank situated in an underground area (such as a basement, cellar, mineworking, drift, shaft, or tunnel) if the storage tank is situated upon or above the surface of the floor.

The term "underground storage tank" or "UST" does not include any pipes connected to any tank which is described in paragraphs (a) through (i) of this definition.

Upgrade means the addition or retrofit of some systems such as cathodic protection, lining, or spill and overfill controls to improve the ability of an underground storage tank system to prevent the release of product.

UST system or *Tank system* means an underground storage tank, connected underground piping, underground ancillary equipment, and containment system, if any.

Wastewater treatment tank means a tank that is designed to receive and treat an influent wastewater through physical, chemical, or biological methods.

Subpart B—UST Systems: Design, Construction, Installation and Notification

§ 280.20 Performance standards for new UST systems.

In order to prevent releases due to structural failure, corrosion, or spills and overfills for as long as the UST system is used to store regulated substances, all owners and operators of new UST systems must meet the following requirements.

(a) *Tanks.* Each tank must be properly designed and constructed, and any portion underground that routinely contains product must be protected from corrosion, in accordance with a code of practice developed by a nationally recognized association or independent testing laboratory as specified below:

(1) The tank is constructed of fiberglass-reinforced plastic; or

NOTE: The following industry codes may be used to comply with paragraph (a)(1) of this section: Underwriters Laboratories Standard 1316, "Standard for Glass- Fiber-Reinforced Plastic Underground Storage Tanks for Petroleum Products"; Underwriter's Laboratories of Canada CAN4-S615-M83, "Standard for Reinforced Plastic Underground Tanks for Petroleum Products"; or American Society of Testing and Materials Standard D4021-86, "Standard Specification for Glass-Fiber-Reinforced Polyester Underground Petroleum Storage Tanks."

(2) The tank is constructed of steel and cathodically protected in the following manner:

(i) The tank is coated with a suitable dielectric material;

(ii) Field-installed cathodic protection systems are designed by a corrosion expert;

(iii) Impressed current systems are designed to allow determination of current operating status as required in § 280.31(c); and

280

112.7(e)(2)

(iv) Cathodic protection systems are operated and maintained in accordance with § 280.31 or according to guidelines established by the implementing agency; or

NOTE: The following codes and standards may be used to comply with paragraph (a)(2) of this section:

(A) Steel Tank Institute "Specification for STI-P3 System of External Corrosion Protection of Underground Steel Storage Tanks";

(B) Underwriters Laboratories Standard 1746, "Corrosion Protection Systems for Underground Storage Tanks";

(C) Underwriters Laboratories of Canada CAN4-S603-M85, "Standard for Steel Underground Tanks for Flammable and Combustible Liquids," and CAN4-G03.1-M85, "Standard for Galvanic Corrosion Protection Systems for Underground Tanks for Flammable and Combustible Liquids," and CAN4-S631-M84, "Isolating Bushings for Steel Underground Tanks Protected with Coatings and Galvanic Systems"; or

(D) National Association of Corrosion Engineers Standard RP-02-85, "Control of External Corrosion on Metallic Buried, Partially Buried, or Submerged Liquid Storage Systems," and Underwriters Laboratories Standard 58, "Standard for Steel Underground Tanks for Flammable and Combustible Liquids."

(3) The tank is constructed of a steel-fiberglass-reinforced-plastic composite; or

NOTE: The following industry codes may be used to comply with paragraph (a)(3) of this section: Underwriters Laboratories Standard 1746, "Corrosion Protection Systems for Underground Storage Tanks," or the Association for Composite Tanks ACT-100, "Specification for the Fabrication of FRP Clad Underground Storage Tanks."

(4) The tank is constructed of metal without additional corrosion protection measures provided that:

(i) The tank is installed at a site that is determined by a corrosion expert not to be corrosive enough to cause it to have a release due to corrosion during its operating life; and

(ii) Owners and operators maintain records that demonstrate compliance with the requirements of paragraphs (a)(4)(i) for the remaining life of the tank; or

(5) The tank construction and corrosion protection are determined by the implementing agency to be designed to prevent the release or threatened release of any stored regulated substance in a manner that is no less protective of human health and the environment than paragraphs (a) (1) through (4) of this section.

(b) *Piping.* The piping that routinely contains regulated substances and is in contact with the ground must be properly designed, constructed, and protected from corrosion in accordance with a code of practice developed by a nationally recognized association or independent testing laboratory as specified below:

112

280

(1) The piping is constructed of fiberglass-reinforced plastic; or

NOTE: The following codes and standards may be used to comply with paragraph (b)(1) of this section:

(A) Underwriters Laboratories Subject 971, "UL Listed Non-Metal Pipe";

(B) Underwriters Laboratories Standard 567, "Pipe Connectors for Flammable and Combustible and LP Gas";

(C) Underwriters Laboratories of Canada Guide ULC-107, "Glass Fiber Reinforced Plastic Pipe and Fittings for Flammable Liquids"; and

(D) Underwriters Laboratories of Canada Standard CAN 4-S633-M81, "Flexible Underground Hose Connectors."

MUST of UST

(2) The piping is constructed of steel and cathodically protected in the following manner:

(i) The piping is coated with a suitable dielectric material;

(ii) Field-installed cathodic protection systems are designed by a corrosion expert;

(iii) Impressed current systems are designed to allow determination of current operating status as required in § 280.31(c); and

Subpart D 280.

(iv) Cathodic protection systems are operated and maintained in accordance with § 280.31 or guidelines established by the implementing agency; or

NOTE: The following codes and standards may be used to comply with paragraph (b)(2) of this section:

(A) National Fire Protection Association Standard 30, "Flammable and Combustible Liquids Code";

(B) American Petroleum Institute Publication 1615, "Installation of Underground Petroleum Storage Systems";

(C) American Petroleum Institute Publication 1632, "Cathodic Protection of Underground Petroleum Storage Tanks and Piping Systems"; and

→ UST system does not apply to

(D) National Association of Corrosion Engineers Standard RP-01-69, "Control of External Corrosion on Submerged Metallic Piping Systems."

(3) The piping is constructed of metal without additional corrosion protection measures provided that:

(i) The piping is installed at a site that is determined by a corrosion expert to not be corrosive enough to cause it to have a release due to corrosion during its operating life; and

(ii) Owners and operators maintain records that demonstrate compliance with the requirements of paragraph (b)(3)(i) of this section for the remaining life of the piping; or

NOTE: National Fire Protection Association Standard 30, "Flammable and Combustible Liquids Code"; and National Association of Corrosion Engineers Standard RP-01-69, "Control of External Corrosion on Submerged Metallic Piping Systems," may be used to comply with paragraph (b)(3) of this section.

(4) The piping construction and corrosion protection are determined by the implementing agency to be designed to prevent the release or threatened release of any stored regulated substance in a manner that is no less protective of human health and the environment than the requirements in paragraphs (b) (1) through (3) of this section.

(c) *Spill and overfill prevention equipment.* (1) Except as provided in paragraph (c)(2) of this section, to prevent spilling and overfilling associated with product transfer to the UST system, owners and operators must use the following spill and overfill prevention equipment:

(i) Spill prevention equipment that will prevent release of product to the environment when the transfer hose is detached from the fill pipe (for example, a spill catchment basin); and

(ii) Overfill prevention equipment that will:

(A) Automatically shut off flow into the tank when the tank is no more than 95 percent full; or

(B) Alert the transfer operator when the tank is no more than 90 percent full by restricting the flow into the tank or triggering a high-level alarm; or

(C) Restrict flow 30 minutes prior to overfilling, alert the operator with a high level alarm one minute before overfilling, or automatically shut off flow into the tank so that none of the fittings located on top of the tank are exposed to product due to overfilling.

(2) Owners and operators are not required to use the spill and overfill prevention equipment specified in paragraph (c)(1) of this section if:

(i) Alternative equipment is used that is determined by the implementing agency to be no less protective of human health and the environment than the equipment specified in paragraph (c)(1) (i) or (ii) of this section; or

(ii) The UST system is filled by transfers of no more than 25 gallons at one time.

(d) *Installation.* All tanks and piping must be properly installed in accordance with a code of practice developed by a nationally recognized association or independent testing laboratory and in accordance with the manufacturer's instructions.

NOTE: Tank and piping system installation practices and procedures described in the following codes may be used to comply with the requirements of paragraph (d) of this section:

(i) American Petroleum Institute Publication 1615, "Installation of Underground Petroleum Storage System"; or

(ii) Petroleum Equipment Institute Publication RP100, "Recommended Practices for Installation of Underground Liquid Storage Systems"; or

(iii) American National Standards Institute Standard B31.3, "Petroleum Refinery Piping," and American National Standards Institute Standard B31.4 "Liquid Petroleum Transportation Piping System."

(e) *Certification of installation.* All owners and operators must ensure that one or more of the following methods of certification, testing, or inspection is used to demonstrate compliance with paragraph (d) of this section by providing a certification of compliance on the UST notification form in accordance with § 280.22.

(1) The installer has been certified by the tank and piping manufacturers; or

(2) The installer has been certified or licensed by the implementing agency; or

(3) The installation has been inspected and certified by a registered professional engineer with education

and experience in UST system installation; or

(4) The installation has been inspected and approved by the implementing agency; or

(5) All work listed in the manufacturer's installation checklists has been completed; or

(6) The owner and operator have complied with another method for ensuring compliance with paragraph (d) of this section that is determined by the implementing agency to be no less protective of human health and the environment.

[53 FR 37194, Sept. 23, 1988, as amended at 56 FR 38344, Aug. 13, 1991]

§ 280.21 Upgrading of existing UST systems.

(a) *Alternatives allowed.* Not later than December 22, 1998, all existing UST systems must comply with one of the following requirements:

(1) New UST system performance standards under § 280.20;

(2) The upgrading requirements in paragraphs (b) through (d) of this section; or

(3) Closure requirements under subpart G of this part, including applicable requirements for corrective action under subpart F.

(b) *Tank upgrading requirements.* Steel tanks must be upgraded to meet one of the following requirements in accordance with a code of practice developed by a nationally recognized association or independent testing laboratory:

(1) *Interior lining.* A tank may be upgraded by internal lining if:

(i) The lining is installed in accordance with the requirements of § 280.33, and

(ii) Within 10 years after lining, and every 5 years thereafter, the lined tank is internally inspected and found to be structurally sound with the lining still performing in accordance with original design specifications.

(2) *Cathodic protection.* A tank may be upgraded by cathodic protection if the cathodic protection system meets the requirements of § 280.20(a)(2) (ii), (iii), and (iv) and the integrity of the tank is ensured using one of the following methods:

(i) The tank is internally inspected and assessed to ensure that the tank is structurally sound and free of corrosion holes prior to installing the cathodic protection system; or

(ii) The tank has been installed for less than 10 years and is monitored monthly for releases in accordance with § 280.43 (d) through (h); or

(iii) The tank has been installed for less than 10 years and is assessed for corrosion holes by conducting two (2) tightness tests that meet the requirements of § 280.43(c). The first tightness test must be conducted prior to installing the cathodic protection system. The second tightness test must be conducted between three (3) and six (6) months following the first operation of the cathodic protection system; or

(iv) The tank is assessed for corrosion holes by a method that is determined by the implementing agency to prevent releases in a manner that is no less protective of human health and the environment than paragraphs (b)(2) (i) through (iii) of this section.

(3) *Internal lining combined with cathodic protection.* A tank may be upgraded by both internal lining and cathodic protection if:

(i) The lining is installed in accordance with the requirements of § 280.33; and

(ii) The cathodic protection system meets the requirements of § 280.20(a)(2) (ii), (iii), and (iv).

NOTE: The following codes and standards may be used to comply with this section:

(A) American Petroleum Institute Publication 1631, "Recommended Practice for the Interior Lining of Existing Steel Underground Storage Tanks";

(B) National Leak Prevention Association Standard 631, "Spill Prevention, Minimum 10 Year Life Extension of Existing Steel Underground Tanks by Lining Without the Addition of Cathodic Protection";

(C) National Association of Corrosion Engineers Standard RP-02-85, "Control of External Corrosion on Metallic Buried, Partially Buried, or Submerged Liquid Storage Systems"; and

(D) American Petroleum Institute Publication 1632, "Cathodic Protection of Underground Petroleum Storage Tanks and Piping Systems."

(c) *Piping upgrading requirements.* Metal piping that routinely contains regulated substances and is in contact with the ground must be cathodically protected in accordance with a code of practice developed by a nationally recognized association or independent testing laboratory and must meet the requirements of § 280.20(b)(2) (ii), (iii), and (iv).

NOTE: The codes and standards listed in the note following § 280.20(b)(2) may be used to comply with this requirement.

(d) *Spill and overfill prevention equipment.* To prevent spilling and overfilling associated with product transfer to the UST system, all existing UST systems must comply with new UST system spill and overfill prevention equipment requirements specified in § 280.20(c).

§ 280.22 Notification requirements.

(a) Any owner who brings an underground storage tank system into use after May 8, 1986, must within 30 days of bringing such tank into use, submit, in the form prescribed in appendix I of this part, a notice of existence of such tank system to the state or local agency or department designated in appendix II of this part to receive such notice.

NOTE: Owners and operators of UST systems that were in the ground on or after May 8, 1986, unless taken out of operation on or before January 1, 1974, were required to notify the designated state or local agency in accordance with the Hazardous and Solid Waste Amendments of 1984, Pub. L. 98-616, on a form published by EPA on November 8, 1985 (50 FR 46602) unless notice was given pursuant to section 103(c) of CERCLA. Owners and operators who have not complied with the notification requirements may use portions I through VI of the notification form contained in appendix I of this part.

(b) In states where state law, regulations, or procedures require owners to use forms that differ from those set forth in appendix I of this part to fulfill the requirements of this section, the state forms may be submitted in lieu of the forms set forth in Appendix I of this part. If a state requires that its form be used in lieu of the form presented in this regulation, such form must meet the requirements of section 9002.

(c) Owners required to submit notices under paragraph (a) of this section must provide notices to the appropriate agencies or departments identified in appendix II of this part for each tank they own. Owners may provide notice for several tanks using one notification form, but owners who own tanks located at more than one place of operation must file a separate notification form for each separate place of operation.

(d) Notices required to be submitted under paragraph (a) of this section must provide all of the information in sections I through VI of the prescribed form (or appropriate state form) for each tank for which notice must be given. Notices for tanks installed after December 22, 1988 must also provide all of the information in section VII of the prescribed form (or appropriate state form) for each tank for which notice must be given.

(e) All owners and operators of new UST systems must certify in the notification form compliance with the following requirements:

(1) Installation of tanks and piping under § 280.20(e);

(2) Cathodic protection of steel tanks and piping under § 280.20 (a) and (b);

(3) Financial responsibility under subpart H of this part; and

(4) Release detection under §§ 280.41 and 280.42.

(f) All owners and operators of new UST systems must ensure that the installer certifies in the notification form that the methods used to install the tanks and piping complies with the requirements in § 280.20(d).

(g) Beginning October 24, 1988, any person who sells a tank intended to be used as an underground storage tank must notify the purchaser of such tank of the owner's notification obligations under paragraph (a) of this section. The form provided in appendix III of this part may be used to comply with this requirement.

Subpart C—General Operating Requirements

§ 280.30 Spill and overfill control.

(a) Owners and operators must ensure that releases due to spilling or overfilling do not occur. The owner and operator must ensure that the volume available in the tank is greater than the volume of product to be transferred to the tank before the transfer is made and that the transfer operation is monitored constantly to prevent overfilling and spilling.

NOTE: The transfer procedures described in National Fire Protection Association Publication 385 may be used to comply with paragraph (a) of this section. Further guidance on spill and overfill prevention appears in American Petroleum Institute Publication 1621, "Recommended Practice for Bulk Liquid Stock Control at Retail Outlets," and National Fire Protection Association Standard 30, "Flammable and Combustible Liquids Code."

(b) The owner and operator must report, investigate, and clean up any spills and overfills in accordance with §280.53.

§280.31 Operation and maintenance of corrosion protection.

All owners and operators of steel UST systems with corrosion protection must comply with the following requirements to ensure that releases due to corrosion are prevented for as long as the UST system is used to store regulated substances:

(a) All corrosion protection systems must be operated and maintained to continuously provide corrosion protection to the metal components of that portion of the tank and piping that routinely contain regulated substances and are in contact with the ground.

(b) All UST systems equipped with cathodic protection systems must be inspected for proper operation by a qualified cathodic protection tester in accordance with the following requirements:

(1) *Frequency.* All cathodic protection systems must be tested within 6 months of installation and at least every 3 years thereafter or according to another reasonable time frame established by the implementing agency; and

(2) *Inspection criteria.* The criteria that are used to determine that cathodic protection is adequate as required by this section must be in accordance with a code of practice developed by a nationally recognized association.

NOTE: National Association of Corrosion Engineers Standard RP-02-85, "Control of External Corrosion on Metallic Buried, Partially Buried, or Submerged Liquid Storage Systems," may be used to comply with paragraph (b)(2) of this section.

(c) UST systems with impressed current cathodic protection systems must also be inspected every 60 days to ensure the equipment is running properly.

(d) For UST systems using cathodic protection, records of the operation of the cathodic protection must be maintained (in accordance with §280.34) to demonstrate compliance with the performance standards in this section. These records must provide the following:

(1) The results of the last three inspections required in paragraph (c) of this section; and

(2) The results of testing from the last two inspections required in paragraph (b) of this section.

§280.32 Compatibility.

Owners and operators must use an UST system made of or lined with materials that are compatible with the substance stored in the UST system.

NOTE: Owners and operators storing alcohol blends may use the following codes to comply with the requirements of this section:

(a) American Petroleum Institute Publication 1626, "Storing and Handling Ethanol and Gasoline-Ethanol Blends at Distribution Terminals and Service Stations"; and

(b) American Petroleum Institute Publication 1627, "Storage and Handling of Gasoline-Methanol/Cosolvent Blends at Distribution Terminals and Service Stations."

§280.33 Repairs allowed.

Owners and operators of UST systems must ensure that repairs will prevent releases due to structural failure or corrosion as long as the UST system is used to store regulated substances. The repairs must meet the following requirements:

(a) Repairs to UST systems must be properly conducted in accordance with a code of practice developed by a nationally recognized association or an independent testing laboratory.

NOTE: The following codes and standards may be used to comply with paragraph (a) of this section: National Fire Protection Association Standard 30, "Flammable and Combustible Liquids Code"; American Petroleum Institute Publication 2200, "Repairing Crude Oil, Liquified Petroleum Gas, and Product Pipelines"; American Petroleum Institute Publication 1631, "Recommended Practice for the Interior Lining of Existing Steel Underground Storage Tanks"; and National Leak Prevention Association Standard 631, "Spill Prevention, Minimum 10 Year Life Extension of Existing Steel Underground Tanks by Lining Without the Addition of Cathodic Protection."

(b) Repairs to fiberglass-reinforced plastic tanks may be made by the manufacturer's authorized representatives or in accordance with a code of practice developed by a nationally recognized association or an independent testing laboratory.

(c) Metal pipe sections and fittings that have released product as a result of corrosion or other damage must be replaced. Fiberglass pipes and fittings may be repaired in accordance with the manufacturer's specifications.

(d) Repaired tanks and piping must be tightness tested in accordance with § 280.43(c) and § 280.44(b) within 30 days following the date of the completion of the repair except as provided in paragraphs (d) (1) through (3), of this section:

(1) The repaired tank is internally inspected in accordance with a code of practice developed by a nationally recognized association or an independent testing laboratory; or

(2) The repaired portion of the UST system is monitored monthly for releases in accordance with a method specified in § 280.43 (d) through (h); or

(3) Another test method is used that is determined by the implementing agency to be no less protective of human health and the environment than those listed above.

(e) Within 6 months following the repair of any cathodically protected UST system, the cathodic protection system must be tested in accordance with § 280.31 (b) and (c) to ensure that it is operating properly.

(f) UST system owners and operators must maintain records of each repair for the remaining operating life of the UST system that demonstrate compliance with the requirements of this section.

§ 280.34 Reporting and recordkeeping.

Owners and operators of UST systems must cooperate fully with inspections, monitoring and testing conducted by the implementing agency, as well as requests for document submission, testing, and monitoring by the owner or operator pursuant to section 9005 of Subtitle I of the Resource Conservation and Recovery Act, as amended.

(a) *Reporting.* Owners and operators must submit the following information to the implementing agency:

(1) Notification for all UST systems (§ 280.22), which includes certification of installation for new UST systems (§ 280.20(e)),

(2) Reports of all releases including suspected releases (§ 280.50), spills and overfills (§ 280.53), and confirmed releases (§ 280.61);

(3) Corrective actions planned or taken including initial abatement measures (§ 280.62), initial site characterization (§ 280.63), free product removal (§ 280.64), investigation of soil and ground-water cleanup (§ 280.65), and corrective action plan (§ 280.66); and

(4) A notification before permanent closure or change-in-service (§ 280.71).

(b) *Recordkeeping.* Owners and operators must maintain the following information:

(1) A corrosion expert's analysis of site corrosion potential if corrosion protection equipment is not used (§ 280.20(a)(4); § 280.20(b)(3)).

(2) Documentation of operation of corrosion protection equipment (§ 280.31);

(3) Documentation of UST system repairs (§ 280.33(f));

(4) Recent compliance with release detection requirements (§ 280.45); and

(5) Results of the site investigation conducted at permanent closure (§ 280.74).

(c) *Availability and Maintenance of Records.* Owners and operators must keep the records required either:

(1) At the UST site and immediately available for inspection by the implementing agency; or

(2) At a readily available alternative site and be provided for inspection to the implementing agency upon request.

(3) In the case of permanent closure records required under § 280.74, owners and operators are also provided with the additional alternative of mailing closure records to the implementing agency if they cannot be kept at the

site or an alternative site as indicated above.

Subpart D—Release Detection

§280.40 General requirements for all UST systems.

(a) Owners and operators of new and existing UST systems must provide a method, or combination of methods, of release detection that:

(1) Can detect a release from any portion of the tank and the connected underground piping that routinely contains product;

(2) Is installed, calibrated, operated, and maintained in accordance with the manufacturer's instructions, including routine maintenance and service checks for operability or running condition; and

(3) Meets the performance requirements in §280.43 or 280.44, with any performance claims and their manner of determination described in writing by the equipment manufacturer or installer. In addition, methods used after the date shown in the following table corresponding with the specified method except for methods permanently installed prior to that date, must be capable of detecting the leak rate or quantity specified for that method in the corresponding section of the rule (also shown in the table) with a probability of detection (Pd) of 0.95 and a probability of false alarm (Pfa) of 0.05.

Method	Section	Date after which Pd/Pfa must be demonstrated
Manual Tank Gauging.	280.43(b)	December 22, 1990.
Tank Tightness Testing.	280.43(c)	December 22, 1990.
Automatic Tank Gauging.	280.43(d)	December 22, 1990.
Automatic Line Leak Detectors.	280.44(a)	September 22, 1991.
Line Tightness Testing.	280.44(b)	December 22, 1990.

(b) When a release detection method operated in accordance with the performance standards in §280.43 and §280.44 indicates a release may have occurred, owners and operators must notify the implementing agency in accordance with subpart E.

(c) Owners and operators of all UST systems must comply with the release detection requirements of this subpart by December 22 of the year listed in the following table:

SCHEDULE FOR PHASE-IN OF RELEASE DETECTION

Year system was installed	Year when release detection is required (by December 22 of the year indicated)				
	1989	1990	1991	1992	1993
Before 1965 or date unknown.	RD	P			
1965–69 ...		P/RD			
1970–74 ...		P	RD		
1975–79 ...		P		RD	
1980–88 ...		P			RD
New tanks (after December 22) immediately upon installation.					

P=Must begin release detection for all pressurized piping as defined in §280.41(b)(1).

RD=Must begin release detection for tanks and suction piping in accordance with §280.41(a), §280.41(b)(2), and §280.42.

(d) Any existing UST system that cannot apply a method of release detection that complies with the requirements of this subpart must complete the closure procedures in subpart G by the date on which release detection is required for that UST system under paragraph (c) of this section.

[53 FR 37194, Sept. 23, 1988, as amended at 55 FR 17753, Apr. 27, 1990; 55 FR 23738, June 12, 1990; 56 FR 26, Jan. 2, 1991]

§280.41 Requirements for petroleum UST systems.

Owners and operators of petroleum UST systems must provide release detection for tanks and piping as follows:

(a) *Tanks.* Tanks must be monitored at least every 30 days for releases using one of the methods listed in §280.43 (d) through (h) except that:

(1) UST systems that meet the performance standards in §280.20 or §280.21, and the monthly inventory control requirements in §280.43 (a) or (b), may use tank tightness testing (conducted in accordance with §280.43(c)) at least every 5 years until December 22, 1998, or until 10 years after the tank is installed or upgraded under §280.21(b), whichever is later;

(2) UST systems that do not meet the performance standards in §280.20 or §280.21 may use monthly inventory controls (conducted in accordance with §280.43(a) or (b)) and annual tank tightness testing (conducted in accordance with §280.43(c)) until December 22, 1998

when the tank must be upgraded under § 280.21 or permanently closed under § 280.71; and

(3) Tanks with capacity of 550 gallons or less may use weekly tank gauging (conducted in accordance with § 280.43(b)).

(b) *Piping*. Underground piping that routinely contains regulated substances must be monitored for releases in a manner that meets one of the following requirements:

(1) *Pressurized piping*. Underground piping that conveys regulated substances under pressure must:

(i) Be equipped with an automatic line leak detector conducted in accordance with § 280.44(a); and

(ii) Have an annual line tightness test conducted in accordance with § 280.44(b) or have monthly monitoring conducted in accordance with § 280.44(c).

(2) *Suction piping*. Underground piping that conveys regulated substances under suction must either have a line tightness test conducted at least every 3 years and in accordance with § 280.44(b), or use a monthly monitoring method conduct in accordance with § 280.44(c). No release detection is required for suction piping that is designed and constructed to meet the following standards:

(i) The below-grade piping operates at less than atmospheric pressure;

(ii) The below-grade piping is sloped so that the contents of the pipe will drain back into the storage tank if the suction is released;

(iii) Only one check valve is included in each suction line;

(iv) The check valve is located directly below and as close as practical to the suction pump; and

(v) A method is provided that allows compliance with paragraphs (b)(2) (ii)–(iv) of this section to be readily determined.

§ 280.42 Requirements for hazardous substance UST systems.

Owners and operators of hazardous substance UST systems must provide release detection that meets the following requirements:

(a) Release detection at existing UST systems must meet the requirements for petroleum UST systems in § 280.41. By December 22, 1998, all existing hazardous substance UST systems must meet the release detection requirements for new systems in paragraph (b) of this section.

(b) Release detection at new hazardous substance UST systems must meet the following requirements:

(1) Secondary containment systems must be designed, constructed and installed to:

(i) Contain regulated substances released from the tank system until they are detected and removed;

(ii) Prevent the release of regulated substances to the environment at any time during the operational life of the UST system; and

(iii) Be checked for evidence of a release at least every 30 days.

NOTE.—The provisions of 40 CFR 265.193, Containment and Detection of Releases, may be used to comply with these requirements.

(2) Double-walled tanks must be designed, constructed, and installed to:

(i) Contain a release from any portion of the inner tank within the outer wall; and

(ii) Detect the failure of the inner wall.

(3) External liners (including vaults) must be designed, constructed, and installed to:

(i) Contain 100 percent of the capacity of the largest tank within its boundary;

(ii) Prevent the interference of precipitation or ground-water intrusion with the ability to contain or detect a release of regulated substances; and

(iii) Surround the tank completely (i.e., it is capable of preventing lateral as well as vertical migration of regulated substances).

(4) Underground piping must be equipped with secondary containment that satisfies the requirements of paragraph (b)(1) of this section (e.g., trench liners, jacketing of double-walled pipe). In addition, underground piping that conveys regulated substances under pressure must be equipped with an automatic line leak detector in accordance with § 280.44(a).

(5) Other methods of release detection may be used if owners and operators:

(i) Demonstrate to the implementing agency that an alternate method can detect a release of the stored substance as effectively as any of the methods allowed in §§280.43(b) through (h) can detect a release of petroleum;

(ii) Provide information to the implementing agency on effective corrective action technologies, health risks, and chemical and physical properties of the stored substance, and the characteristics of the UST site; and,

(iii) Obtain approval from the implementing agency to use the alternate release detection method before the installation and operation of the new UST system.

§280.43 Methods of release detection for tanks.

Each method of release detection for tanks used to meet the requirements of §280.41 must be conducted in accordance with the following:

(a) *Inventory control.* Product inventory control (or another test of equivalent performance) must be conducted monthly to detect a release of at least 1.0 percent of flow-through plus 130 gallons on a monthly basis in the following manner:

(1) Inventory volume measurements for regulated substance inputs, withdrawals, and the amount still remaining in the tank are recorded each operating day;

(2) The equipment used is capable of measuring the level of product over the full range of the tank's height to the nearest one-eighth of an inch;

(3) The regulated substance inputs are reconciled with delivery receipts by measurement of the tank inventory volume before and after delivery;

(4) Deliveries are made through a drop tube that extends to within one foot of the tank bottom;

(5) Product dispensing is metered and recorded within the local standards for meter calibration or an accuracy of 6 cubic inches for every 5 gallons of product withdrawn; and

(6) The measurement of any water level in the bottom of the tank is made to the nearest one-eighth of an inch at least once a month.

NOTE: Practices described in the American Petroleum Institute Publication 1621, "Recommended Practice for Bulk Liquid Stock Control at Retail Outlets," may be used, where applicable, as guidance in meeting the requirements of this paragraph.

(b) *Manual tank gauging.* Manual tank gauging must meet the following requirements:

(1) Tank liquid level measurements are taken at the beginning and ending of a period of at least 36 hours during which no liquid is added to or removed from the tank;

(2) Level measurements are based on an average of two consecutive stick readings at both the beginning and ending of the period;

(3) The equipment used is capable of measuring the level of product over the full range of the tank's height to the nearest one-eighth of an inch;

(4) A leak is suspected and subject to the requirements of subpart E if the variation between beginning and ending measurements exceeds the weekly or monthly standards in the following table:

Nominal tank capacity	Weekly standard (one test)	Monthly standard (average of four tests)
550 gallons or less.	10 gallons	5 gallons.
551–1,000 gallons.	13 gallons	7 gallons.
1,001–2,000 gallons.	26 gallons	13 gallons.

(5) Only tanks of 550 gallons or less nominal capacity may use this as the sole method of release detection. Tanks of 551 to 2,000 gallons may use the method in place of manual inventory control in §280.43(a). Tanks of greater than 2,000 gallons nominal capacity may not use this method to meet the requirements of this subpart.

(c) *Tank tightness testing.* Tank tightness testing (or another test of equivalent performance) must be capable of detecting a 0.1 gallon per hour leak rate from any portion of the tank that routinely contains product while accounting for the effects of thermal expansion or contraction of the product, vapor pockets, tank deformation, evaporation or condensation, and the location of the water table.

(d) *Automatic tank gauging.* Equipment for automatic tank gauging that tests for the loss of product and conducts inventory control must meet the following requirements:

(1) The automatic product level monitor test can detect a 0.2 gallon per hour leak rate from any portion of the tank that routinely contains product; and

(2) Inventory control (or another test of equivalent performance) is conducted in accordance with the requirements of § 280.43(a).

(e) *Vapor monitoring.* Testing or monitoring for vapors within the soil gas of the excavation zone must meet the following requirements:

(1) The materials used as backfill are sufficiently porous (e.g., gravel, sand, crushed rock) to readily allow diffusion of vapors from releases into the excavation area;

(2) The stored regulated substance, or a tracer compound placed in the tank system, is sufficiently volatile (e.g., gasoline) to result in a vapor level that is detectable by the monitoring devices located in the excavation zone in the event of a release from the tank;

(3) The measurement of vapors by the monitoring device is not rendered inoperative by the ground water, rainfall, or soil moisture or other known interferences so that a release could go undetected for more than 30 days;

(4) The level of background contamination in the excavation zone will not interfere with the method used to detect releases from the tank;

(5) The vapor monitors are designed and operated to detect any significant increase in concentration above background of the regulated substance stored in the tank system, a component or components of that substance, or a tracer compound placed in the tank system;

(6) In the UST excavation zone, the site is assessed to ensure compliance with the requirements in paragraphs (e) (1) through (4) of this section and to establish the number and positioning of monitoring wells that will detect releases within the excavation zone from any portion of the tank that routinely contains product; and

(7) Monitoring wells are clearly marked and secured to avoid unauthorized access and tampering.

(f) *Ground-water monitoring.* Testing or monitoring for liquids on the ground water must meet the following requirements:

(1) The regulated substance stored is immiscible in water and has a specific gravity of less than one;

(2) Ground water is never more than 20 feet from the ground surface and the hydraulic conductivity of the soil(s) between the UST system and the monitoring wells or devices is not less than 0.01 cm/sec (e.g., the soil should consist of gravels, coarse to medium sands, coarse silts or other permeable materials);

(3) The slotted portion of the monitoring well casing must be designed to prevent migration of natural soils or filter pack into the well and to allow entry of regulated substance on the water table into the well under both high and low ground-water conditions;

(4) Monitoring wells shall be sealed from the ground surface to the top of the filter pack;

(5) Monitoring wells or devices intercept the excavation zone or are as close to it as is technically feasible;

(6) The continuous monitoring devices or manual methods used can detect the presence of at least one-eighth of an inch of free product on top of the ground water in the monitoring wells;

(7) Within and immediately below the UST system excavation zone, the site is assessed to ensure compliance with the requirements in paragraphs (f) (1) through (5) of this section and to establish the number and positioning of monitoring wells or devices that will detect releases from any portion of the tank that routinely contains product; and

(8) Monitoring wells are clearly marked and secured to avoid unauthorized access and tampering.

(g) *Interstitial monitoring.* Interstitial monitoring between the UST system and a secondary barrier immediately around or beneath it may be used, but only if the system is designed, constructed and installed to detect a leak from any portion of the tank that routinely contains product and also meets one of the following requirements:

(1) For double-walled UST systems, the sampling or testing method can detect a release through the inner wall in any portion of the tank that routinely contains product;

NOTE: The provisions outlined in the Steel Tank Institute's "Standard for Dual Wall

Underground Storage Tanks" may be used as guidance for aspects of the design and construction of underground steel double-walled tanks.

(2) For UST systems with a secondary barrier within the excavation zone, the sampling or testing method used can detect a release between the UST system and the secondary barrier;

(i) The secondary barrier around or beneath the UST system consists of artificially constructed material that is sufficiently thick and impermeable (at least 10^{-6} cm/sec for the regulated substance stored) to direct a release to the monitoring point and permit its detection;

(ii) The barrier is compatible with the regulated substance stored so that a release from the UST system will not cause a deterioration of the barrier allowing a release to pass through undetected;

(iii) For cathodically protected tanks, the secondary barrier must be installed so that it does not interfere with the proper operation of the cathodic protection system;

(iv) The ground water, soil moisture, or rainfall will not render the testing or sampling method used inoperative so that a release could go undetected for more than 30 days;

(v) The site is assessed to ensure that the secondary barrier is always above the ground water and not in a 25-year flood plain, unless the barrier and monitoring designs are for use under such conditions; and,

(vi) Monitoring wells are clearly marked and secured to avoid unauthorized access and tampering.

(3) For tanks with an internally fitted liner, an automated device can detect a release between the inner wall of the tank and the liner, and the liner is compatible with the substance stored.

(h) *Other methods.* Any other type of release detection method, or combination of methods, can be used if:

(1) It can detect a 0.2 gallon per hour leak rate or a release of 150 gallons within a month with a probability of detection of 0.95 and a probability of false alarm of 0.05; or

(2) The implementing agency may approve another method if the owner and operator can demonstrate that the method can detect a release as effectively as any of the methods allowed in paragraphs (c) through (h) of this section. In comparing methods, the implementing agency shall consider the size of release that the method can detect and the frequency and reliability with which it can be detected. If the method is approved, the owner and operator must comply with any conditions imposed by the implementing agency on its use to ensure the protection of human health and the environment.

§ 280.44 Methods of release detection for piping.

Each method of release detection for piping used to meet the requirements of § 280.41 must be conducted in accordance with the following:

(a) *Automatic line leak detectors.* Methods which alert the operator to the presence of a leak by restricting or shutting off the flow of regulated substances through piping or triggering an audible or visual alarm may be used only if they detect leaks of 3 gallons per hour at 10 pounds per square inch line pressure within 1 hour. An annual test of the operation of the leak detector must be conducted in accordance with the manufacturer's requirements.

(b) *Line tightness testing.* A periodic test of piping may be conducted only if it can detect a 0.1 gallon per hour leak rate at one and one-half times the operating pressure.

(c) *Applicable tank methods.* Any of the methods in § 280.43 (e) through (h) may be used if they are designed to detect a release from any portion of the underground piping that routinely contains regulated substances.

§ 280.45 Release detection recordkeeping.

All UST system owners and operators must maintain records in accordance with § 280.34 demonstrating compliance with all applicable requirements of this subpart. These records must include the following:

(a) All written performance claims pertaining to any release detection system used, and the manner in which these claims have been justified or tested by the equipment manufacturer or installer, must be maintained for 5 years, or for another reasonable period of time determined by the implementing agency, from the date of installation;

(b) The results of any sampling, testing, or monitoring must be maintained for at least 1 year, or for another reasonable period of time determined by the implementing agency, except that the results of tank tightness testing conducted in accordance with § 280.43(c) must be retained until the next test is conducted; and

(c) Written documentation of all calibration, maintenance, and repair of release detection equipment permanently located on-site must be maintained for at least one year after the servicing work is completed, or for another reasonable time period determined by the implementing agency. Any schedules of required calibration and maintenance provided by the release detection equipment manufacturer must be retained for 5 years from the date of installation.

Subpart E—Release Reporting, Investigation, and Confirmation

§ 280.50 Reporting of suspected releases.

Owners and operators of UST systems must report to the implementing agency within 24 hours, or another reasonable time period specified by the implementing agency, and follow the procedures in § 280.52 for any of the following conditions:

(a) The discovery by owners and operators or others of released regulated substances at the UST site or in the surrounding area (such as the presence of free product or vapors in soils, basements, sewer and utility lines, and nearby surface water).

(b) Unusual operating conditions observed by owners and operators (such as the erratic behavior of product dispensing equipment, the sudden loss of product from the UST system, or an unexplained presence of water in the tank), unless system equipment is found to be defective but not leaking, and is immediately repaired or replaced; and,

(c) Monitoring results from a release detection method required under § 280.41 and § 280.42 that indicate a release may have occurred unless:

(1) The monitoring device is found to be defective, and is immediately repaired, recalibrated or replaced, and additional monitoring does not confirm the initial result; or

(2) In the case of inventory control, a second month of data does not confirm the initial result.

§ 280.51 Investigation due to off-site impacts.

When required by the implementing agency, owners and operators of UST systems must follow the procedures in § 280.52 to determine if the UST system is the source of off-site impacts. These impacts include the discovery of regulated substances (such as the presence of free product or vapors in soils, basements, sewer and utility lines, and nearby surface and drinking waters) that has been observed by the implementing agency or brought to its attention by another party.

§ 280.52 Release investigation and confirmation steps.

Unless corrective action is initiated in accordance with subpart F, owners and operators must immediately investigate and confirm all suspected releases of regulated substances requiring reporting under § 280.50 within 7 days, or another reasonable time period specified by the implementing agency, using either the following steps or another procedure approved by the implementing agency:

(a) *System test.* Owners and operators must conduct tests (according to the requirements for tightness testing in § 280.43(c) and § 280.44(b)) that determine whether a leak exists in that portion of the tank that routinely contains product, or the attached delivery piping, or both.

(1) Owners and operators must repair, replace or upgrade the UST system, and begin corrective action in accordance with subpart F if the test results for the system, tank, or delivery piping indicate that a leak exists.

(2) Further investigation is not required if the test results for the system, tank, and delivery piping do not indicate that a leak exists and if environmental contamination is not the basis for suspecting a release.

(3) Owners and operators must conduct a site check as described in paragraph (b) of this section if the test results for the system, tank, and delivery piping do not indicate that a leak exists but environmental contamination is the basis for suspecting a release.

(b) *Site check.* Owners and operators must measure for the presence of a release where contamination is most likely to be present at the UST site. In selecting sample types, sample locations, and measurement methods, owners and operators must consider the nature of the stored substance, the type of initial alarm or cause for suspicion, the type of backfill, the depth of ground water, and other factors appropriate for identifying the presence and source of the release.

(1) If the test results for the excavation zone or the UST site indicate that a release has occurred, owners and operators must begin corrective action in accordance with subpart F;

(2) If the test results for the excavation zone or the UST site do not indicate that a release has occurred, further investigation is not required.

§280.53 Reporting and cleanup of spills and overfills.

(a) Owners and operators of UST systems must contain and immediately clean up a spill or overfill and report to the implementing agency within 24 hours, or another reasonable time period specified by the implementing agency, and begin corrective action in accordance with subpart F in the following cases:

(1) Spill or overfill of petroleum that results in a release to the environment that exceeds 25 gallons or another reasonable amount specified by the implementing agency, or that causes a sheen on nearby surface water; and

(2) Spill or overfill of a hazardous substance that results in a release to the environment that equals or exceeds its reportable quantity under CERCLA (40 CFR part 302).

(b) Owners and operators of UST systems must contain and immediately clean up a spill or overfill of petroleum that is less than 25 gallons or another reasonable amount specified by the implementing agency, and a spill or overfill of a hazardous substance that is less than the reportable quantity. If cleanup cannot be accomplished within 24 hours, or another reasonable time period established by the implementing agency, owners and operators must immediately notify the implementing agency.

NOTE: Pursuant to §§302.6 and 355.40, a release of a hazardous substance equal to or in excess of its reportable quantity must also be reported immediately (rather than within 24 hours) to the National Response Center under sections 102 and 103 of the Comprehensive Environmental Response, Compensation, and Liability Act of 1980 and to appropriate state and local authorities under Title III of the Superfund Amendments and Reauthorization Act of 1986.

Subpart F—Release Response and Corrective Action for UST Systems Containing Petroleum or Hazardous Substances

§280.60 General.

Owners and operators of petroleum or hazardous substance UST systems must, in response to a confirmed release from the UST system, comply with the requirements of this subpart except for USTs excluded under §280.10(b) and UST systems subject to RCRA Subtitle C corrective action requirements under section 3004(u) of the Resource Conservation and Recovery Act, as amended.

§280.61 Initial response.

Upon confirmation of a release in accordance with §280.52 or after a release from the UST system is identified in any other manner, owners and operators must perform the following initial response actions within 24 hours of a release or within another reasonable period of time determined by the implementing agency:

(a) Report the release to the implementing agency (e.g., by telephone or electronic mail);

(b) Take immediate action to prevent any further release of the regulated substance into the environment; and

(c) Identify and mitigate fire, explosion, and vapor hazards.

§ 280.62 Initial abatement measures and site check.

(a) Unless directed to do otherwise by the implementing agency, owners and operators must perform the following abatement measures:

(1) Remove as much of the regulated substance from the UST system as is necessary to prevent further release to the environment;

(2) Visually inspect any aboveground releases or exposed belowground releases and prevent further migration of the released substance into surrounding soils and ground water;

(3) Continue to monitor and mitigate any additional fire and safety hazards posed by vapors or free product that have migrated from the UST excavation zone and entered into subsurface structures (such as sewers or basements);

(4) Remedy hazards posed by contaminated soils that are excavated or exposed as a result of release confirmation, site investigation, abatement, or corrective action activities. If these remedies include treatment or disposal of soils, the owner and operator must comply with applicable State and local requirements;

(5) Measure for the presence of a release where contamination is most likely to be present at the UST site, unless the presence and source of the release have been confirmed in accordance with the site check required by § 280.52(b) or the closure site assessment of § 280.72(a). In selecting sample types, sample locations, and measurement methods, the owner and operator must consider the nature of the stored substance, the type of backfill, depth to ground water and other factors as appropriate for identifying the presence and source of the release; and

(6) Investigate to determine the possible presence of free product, and begin free product removal as soon as practicable and in accordance with § 280.64.

(b) Within 20 days after release confirmation, or within another reasonable period of time determined by the implementing agency, owners and operators must submit a report to the implementing agency summarizing the initial abatement steps taken under paragraph (a) of this section and any resulting information or data.

§ 280.63 Initial site characterization.

(a) Unless directed to do otherwise by the implementing agency, owners and operators must assemble information about the site and the nature of the release, including information gained while confirming the release or completing the initial abatement measures in §§ 280.60 and 280.61. This information must include, but is not necessarily limited to the following:

(1) Data on the nature and estimated quantity of release;

(2) Data from available sources and/or site investigations concerning the following factors: surrounding populations, water quality, use and approximate locations of wells potentially affected by the release, subsurface soil conditions, locations of subsurface sewers, climatological conditions, and land use;

(3) Results of the site check required under § 280.62(a)(5); and

(4) Results of the free product investigations required under § 280.62(a)(6), to be used by owners and operators to determine whether free product must be recovered under § 280.64.

(b) Within 45 days of release confirmation or another reasonable period of time determined by the implementing agency, owners and operators must submit the information collected in compliance with paragraph (a) of this section to the implementing agency in a manner that demonstrates its applicability and technical adequacy, or in a format and according to the schedule required by the implementing agency.

§ 280.64 Free product removal.

At sites where investigations under § 280.62(a)(6) indicate the presence of free product, owners and operators must remove free product to the maximum extent practicable as determined by the implementing agency while continuing, as necessary, any actions initiated under §§ 280.61 through 280.63, or preparing for actions required under §§ 280.65 through 280.66. In meeting the requirements of this section, owners and operators must:

(a) Conduct free product removal in a manner that minimizes the spread of contamination into previously uncontaminated zones by using recovery and disposal techniques appropriate to the hydrogeologic conditions at the site, and that properly treats, discharges or disposes of recovery by-products in compliance with applicable local, State and Federal regulations;

(b) Use abatement of free product migration as a minimum objective for the design of the free product removal system;

(c) Handle any flammable products in a safe and competent manner to prevent fires or explosions; and

(d) Unless directed to do otherwise by the implementing agency, prepare and submit to the implementing agency, within 45 days after confirming a release, a free product removal report that provides at least the following information:

(1) The name of the person(s) responsible for implementing the free product removal measures;

(2) The estimated quantity, type, and thickness of free product observed or measured in wells, boreholes, and excavations;

(3) The type of free product recovery system used;

(4) Whether any discharge will take place on-site or off-site during the recovery operation and where this discharge will be located;

(5) The type of treatment applied to, and the effluent quality expected from, any discharge;

(6) The steps that have been or are being taken to obtain necessary permits for any discharge; and

(7) The disposition of the recovered free product.

§ 280.65 Investigations for soil and ground-water cleanup.

(a) In order to determine the full extent and location of soils contaminated by the release and the presence and concentrations of dissolved product contamination in the ground water, owners and operators must conduct investigations of the release, the release site, and the surrounding area possibly affected by the release if any of the following conditions exist:

(1) There is evidence that ground-water wells have been affected by the release (e.g., as found during release confirmation or previous corrective action measures);

(2) Free product is found to need recovery in compliance with § 280.64;

(3) There is evidence that contaminated soils may be in contact with ground water (e.g., as found during conduct of the initial response measures or investigations required under §§ 280.60 through 280.64); and

(4) The implementing agency requests an investigation, based on the potential effects of contaminated soil or ground water on nearby surface water and ground-water resources.

(b) Owners and operators must submit the information collected under paragraph (a) of this section as soon as practicable or in accordance with a schedule established by the implementing agency.

§ 280.66 Corrective action plan.

(a) At any point after reviewing the information submitted in compliance with §§ 280.61 through 280.63, the implementing agency may require owners and operators to submit additional information or to develop and submit a corrective action plan for responding to contaminated soils and ground water. If a plan is required, owners and operators must submit the plan according to a schedule and format established by the implementing agency. Alternatively, owners and operators may, after fulfilling the requirements of §§ 280.61 through 280.63, choose to submit a corrective action plan for responding to contaminated soil and ground water. In either case, owners and operators are responsible for submitting a plan that provides for adequate protection of human health and the environment as determined by the implementing agency, and must modify their plan as necessary to meet this standard.

(b) The implementing agency will approve the corrective action plan only after ensuring that implementation of the plan will adequately protect human health, safety, and the environment. In making this determination, the implementing agency should consider the following factors as appropriate:

(1) The physical and chemical characteristics of the regulated substance, in-

cluding its toxicity, persistence, and potential for migration;

(2) The hydrogeologic characteristics of the facility and the surrounding area;

(3) The proximity, quality, and current and future uses of nearby surface water and ground water;

(4) The potential effects of residual contamination on nearby surface water and ground water;

(5) An exposure assessment; and

(6) Any information assembled in compliance with this subpart.

(c) Upon approval of the corrective action plan or as directed by the implementing agency, owners and operators must implement the plan, including modifications to the plan made by the implementing agency. They must monitor, evaluate, and report the results of implementing the plan in accordance with a schedule and in a format established by the implementing agency.

(d) Owners and operators may, in the interest of minimizing environmental contamination and promoting more effective cleanup, begin cleanup of soil and ground water before the corrective action plan is approved provided that they:

(1) Notify the implementing agency of their intention to begin cleanup;

(2) Comply with any conditions imposed by the implementing agency, including halting cleanup or mitigating adverse consequences from cleanup activities; and

(3) Incorporate these self-initiated cleanup measures in the corrective action plan that is submitted to the implementing agency for approval.

§ 280.67 Public participation.

(a) For each confirmed release that requires a corrective action plan, the implementing agency must provide notice to the public by means designed to reach those members of the public directly affected by the release and the planned corrective action. This notice may include, but is not limited to, public notice in local newspapers, block advertisements, public service announcements, publication in a state register, letters to individual households, or personal contacts by field staff.

(b) The implementing agency must ensure that site release information and decisions concerning the corrective action plan are made available to the public for inspection upon request.

(c) Before approving a corrective action plan, the implementing agency may hold a public meeting to consider comments on the proposed corrective action plan if there is sufficient public interest, or for any other reason.

(d) The implementing agency must give public notice that complies with paragraph (a) of this section if implementation of an approved corrective action plan does not achieve the established cleanup levels in the plan and termination of that plan is under consideration by the implementing agency.

Subpart G—Out-of-Service UST Systems and Closure

§ 280.70 Temporary closure.

(a) When an UST system is temporarily closed, owners and operators must continue operation and maintenance of corrosion protection in accordance with § 280.31, and any release detection in accordance with subpart D. Subparts E and F must be complied with if a release is suspected or confirmed. However, release detection is not required as long as the UST system is empty. The UST system is empty when all materials have been removed using commonly employed practices so that no more than 2.5 centimeters (one inch) of residue, or 0.3 percent by weight of the total capacity of the UST system, remain in the system.

(b) When an UST system is temporarily closed for 3 months or more, owners and operators must also comply with the following requirements:

(1) Leave vent lines open and functioning; and

(2) Cap and secure all other lines, pumps, manways, and ancillary equipment.

(c) When an UST system is temporarily closed for more than 12 months, owners and operators must permanently close the UST system if it does not meet either performance standards in § 280.20 for new UST systems or the upgrading requirements in § 280.21, *except that* the spill and overfill equipment requirements do not have to be met. Owners and operators must permanently close the substandard UST systems at the end of this 12-month period in accordance with §§ 280.71–280.74, *unless* the implementing agency provides an extension of the 12-month temporary closure period. Owners and operators must complete a site assessment in accordance with § 280.72 before such an extension can be applied for.

§ 280.71 Permanent closure and changes-in-service.

(a) At least 30 days before beginning either permanent closure or a change-in-service under paragraphs (b) and (c) of this section, or within another reasonable time period determined by the implementing agency, owners and operators must notify the implementing agency of their intent to permanently close or make the change-in-service, *unless* such action is in response to corrective action. The required assessment of the excavation zone under § 280.72 must be performed after notifying the implementing agency but before completion of the permanent closure or a change-in-service.

(b) To permanently close a tank, owners and operators must empty and clean it by removing all liquids and accumulated sludges. All tanks taken out of service permanently must also be either removed from the ground or filled with an inert solid material.

(c) Continued use of an UST system to store a non-regulated substance is considered a change-in-service. Before a change-in-service, owners and operators must empty and clean the tank by removing all liquid and accumulated sludge and conduct a site assessment in accordance with § 280.72.

NOTE: The following cleaning and closure procedures may be used to comply with this section:

(A) American Petroleum Institute Recommended Practice 1604, "Removal and Disposal of Used Underground Petroleum Storage Tanks";

(B) American Petroleum Institute Publication 2015, "Cleaning Petroleum Storage Tanks";

(C) American Petroleum Institute Recommended Practice 1631, "Interior Lining of Underground Storage Tanks," may be used as guidance for compliance with this section; and

(D) The National Institute for Occupational Safety and Health "Criteria for a Recommended Standard * * * Working in Confined Space" may be used as guidance for conducting safe closure procedures at some hazardous substance tanks.

§ 280.72 Assessing the site at closure or change-in-service.

(a) Before permanent closure or a change-in-service is completed, owners and operators must measure for the presence of a release where contamination is most likely to be present at the UST site. In selecting sample types, sample locations, and measurement methods, owners and operators must consider the method of closure, the nature of the stored substance, the type of backfill, the depth to ground water, and other factors appropriate for identifying the presence of a release. The requirements of this section are satisfied if one of the external release detection methods allowed in § 280.43 (e) and (f) is operating in accordance with the requirements in § 280.43 at the time of closure, and indicates no release has occurred.

(b) If contaminated soils, contaminated ground water, or free product as a liquid or vapor is discovered under paragraph (a) of this section, or by any other manner, owners and operators must begin corrective action in accordance with subpart F.

§ 280.73 Applicability to previously closed UST systems.

When directed by the implementing agency, the owner and operator of an UST system permanently closed before December 22, 1988 must assess the excavation zone and close the UST system in accordance with this subpart if releases from the UST may, in the judgment of the implementing agency, pose a current or potential threat to human health and the environment.

§ 280.74 Closure records.

Owners and operators must maintain records in accordance with § 280.34 that are capable of demonstrating compliance with closure requirements under this subpart. The results of the excavation zone assessment required in § 280.72 must be maintained for at least 3 years after completion of permanent closure or change-in-service in one of the following ways:

(a) By the owners and operators who took the UST system out of service;

(b) By the current owners and operators of the UST system site; or

(c) By mailing these records to the implementing agency if they cannot be maintained at the closed facility.

Subpart H—Financial Responsibility

SOURCE: 53 FR 43370, Oct. 26, 1988, unless otherwise noted.

§ 280.90 Applicability.

(a) This subpart applies to owners and operators of all petroleum underground storage tank (UST) systems except as otherwise provided in this section.

(b) Owners and operators of petroleum UST systems are subject to these requirements if they are in operation on or after the date for compliance established in § 280.91.

(c) State and Federal government entities whose debts and liabilities are the debts and liabilities of a state or the United States are exempt from the requirements of this subpart.

(d) The requirements of this subpart do not apply to owners and operators of any UST system described in § 280.10 (b) or (c).

(e) If the owner and operator of a petroleum underground storage tank are separate persons, only one person is required to demonstrate financial responsibility; however, both parties are liable in event of noncompliance. Regardless of which party complies, the date set for compliance at a particular facility is determined by the characteristics of the owner as set forth in § 280.91.

§ 280.91 Compliance dates.

Owners of petroleum underground storage tanks are required to comply with the requirements of this subpart by the following dates:

(a) All petroleum marketing firms owning 1,000 or more USTs and all other UST owners that report a tangible net worth of $20 million or more to the U.S. Securities and Exchange Commission (SEC), Dun and Bradstreet, the Energy Information Administration, or the Rural Electrification Administration; January 24, 1989, except that compliance with § 280.94(b) is required by: July 24, 1989.

(b) All petroleum marketing firms owning 100–999 USTs; October 26, 1989.

(c) All petroleum marketing firms owning 13–99 USTs at more than one facility; April 26, 1991.

(d) All petroleum UST owners not described in paragraphs (a), (b), or (c) of this section, excluding local government entities; December 31, 1993.

(e) All local government entities; one year from the date of promulgation of additional mechanisms for use by local government entities to comply with financial responsibility requirements for underground storage tanks containing petroleum.

[53 FR 43370, Oct. 26, 1988, as amended at 54 FR 5452, Feb. 3, 1989; 55 FR 18567, May 2, 1990; 55 FR 46025, Oct. 31, 1990; 56 FR 66373, Dec. 23, 1991]

§ 280.92 Definition of terms.

When used in this subpart, the following terms shall have the meanings given below:

Accidental release means any sudden or nonsudden release of petroleum from an underground storage tank that results in a need for corrective action and/or compensation for bodily injury or property damage neither expected nor intended by the tank owner or operator.

Bodily injury shall have the meaning given to this term by applicable state law; however, this term shall not include those liabilities which, consistent with standard insurance industry practices, are excluded from coverage in liability insurance policies for bodily injury.

Chief Financial Officer, in the case of local government owners and operators, means the individual with the overall authority and responsibility for the collection, disbursement, and use of funds by the local government.

Controlling interest means direct ownership of at least 50 percent of the voting stock of another entity.

Director of the Implementing Agency means the EPA Regional Administrator, or, in the case of a state with a program approved under section 9004, the Director of the designated state or local agency responsible for carrying out an approved UST program.

Financial reporting year means the latest consecutive twelve-month period for which any of the following reports used to support a financial test is prepared:

(1) a 10-K report submitted to the SEC;

(2) an annual report of tangible net worth submitted to Dun and Bradstreet; or

(3) annual reports submitted to the Energy Information Administration or the Rural Electrification Administration.

"Financial reporting year" may thus comprise a fiscal or a calendar year period.

Legal defense cost is any expense that an owner or operator or provider of financial assurance incurs in defending against claims or actions brought,

(1) By EPA or a state to require corrective action or to recover the costs of corrective action;

(2) By or on behalf of a third party for bodily injury or property damage caused by an accidental release; or

(3) By any person to enforce the terms of a financial assurance mechanism.

Local government shall have the meaning given this term by applicable state law and includes Indian tribes. The term is generally intended to include: (1) Counties, municipalities, townships, separately chartered and operated special districts (including local government public transit systems and redevelopment authorities), and independent school districts authorized as governmental bodies by state charter or constitution; and (2) Special districts and independent school districts established by counties, municipalities, townships, and other general purpose governments to provide essential services.

Occurrence means an accident, including continuous or repeated exposure to conditions, which results in a release from an underground storage tank.

NOTE: This definition is intended to assist in the understanding of these regulations and is not intended either to limit the meaning of "occurrence" in a way that conflicts with standard insurance usage or to prevent the use of other standard insurance terms in place of "occurrence."

Owner or operator, when the owner or operator are separate parties, refers to the party that is obtaining or has obtained financial assurances.

Petroleum marketing facilities include all facilities at which petroleum is produced or refined and all facilities from which petroleum is sold or transferred to other petroleum marketers or to the public.

Petroleum marketing firms are all firms owning petroleum marketing facilities. Firms owning other types of facilities with USTs as well as petroleum marketing facilities are considered to be petroleum marketing firms.

Property damage shall have the meaning given this term by applicable state law. This term shall not include those liabilities which, consistent with standard insurance industry practices, are excluded from coverage in liability insurance policies for property damage. However, such exclusions for property damage shall not include corrective action associated with releases from tanks which are covered by the policy.

Provider of financial assurance means an entity that provides financial assurance to an owner or operator of an underground storage tank through one of the mechanisms listed in §§ 280.95–280.103, including a guarantor, insurer, risk retention group, surety, issuer of a letter of credit, issuer of a state-required mechanism, or a state.

Substantial business relationship means the extent of a business relationship necessary under applicable state law to make a guarantee contract issued incident to that relationship valid and enforceable. A guarantee contract is issued "incident to that relationship" if it arises from and depends on existing economic transactions between the guarantor and the owner or operator.

Substantial governmental relationship means the extent of a governmental relationship necessary under applicable state law to make an added guarantee contract issued incident to that relationship valid and enforceable. A guarantee contract is issued "incident to that relationship" if it arises from a clear commonality of interest in the event of an UST release such as coterminous boundaries, overlapping constituencies, common ground-water aquifer, or other relationship other than monetary compensation that provides a motivation for the guarantor to provide a guarantee.

Tangible net worth means the tangible assets that remain after deducting liabilities; such assets do not include intangibles such as goodwill and rights to patents or royalties. For purposes of this definition, "assets" means all existing and all probable future economic benefits obtained or controlled by a particular entity as a result of past transactions.

Termination under § 280.97(b)(1) and § 280.97(b)(2) means only those changes that could result in a gap in coverage as where the insured has not obtained substitute coverage or has obtained substitute coverage with a different retroactive date than the retroactive date of the original policy.

[53 FR 43370, Oct. 26, 1988, as amended at 54 FR 47081, Nov. 9, 1989; 58 FR 9050, Feb. 18, 1993]

§ 280.93 Amount and scope of required financial responsibility.

(a) Owners or operators of petroleum underground storage tanks must demonstrate financial responsibility for taking corrective action and for compensating third parties for bodily injury and property damage caused by accidental releases arising from the operation of petroleum underground storage tanks in at least the following per-occurrence amounts:

(1) For owners or operators of petroleum underground storage tanks that are located at petroleum marketing facilities, or that handle an average of more than 10,000 gallons of petroleum per month based on annual throughput for the previous calendar year; $1 million.

(2) For all other owners or operators of petroleum underground storage tanks; $500,000.

(b) Owners or operators of petroleum underground storage tanks must demonstrate financial responsibility for taking corrective action and for compensating third parties for bodily injury and property damage caused by accidental releases arising from the operation of petroleum underground storage tanks in at least the following annual aggregate amounts:

(1) For owners or operators of 1 to 100 petroleum underground storage tanks, $1 million; and

(2) For owners or operators of 101 or more petroleum underground storage tanks, $2 million.

(c) For the purposes of paragraphs (b) and (f) of this section, only, "a petroleum underground storage tank" means a single containment unit and does not mean combinations of single containment units.

(d) Except as provided in paragraph (e) of this section, if the owner or operator uses separate mechanisms or separate combinations of mechanisms to demonstrate financial responsibility for:

(1) Taking corrective action;

(2) Compensating third parties for bodily injury and property damage caused by sudden accidental releases; or

(3) Compensating third parties for bodily injury and property damage caused by nonsudden accidental releases, the amount of assurance provided by each mechanism or combination of mechanisms must be in the full amount specified in paragraphs (a) and (b) of this section.

(e) If an owner or operator uses separate mechanisms or separate combinations of mechanisms to demonstrate financial responsibility for different petroleum underground storage tanks, the annual aggregate required shall be based on the number of tanks covered by each such separate mechanism or combination of mechanisms.

(f) Owners or operators shall review the amount of aggregate assurance provided whenever additional petroleum underground storage tanks are acquired or installed. If the number of petroleum underground storage tanks for which assurance must be provided exceeds 100, the owner or operator shall demonstrate financial responsibility in the amount of at least $2 million of annual aggregate assurance by the anniversary of the date on which the mechanism demonstrating financial responsibility became effective. If assurance is being demonstrated by a combination of mechanisms, the owner or operator shall demonstrate financial responsibility in the amount of at least $2 million of annual aggregate assurance by the first-occurring effective date anniversary of any one of the mechanisms combined (other than a financial test or guarantee) to provide assurance.

(g) The amounts of assurance required under this section exclude legal defense costs.

(h) The required per-occurrence and annual aggregate coverage amounts do not in any way limit the liability of the owner or operator.

§280.94 Allowable mechanisms and combinations of mechanisms.

(a) Subject to the limitations of paragraphs (b) and (c) of this section,

(1) An owner or operator, including a local government owner or operator, may use any one or combination of the mechanisms listed in §§280.95 through 280.103 to demonstrate financial responsibility under this subpart for one or more underground storage tanks, and

(2) A local government owner or operator may use any one or combination of the mechanisms listed in §§280.104 through 280.107 to demonstrate financial responsibility under this subpart for one or more underground storage tanks.

(b) An owner or operator may use a guarantee under §280.96 or surety bond under §280.98 to establish financial responsibility only if the Attorney(s) General of the state(s) in which the underground storage tanks are located has (have) submitted a written statement to the implementing agency that a guarantee or surety bond executed as described in this section is a legally valid and enforceable obligation in that state.

(c) An owner or operator may use self-insurance in combination with a guarantee only if, for the purpose of meeting the requirements of the financial test under this rule, the financial statements of the owner or operator are not consolidated with the financial statements of the guarantor.

[53 FR 43370, Oct. 26, 1988, as amended at 58 FR 9051, Feb. 18, 1993]

§280.95 Financial test of self-insurance.

(a) An owner or operator, and/or guarantor, may satisfy the requirements of §280.93 by passing a financial test as specified in this section. To pass the financial test of self-insurance, the owner or operator, and/or guarantor must meet the criteria of paragraph (b) or (c) of this section based on year-end financial statements for the latest completed fiscal year.

(b)(1) The owner or operator, and/or guarantor, must have a tangible net worth of at least ten times:

(i) The total of the applicable aggregate amount required by §280.93, based on the number of underground storage tanks for which a financial test is used to demonstrate financial responsibility to EPA under this section or to a state implementing agency under a state program approved by EPA under 40 CFR part 281;

(ii) The sum of the corrective action cost estimates, the current closure and post-closure care cost estimates, and amount of liability coverage for which a financial test is used to demonstrate financial responsibility to EPA under 40 CFR 264.101, 264.143, 264.145, 265.143, 165.145, 264.147, and 265.147 or to a state implementing agency under a state program authorized by EPA under 40 CFR part 271; and

(iii) The sum of current plugging and abandonment cost estimates for which a financial test is used to demonstrate financial responsibility to EPA under 40 CFR 144.63 or to a state implementing agency under a state program authorized by EPA under 40 CFR part 145.

(2) The owner or operator, and/or guarantor, must have a tangible net worth of at least $10 million.

(3) The owner or operator, and/or guarantor, must have a letter signed by the chief financial officer worded as specified in paragraph (d) of this section.

(4) The owner or operator, and/or guarantor, must either:

(i) File financial statements annually with the U.S. Securities and Exchange Commission, the Energy Information Administration, or the Rural Electrification Administration; or

(ii) Report annually the firm's tangible net worth to Dun and Bradstreet, and Dun and Bradstreet must have assigned the firm a financial strength rating of 4A or 5A.

(5) The firm's year-end financial statements, if independently audited, cannot include an adverse auditor's opinion, a disclaimer of opinion, or a "going concern" qualification.

(c)(1) The owner or operator, and/or guarantor must meet the financial test requirements of 40 CFR 264.147(f)(1), substituting the appropriate amounts specified in § 280.93 (b)(1) and (b)(2) for the "amount of liability coverage" each time specified in that section.

(2) The fiscal year-end financial statements of the owner or operator, and/or guarantor, must be examined by an independent certified public accountant and be accompanied by the accountant's report of the examination.

(3) The firm's year-end financial statements cannot include an adverse auditor's opinion, a disclaimer of opinion, or a "going concern" qualification.

(4) The owner or operator, and/or guarantor, must have a letter signed by the chief financial officer, worded as specified in paragraph (d) of this section.

(5) If the financial statements of the owner or operator, and/or guarantor, are not submitted annually to the U.S. Securities and Exchange Commission, the Energy Information Administration or the Rural Electrification Administration, the owner or operator, and/or guarantor, must obtain a special report by an independent certified public accountant stating that:

(i) He has compared the data that the letter form the chief financial officer specifies as having been derived from the latest year-end financial statements of the owner or operator, and/or guarantor, with the amounts in such financial statements; and

(ii) In connection with that comparison, no matters came to his attention which caused him to believe that the specified data should be adjusted.

(d) To demonstrate that it meets the financial test under paragraph (b) or (c) of this section, the chief financial officer of the owner or operator, or guarantor, must sign, within 120 days of the close of each financial reporting year, as defined by the twelve-month period for which financial statements used to support the financial test are prepared, a letter worded exactly as follows, except that the instructions in brackets are to be replaced by the relevant information and the brackets deleted:

LETTER FROM CHIEF FINANCIAL OFFICER

I am the chief financial officer of [insert: name and address of the owner or operator, or guarantor]. This letter is in support of the use of [insert: "the financial test of self-insurance," and/or "guarantee"] to demonstrate financial responsibility for [insert: "taking corrective action" and/or "compensating third parties for bodily injury and property damage"] caused by [insert: "suddent accidential releases" and/or "nonsudden accidential releases"] in the amount of at least [insert: dollar amount] per occurrence and [insert: dollar amount] annual aggregate arising from operating (an) underground storage tank(s).

Underground storage tanks at the following facilities are assured by this financial test or a financial test under an authorized State program by this [insert: "owner or operator," and/or "guarantor"]: [List for each facility: the name and address of the facility where tanks assured by this financial test are located, and whether tanks are assured by this financial test or a financial test under a State program approved under 40 CFR part 281. If separate mechanisms or combinations of mechanisms are being used to assure any of the tanks at this facility, list each tank assured by this financial test or a financial test under a State program authorized under 40 CFR part 281 by the tank identification number provided in the notification submitted pursuant to 40 CFR 280.22 or the corresponding State requirements.]

A [insert: "financial test," and/or "guarantee"] is also used by this [insert: "owner or operator," or "guarantor"] to demonstrate evidence of financial responsibility in the following amounts under other EPA regulations or state programs authorized by EPA under 40 CFR parts 271 and 145:

EPA Regulations	*Amount*
Closure (§§ 264.143 and 265.143)	$______
Post-Closure Care (§§ 264.145 and 265.145)	$______
Liability Coverage (§§ 264.147 and 265.147)	$______
Corrective Action (§§ 264.101(b)) ...	$______
Plugging and Abandonment (§ 144.63)	$______
Closure	$______
Post-Closure Care	$______
Liabilitly Coverage	$______
Corrective Action	$______
Plugging and Abandonment	$______
Total	$______

This [insert: "owner or operator," or "guarantor"] has not received an adverse opinion, a disclaimer of opinion, or a "going concern" qualification from an independent auditor on his financial statements for the latest completed fiscal year.

[Fill in the information for Alternative I if the criteria of paragraph (b) of §280.95 are being used to demonstrate compliance with the financial test requirements. Fill in the information for Alternative II if the criteria of paragraph (c) of §280.95 are being used to demonstrate compliance with the financial test requirements.]

Alternative I

1.	Amount of annual UST aggregate coverage being assured by a financial test, and/or guarantee.	$______	
2.	Amount of corrective action, closure and post-closure care costs, liability coverage, and plugging and abandonment costs covered by a financial test, and/or guarantee.	$______	
3.	Sum of lines 1 and 2	$______	
4.	Total tangible assets	$______	
5.	Total liabilities [if any of the amount reported on line 3 is included in total liabilities, you may deduct that amount from this line and add that amount to line 6].	$______	
6.	Tangible net worth [subtract line 5 from line 4].	$______	
		Yes	No
7.	Is line 6 at least $10 million?	___	_
8.	Is line 6 at least 10 times line 3?.	___	_
9.	Have financial statements for the latest fiscal year been filed with the Securities and Exchange Commission?.	___	_
10.	Have financial statements for the latest fiscal year been filed with the Energy Information Administration?.	___	_
11.	Have financial statements for the lastest fiscal year been filed with the Rural Electrification Administration?.	___	_
12.	Has financial information been provided to Dun and Bradstreet, and has Dun and Bradstreet provided a financial strength rating of 4A or 5A? [Answer "Yes" only if both criteria have been met.].	___	_

Alternative I—Continued

Alternative II

1.	Amount of annual UST aggregate coverage being assured by a test, and/or guarantee.	$______	
2.	Amount of corrective action, closure and post-closure care costs, liability coverage, and plugging and abandonment costs covered by a financial test, and/or guarantee.	$______	
3.	Sum of lines 1 and 2	$______	
4.	Total tangible assets	$______	
5.	Total liabilities [if any of the amount reported on line 3 is included in total liabilities, you may deduct that amount from this line and add that amount to line 6].	$______	
6.	Tangible net worth [subtract line 5 from line 4].	$______	
7.	Total assets in the U.S. [required only if less than 90 percent of assets are located in the U.S.].	$______	
		Yes	No
8.	Is line 6 at least $10 million?	$___	_
9.	Is line 6 at least 6 times line 3?.	___	_
10.	Are at least 90 percent of assets located in the U.S.? [If "No," complete line 11.].	___	_
11.	Is line 7 at least 6 times line 3?.	___	_

[Fill in either lines 12–15 or lines 16–18:]

12.	Current assets	$______	
13.	Current liabilities	______	
14.	Net working capital [subtract line 13 from line 12].	______	
		Yes	No
15.	Is line 14 at least 6 times line 3?.	___	_
16.	Current bond rating of most recent bond issue.	___	_
17.	Name of rating service	___	_
18.	Date of maturity of bond	___	_
19.	Have financial statements for the latest fiscal year been filed with the SEC, the Energy Information Administration, or the Rural Electrification Administration?.	___	_

[If "No," please attach a report from an independent certified public accountant certifying that there are no material differences between the data as reported in lines 4–18 above and the financial statements for the latest fiscal year.]

[For both Alternative I and Alternative II complete the certification with this statement.]

I hereby certify that the wording of this letter is identical to the wording specified in 40 CFR part 280.95(d) as such regulations were constituted on the date shown immediately below.

[Signature]

[Name]

[Title]

[Date]

(e) If an owner or operator using the test to provide financial assurance finds that he or she no longer meets the requirements of the financial test based on the year-end financial statements, the owner or operator must obtain alternative coverage within 150 days of the end of the year for which financial statements have been prepared.

(f) The Director of the implementing agency may require reports of financial condition at any time from the owner or operator, and/or guarantor. If the Director finds, on the basis of such reports or other information, that the owner or operator, and/or guarantor, no longer meets the financial test requirements of § 280.95(b) or (c) and (d), the owner or operator must obtain alternate coverage within 30 days after notification of such a finding.

(g) If the owner or operator fails to obtain alternate assurance within 150 days of finding that he or she no longer meets the requirements of the financial test based on the year-end financial statements, or within 30 days of notification by the Director of the implementing agency that he or she no longer meets the requirements of the financial test, the owner or operator must notify the Director of such failure within 10 days.

§ 280.96 Guarantee.

(a) An owner or operator may satisfy the requirements of § 280.93 by obtaining a guarantee that conforms to the requirements of this section. The guarantor must be:

(1) A firm that (i) possesses a controlling interest in the owner or operator; (ii) possesses a controlling interest in a firm described under paragraph (a)(1)(i) of this section; or, (iii) is controlled through stock ownership by a common parent firm that possesses a controlling interest in the owner or operator; or,

(2) A firm engaged in a substantial business relationship with the owner or operator and issuing the guarantee as an act incident to that business relationship.

(b) Within 120 days of the close of each financial reporting year the guarantor must demonstrate that it meets the financial test criteria of § 280.95 based on year-end financial statements for the latest completed financial reporting year by completing the letter from the chief financial officer described in § 280.95(d) and must deliver the letter to the owner or operator. If the guarantor fails to meet the requirements of the financial test at the end of any financial reporting year, within 120 days of the end of that financial reporting year the guarantor shall send by certified mail, before cancellation or nonrenewal of the guarantee, notice to the owner or operator. If the Director of the implementing agency notifies the guarantor that he no longer meets the requirements of the financial test of § 280.95 (b) or (c) and (d), the guarantor must notify the owner or operator within 10 days of receiving such notification from the Director. In both cases, the guarantee will terminate no less than 120 days after the date the owner or operator receives the notification, as evidenced by the return receipt. The owner or operator must obtain alternative coverage as specified in § 280.110(c).

(c) The guarantee must be worded as follows, except that instructions in brackets are to be replaced with the relevant information and the brackets deleted:

GUARANTEE

Guarantee made this [date] by [name of guaranteeing entity], a business entity organized under the laws of the state of [name of state], herein referred to as guarantor, to [the state implementing agency] and to any and all third parties, and obligees, on behalf of [owner or operator] of [business address].

Recitals.

(1) Guarantor meets or exceeds the financial test criteria of 40 CFR 280.95 (b) or (c) and (d) and agrees to comply with the requirements for guarantors as specified in 40 CFR 280.96(b).

(2) [Owner or operator] owns or operates the following underground storage tank(s) covered by this guarantee: [List the number of tanks at each facility and the name(s) and address(es) of the facility(ies) where the tanks are located. If more than one instrument is used to assure different tanks at any one facility, for each tank covered by this instrument, list the tank identification number provided in the notification submitted pursuant to 40 CFR 280.22 or the corresponding state requirement, and the name and address of the facility.] This guarantee satisfies 40 CFR part 280, subpart H requirements for assuring funding for [insert: "taking corrective action" and/or "compensating third parties for bodily injury and property damage caused by" either "sudden accidental releases" or "nonsudden accidental releases" or "accidental releases"; if coverage is different for different tanks or locations, indicate the type of coverage applicable to each tank or location] arising from operating the above-identified underground storage tank(s) in the amount of [insert dollar amount] per occurrence and [insert dollar amount] annual aggregate.

(3) [Insert appropriate phrase: "On behalf of our subsidiary" (if guarantor is corporate parent of the owner or operator); "On behalf of our affiliate" (if guarantor is a related firm of the owner or operator); or "Incident to our business relationship with" (if guarantor is providing the guarantee as an incident to a substantial business relationship with owner or operator)] [owner or operator], guarantor guarantees to [implementing agency] and to any and all third parties that:

In the event that [owner or operator] fails to provide alternative coverage within 60 days after receipt of a notice of cancellation of this guarantee and the [Director of the implementing agency] has determined or suspects that a release has occurred at an underground storage tank covered by this guarantee, the guarantor, upon instructions from the [Director], shall fund a standby trust fund in accordance with the provisions of 40 CFR 280.108, in an amount not to exceed the coverage limits specified above.

In the event that the [Director] determines that [owner or operator] has failed to perform corrective action for releases arising out of the operation of the above-identified tank(s) in accordance with 40 CFR part 280, subpart F, the guarantor upon written instructions from the [Director] shall fund a standby trust in accordance with the provisions of 40 CFR 280.108, in an amount not to exceed the coverage limits specified above.

If [owner or operator] fails to satisfy a judgment or award based on a determination of liability for bodily injury or property damage to third parties caused by ["sudden" and/or "nonsudden"] accidental releases arising from the operation of the above-identified tank(s), or fails to pay an amount agreed to in settlement of a claim arising from or alleged to arise from such injury or damage, the guarantor, upon written instructions from the [Director], shall fund a standby trust in accordance with the provisions of 40 CFR 280.108 to satisfy such judgment(s), award(s), or settlement agreement(s) up to the limits of coverage specified above.

(4) Guarantor agrees that if, at the end of any fiscal year before cancellation of this guarantee, the guarantor fails to meet the financial test criteria of 40 CFR 280.95 (b) or (c) and (d), guarantor shall send within 120 days of such failure, by certified mail, notice to [owner or operator]. The guarantee will terminate 120 days from the date of receipt of the notice by [owner or operator], as evidenced by the return receipt.

(5) Guarantor agrees to notify [owner or operator] by certified mail of a voluntary or involuntary proceeding under Title 11 (Bankruptcy), U.S. Code naming guarantor as debtor, within 10 days after commencement of the proceeding.

(6) Guarantor agrees to remain bound under this guarantee notwithstanding any modification or alteration of any obligation of [owner or operator] pursuant to 40 CFR part 280.

(7) Guarantor agrees to remain bound under this guarantee for so long as [owner or operator] must comply with the applicable financial responsibility requirements of 40 CFR part 280, subpart H for the above-identified tank(s), except that guarantor may cancel this guarantee by sending notice by certified mail to [owner or operator], such cancellation to become effective no earlier than 120 days after receipt of such notice by [owner or operator], as evidenced by the return receipt.

(8) The guarantor's obligation does not apply to any of the following:

(a) Any obligation of [insert owner or operator] under a workers' compensation, disability benefits, or unemployment compensation law or other similar law;

(b) Bodily injury to an employee of [insert owner or operator] arising from, and in the course of, employment by [insert owner or operator];

(c) Bodily injury or property damage arising from the ownership, maintenance, use, or entrustment to others of any aircraft, motor vehicle, or watercraft;

(d) Property damage to any property owned, rented, loaded to, in the care, custody, or control of, or occupied by [insert owner or operator] that is not the direct result of a release from a petroleum underground storage tank;

(e) Bodily damage or property damage for which [insert owner or operator] is obligated to pay damages by reason of the assumption of liability in a contract or agreement other than a contract or agreement entered into to meet the requirements of 40 CFR 280.93.

(9) Guarantor expressly waives notice of acceptance of this guarantee by [the implementing agency], by any or all third parties, or by [owner or operator].

I hereby certify that the wording of this guarantee is identical to the wording specified in 40 CFR 280.96(c) as such regulations were constituted on the effective date shown immediately below.

Effective date: ______

[Name of guarantor]

[Authorized signature for guarantor]

[Name of person signing]

[Title of person signing]

Signature of witness or notary: ______

(d) An owner or operator who uses a guarantee to satisfy the requirements of § 280.93 must establish a standby trust fund when the guarantee is obtained. Under the terms of the guarantee, all amounts paid by the guarantor under the guarantee will be deposited directly into the standby trust fund in accordance with instructions from the Director of the implementing agency under § 280.108. This standby trust fund must meet the requirements specified in § 280.103.

§ 280.97 Insurance and risk retention group coverage.

(a) An owner or operator may satisfy the requirements of § 290.93 by obtaining liability insurance that conforms to the requirements of this section from a qualified insurer or risk retention group. Such insurance may be in the form of a separate insurance policy or an endorsement to an existing insurance policy.

(b) Each insurance policy must be amended by an endorsement worded as specified in paragraph (b)(1) of this section, or evidenced by a certificate of insurance worded as specified in paragraph (b)(2) of this section, except that instructions in brackets must be replaced with the relevant information and the brackets deleted:

(1) Endorsement

Name: [name of each covered location] ______

Address: [address of each covered location] ______

Policy Number: ______

Period of Coverage: [current policy period] ______

Name of [Insurer or Risk Retention Group]: ______

Address of [Insurer or Risk Retention Group]: ______

Name of Insured: ______

Address of Insured: ______

ENDORSEMENT:

1. This endorsement certifies that the policy to which the endorsement is attached provides liability insurance covering the following underground storage tanks:

[List the number of tanks at each facility and the name(s) and address(es) of the facility(ies) where the tanks are located. If more than one instrument is used to assure different tanks at any one facility, for each tank covered by this instrument, list the tank identification number provided in the notification submitted pursuant to 40 CFR 280.22, or the corresponding state requirement, and the name and address of the facility.]

for [insert: "taking corrective action" and/or "compensating third parties for bodily injury and property damage caused by" either "sudden accidental releases" or "nonsudden accidental releases" or "accidental releases"; in accordance with and subject to the limits of liability, exclusions, conditions, and other terms of the policy; if coverage is different for different tanks or locations, indicate the type of coverage applicable to each tank or location] arising from operating the underground storage tank(s) identified above.

The limits of liability are [insert the dollar amount of the "each occurrence" and "annual aggregate" limits of the Insurer's or Group's liability; if the amount of coverage is different for different types of coverage or for different underground storage tanks or locations, indicate the amount of coverage for each type of coverage and/or for each underground storage tank or location], exclusive of legal defense costs, which are subject to a separate limit under the policy. This coverage is provided under [policy number]. The effective date of said policy is [date].

2. The insurance afforded with respect to such occurrences is subject to all of the terms and conditions of the policy; provided, however, that any provisions inconsistent with subsections (a) through (e) of this Paragraph 2 are hereby amended to conform with subsections (a) through (e);

a. Bankruptcy or insolvency of the insured shall not relieve the ["Insurer" or "Group"] of its obligations under the policy to which this endorsement is attached.

b. The ["Insurer" or "Group"] is liable for the payment of amounts within any deductible applicable to the policy to the provider of corrective action or a damaged third-party, with a right of reimbursement by the insured for any such payment made by the ["Insurer" or "Group"]. This provision does not apply with respect to that amount of any deductible for which coverage is demonstrated under another mechanism or combination of mechanisms as specified in 40 CFR 280.95–280.102.

c. Whenever requested by [a Director of an implementing agency], the ["Insurer" or "Group"] agrees to furnish to [the Director] a signed duplicate original of the policy and all endorsements.

d. Cancellation or any other termination of the insurance by the ["Insurer" or "Group"], except for non-payment of premium or misrepresentation by the insured, will be effective only upon written notice and only after the expiration of 60 days after a copy of such written notice is received by the insured. Cancellation for non-payment of premium or misrepresentation by the insured will be effective only upon written notice and only after expiration of a minimum of 10 days after a copy of such written notice is received by the insured.

[Insert for claims-made policies:

e. The insurance covers claims otherwise covered by the policy that are reported to the ["Insurer" or "Group"] within six months of the effective date of cancellation or non-renewal of the policy except where the new or renewed policy has the same retroactive date or a retroactive date earlier than that of the prior policy, and which arise out of any covered occurrence that commenced after the policy retroactive date, if applicable, and prior to such policy renewal or termination date. Claims reported during such extended reporting period are subject to the terms, conditions, limits, including limits of liability, and exclusions of the policy.]

I hereby certify that the wording of this instrument is identical to the wording in 40 CFR 280.97(b)(1) and that the ["Insurer" or "Group"] is ["licensed to transact the business of insurance or eligible to provide insurance as an excess or surplus lines insurer in one or more states"].

[Signature of authorized representative of Insurer or Risk Retention Group]

[Name of person signing]

[Title of person signing], Authorized Representative of [name of Insurer or Risk Retention Group]

[Address of Representative]

(2) Certificate of Insurance

Name: [name of each covered location]

____

____

Address: [address of each covered location]

____

____

Policy Number: __________

Endorsement (if applicable): __________

Period of Coverage: [current policy period]

____

Name of [Insurer or Risk Retention Group]:

____

____

Address of [Insurer or Risk Retention Group]:

____

____

Name of Insured: __________

Address of Insured:

____

____

____

Certification:

1. [Name of Insurer or Risk Retention Group], [the "Insurer" or "Group"], as identified above, hereby certifies that it has issued liability insurance covering the following underground storage tank(s):

[List the number of tanks at each facility and the name(s) and address(es) of the facility(ies) where the tanks are located. If more than one instrument is used to assure different tanks at any one facility, for each tank covered by this instrument, list the tank identification number provided in the notification submitted pursuant to 40 CFR 280.22, or the corresponding state requirement, and the name and address of the facility.]

for [insert: "taking corrective action" and/or "compensating third parties for bodily injury and property damage caused by" either "sudden accidental releases" or "nonsudden accidental releases" or "accidental releases"; in accordance with and subject to the limits of liability, exclusions, conditions, and other terms of the policy; if coverage is different for different tanks or locations, indicate the type of coverage applicable to each tank or location] arising from operating the underground storage tank(s) identified above.

The limits of liability are [insert the dollar amount of the "each occurrence" and "annual aggregate" limits of the Insurer's or Group's liability; if the amount of coverage is different for different types of coverage or for different underground storage tanks or locations, indicate the amount of coverage for each type of coverage and/or for each underground storage tank or location], exclusive of legal defense costs, which are subject to a separate limit under the policy. This coverage is provided under [policy number]. The effective date of said policy is [date].

2. The ["Insurer" or "Group"] further certifies the following with respect to the insurance described in Paragraph 1:

a. Bankruptcy or insolvency of the insured shall not relieve the ["Insurer" or "Group"] of its obligations under the policy to which this certificate applies.

b. The ["Insurer" or "Group"] is liable for the payment of amounts within any deductible applicable to the policy to the provider of corrective action or a damaged third-party, with a right of reimbursement by the insured for any such payment made by the ["Insurer" or "Group"]. This provision does not apply with respect to that amount of any deductible for which coverage is demonstrated under another mechanism or combination of mechanisms as specified in 40 CFR 280.95–280.102.

c. Whenever requested by [a Director of an implementing agency], the ["Insurer" or "Group"] agrees to furnish to [the Director] a signed duplicate original of the policy and all endorsements.

d. Cancellation or any other termination of the insurance by the ["Insurer" or "Group"], except for non-payment of premium or misrepresentation by the insured, will be effective only upon written notice and only after the expiration of 60 days after a copy of such written notice is received by the insured. Cancellation for non-payment of premium or misrepresentation by the insured will be effective only upon written notice and only after expiration of a minimum of 10 days after a copy of such written notice is received by the insured.

[Insert for claims-made policies:

e. The insurance covers claims otherwise covered by the policy that are reported to the ["Insurer" or "Group"] within six months of the effective date of cancellation or non-renewal of the policy except where the new or renewed policy has the same retroactive date or a retroactive date earlier than that of the prior policy, and which arise out of any covered occurrence that commenced after the policy retroactive date, if applicable, and prior to such policy renewal or termination date. Claims reported during such extended reporting period are subject to the terms, conditions, limits, including limits of liability, and exclusions of the policy.]

I hereby certify that the wording of this instrument is identical to the wording in 40 CFR 280.97(b)(2) and that the ["Insurer" or "Group"] is ["licensed to transact the business of insurance, or eligible to provide insurance as an excess or surplus lines insurer, in one or more states"].

[Signature of authorized representative of Insurer]

[Type name]

[Title], Authorized Representative of [name of Insurer or Risk Retention Group]

[Address of Representative]

(c) Each insurance policy must be issued by an insurer or a risk retention group that, at a minimum, is licensed to transact the business of insurance or eligible to provide insurance as an excess or surplus lines insurer in one or more states.

[53 FR 43370, Oct. 26, 1988, as amended at 54 FR 47081, Nov. 9, 1989]

§ 280.98 Surety bond.

(a) An owner or operator may satisfy the requirements of § 280.93 by obtaining a surety bond that conforms to the requirements of this section. The surety company issuing the bond must be among those listed as acceptable sureties on federal bonds in the latest Circular 570 of the U.S. Department of the Treasury.

(b) The surety bond must be worded as follows, except that instructions in brackets must be replaced with the relevant information and the brackets deleted:

PERFORMANCE BOND

Date bond executed: ______

Period of coverage: ______

Principal: [legal name and business address of owner or operator]

______

Type of organization: [insert "individual," "joint venture," "partnership," or "corporation"]

______

State of incorporation (if applicable):

______

Surety(ies): [name(s) and business address(es)]

______

Scope of Coverage: [List the number of tanks at each facility and the name(s) and address(es) of the facility(ies) where the tanks are located. If more than one instrument is used to assure different tanks at any one facility, for each tank covered by this instrument, list the tank identification number provided in the notification submitted pursuant to 40 CFR 280.22, or the corresponding state requirement, and the name and address of the facility. List the coverage guaranteed by the bond: "taking corrective action" and/or "compensating third parties for bodily injury and property damage caused by" either "sudden accidental releases" or "nonsudden accidental releases" or "accidental releases" "arising from operating the underground storage tank"].

Penal sums of bond:

Per occurrence $ ______

Annual aggregate $ ______

Surety's bond number: ______

Know All Persons by These Presents, that we, the Principal and Surety(ies), hereto are firmly bound to [the implementing agency], in the above penal sums for the payment of which we bind ourselves, our heirs, executors, administrators, successors, and assigns jointly and severally; provided that, where the Surety(ies) are corporations acting as co-sureties, we, the Sureties, bind ourselves in such sums jointly and severally only for the purpose of allowing a joint action or actions against any or all of us, and for all other purposes each Surety binds itself, jointly and severally with the Principal, for the payment of such sums only as is set forth opposite the name of such Surety, but if no limit of liability is indicated, the limit of liability shall be the full amount of the penal sums.

Whereas said Principal is required under Subtitle I of the Resource Conservation and Recovery Act (RCRA), as amended, to provide financial assurance for [insert: "taking corrective action" and/or "compensating third parties for bodily injury and property damage caused by" either "sudden accidental releases" or "nonsudden accidental releases" or "accidental releases"; if coverage is different for different tanks or locations, indicate the type of coverage applicable to each tank or location] arising from operating the underground storage tanks identified above, and

Whereas said Principal shall establish a standby trust fund as is required when a surety bond is used to provide such financial assurance;

Now, therefore, the conditions of the obligation are such that if the Principal shall faithfully ["take corrective action, in accordance with 40 CFR part 280, subpart F and the Director of the state implementing agency's instructions for," and/or "compensate injured third parties for bodily injury and property damage caused by" either "sudden" or "nonsudden" or "sudden and nonsudden"] accidental releases arising from operating the tank(s) indentified above, or if the Principal shall provide alternate financial assurance, as specified in 40 CFR part 280, subpart H, within 120 days after the date the notice of cancellation is received by the Principal from the Surety(ies), then this obligation shall be null and void; otherwise it is to remain in full force and effect.

Such obligation does not apply to any of the following:

(a) Any obligation of [insert owner or operator] under a workers' compensation, disability benefits, or unemployment compensation law or other similar law;

(b) Bodily injury to an employee of [insert owner or operator] arising from, and in the course of, employment by [insert owner or operator];

(c) Bodily injury or property damage arising from the ownership, maintenance, use, or entrustment to others of any aircraft, motor vehicle, or watercraft;

(d) Property damage to any property owned, rented, loaned to, in the care, custody, or control of, or occupied by [insert owner or operator] that is not the direct result of a release from a petroleum underground storage tank;

(e) Bodily injury or property damage for which [insert owner or operator] is obligated to pay damages by reason of the assumption of liability in a contract or agreement other than a contract or agreement entered into to meet the requirements of 40 CFR 280.93.

The Surety(ies) shall become liable on this bond obligation only when the Principal has failed to fulfill the conditions described above.

Upon notification by [the Director of the implementing agency] that the Principal has failed to ["take corrective action, in accordance with 40 CFR part 280, subpart F and the Director's instructions," and/or "compensate injured third parties"] as guaranteed by this bond, the Surety(ies) shall either perform ["corrective action in accordance with 40 CFR part 280 and the Director's instructions," and/or "third-party liability compensation"] or place funds in an amount up to the annual aggregate penal sum into the standby trust fund as directed by [the Regional Administrator or the Director] under 40 CFR 280.108.

Upon notification by [the Director] that the Principal has failed to provide alternate financial assurance within 60 days after the date the notice of cancellation is received by the Principal from the Surety(ies) and that [the Director] has determined or suspects that a release has occurred, the Surety(ies) shall place funds in an amount not exceeding the annual aggregate penal sum into the standby trust fund as directed by [the Director] under 40 CFR 280.108.

The Surety(ies) hereby waive(s) notification of amendments to applicable laws, statutes, rules, and regulations and agrees that no such amendment shall in any way alleviate its (their) obligation on this bond.

The liability of the Surety(ies) shall not be discharged by any payment or succession of payments hereunder, unless and until such payment or payments shall amount in the annual aggregate to the penal sum shown on the face of the bond, but in no event shall the obligation of the Surety(ies) hereunder exceed the amount of said annual aggregate penal sum.

The Surety(ies) may cancel the bond by sending notice of cancellation by certified mail to the Principal, provided, however, that cancellation shall not occur during the 120 days beginning on the date of receipt of the notice of cancellation by the Principal, as evidenced by the return receipt.

The Principal may terminate this bond by sending written notice to the Surety(ies).

In Witness Thereof, the Principal and Surety(ies) have executed this Bond and have affixed their seals on the date set forth above.

The persons whose signatures appear below hereby certify that they are authorized to execute this surety bond on behalf of the Principal and Surety(ies) and that the wording of this surety bond is identical to the wording specified in 40 CFR 280.98(b) as such regulations were constituted on the date this bond was executed.

Principal

[Signature(s)]
[Names(s)]
[Title(s)]
[Corporate seal]

Corporate Surety(ies)

[Name and address]
[State of Incorporation: ________
[Liability limit: $________
[Signature(s)]
[Names(s) and title(s)]
[Corporate seal]

[For every co-surety, provide signature(s), corporate seal, and other information in the same manner as for Surety above.]

Bond premium: $________

(c) Under the terms of the bond, the surety will become liable on the bond obligation when the owner or operator fails to perform as guaranteed by the bond. In all cases, the surety's liability is limited to the per-occurrence and annual aggregate penal sums.

(d) The owner or operator who uses a surety bond to satisfy the requirements of § 280.93 must establish a standby trust fund when the surety bond is acquired. Under the terms of the bond, all amounts paid by the surety under the bond will be deposited directly into the standby trust fund in accordance with instructions from the Director under § 280.108. This standby trust fund must meet the requirements specified in § 280.103.

§ 280.99 Letter of credit.

(a) An owner or operator may satisfy the requirements of § 280.93 by obtaining an irrevocable standby letter of credit that conforms to the requirements of this section. The issuing institution must be an entity that has the authority to issue letters of credit in each state where used and whose letter-of-credit operations are regulated and examined by a federal or state agency.

(b) The letter of credit must be worded as follows, except that instructions in brackets are to be replaced with the relevant information and the brackets deleted:

IRREVOCABLE STANDBY LETTER OF CREDIT

[Name and address of issuing institution]
[Name and address of Director(s) of state implementing agency(ies)]

Dear Sir or Madam: We hereby establish our Irrevocable Standby Letter of Credit No. —— in your favor, at the request and for the account of [owner or operator name] of [address] up to the aggregate amount of [in words] U.S. dollars ($[insert dollar amount]), available upon presentation [insert, if more than one Director of a state implementing agency is a beneficiary, "by any one of you"] of

(1) your sight draft, bearing reference to this letter of credit, No. ——, and

(2) your signed statement reading as follows: "I certify that the amount of the draft is payable persuant to regulations issued under authority of Subtitle I of the Resource Conservation and Recovery Act of 1976, as amended."

This letter of credit may be drawn on to cover [insert: "taking corrective action" and/or "compensating third parties for bodily injury and property damage caused by" either "sudden accidental releases" or "nonsudden accidental releases" or "accidental releases"] arising from operating the underground storage tank(s) identified below in the amount of [in words] $[insert dollar amount] per occurrence and [in words] $[insert dollar amount] annual aggregate:

[List the number of tanks at each facility and the name(s) and address(es) of the facility(ies) where the tanks are located. If more than one instrument is used to assure different tanks at any one facility, for each tank covered by this instrument, list the tank identification number provided in the notification submitted pursuant to 40 CFR 280.22, or the corresponding state requirement, and the name and address of the facility.]

The letter of credit may not be drawn on to cover any of the following:

(a) Any obligation of [insert owner or operator] under a workers' compensation, disability benefits, or unemployment compensation law or other similar law;

(b) Bodily injury to an employee of [insert owner or operator] arising from, and in the course of, employment by [insert owner or operator];

(c) Bodily injury or property damage arising from the ownership, maintenance, use, or entrustment to others of any aircraft, motor vehicle, or watercraft;

(d) Property damage to any property owned, rented, loaned to, in the care, custody, or control of, or occupied by [insert owner or operator] that is not the direct result of a release from a petroleum underground storage tank;

(e) Bodily injury or property damage for which [insert owner or operator] is obligated to pay damages by reason of the assumption of liability in a contract or agreement other than a contract or agreement entered into to meet the requirements of 40 CFR 280.93.

This letter of credit is effective as of [date] and shall expire on [date], but such expiration date shall be automatically extended for a period of [at least the length of the original term] on [expiration date] and on each successive expiration date, unless, at least 120 days before the curent expiration date, we notify [owner or operator] by certified mail that we have decided not to extend this letter of credit beyond the current expiration date. In the event that [owner or operator] is so notified, any unused portion of the credit shall be available upon presentation of your sight draft for 120 days after the date of receipt by [owner or operator], as shown on the signed return receipt.

Whenever this letter of credit is drawn on under and in compliance with the terms of this credit, we shall duly honor such draft upon presentation to us, and we shall deposit the amount of the draft directly into the standby trust fund of [owner or operator] in accordance with your instructions.

We certify that the wording of this letter of credit is identical to the wording specified in 40 CFR 280.99(b) as such regulations were constituted on the date shown immediately below.

[Signature(s) and title(s) of official(s) of issuing institution]

[Date]

This credit is subject to [insert "the most recent edition of the Uniform Customs and Practice for Documentary Credits, published by the International Chamber of Commerce," or "the Uniform Commercial Code"].

(c) An owner or operator who uses a letter of credit to satisfy the requirements of §280.93 must also establish a standby trust fund when the letter of credit is acquired. Under the terms of the letter of credit, all amounts paid pursuant to a draft by the Director of the implementing agency will be deposited by the issuing institution directly into the standby trust fund in accordance with instructions from the Director under §280.108. This standby trust fund must meet the requirements specified in §280.103.

(d) The letter of credit must be irrevocable with a term specified by the issuing institution. The letter of credit must provide that credit be automatically renewed for the same term as the original term, unless, at least 120 days before the current expiration date, the issuing institution notifies the owner or operator by certified mail of its decision not to renew the letter of credit. Under the terms of the letter of credit, the 120 days will begin on the date when the owner or operator receives the notice, as evidenced by the return receipt.

§280.100 Use of state-required mechanism.

(a) For underground storage tanks located in a state that does not have an approved program, and where the state requires owners or operators of underground storage tanks to demonstrate financial responsibility for taking corrective action and/or for compensating third parties for bodily injury and property damage, an owner or operator may use a state-required financial mechanism to meet the requirements of §280.93 if the Regional Administrator determines that the state mechanism is at least equivalent to the financial mechanisms specified in this subpart.

(b) The Regional Administrator will evaluate the equivalency of a state-required mechanism principally in terms of: certainty of the availability of funds for taking corrective action and/or for compensating third parties; the amount of funds that will be made available; and the types of costs covered. The Regional Administrator may also consider other factors as is necessary.

(c) The state, an owner or operator, or any other interested party may submit to the Regional Administrator a written petition requesting that one or more of the state-required mechanisms be considered acceptable for meeting the requirements of §280.93. The submission must include copies of the appropriate state statutory and regulatory requirements and must show the amount of funds for corrective action and/or for compensating third parties assured by the mechanism(s). The Regional Administrator may require the petitioner to submit additional information as is deemed necessary to make this determination.

(d) Any petition under this section may be submitted on behalf of all of

341-150 O—93——34

the state's underground storage tank owners and operators.

(e) The Regional Administrator will notify the petitioner of his determination regarding the mechanism's acceptability in lieu of financial mechanisms specified in this subpart. Pending this determination, the owners and operators using such mechanisms will be deemed to be in compliance with the requirements of § 280.93 for underground storage tanks located in the state for the amounts and types of costs covered by such mechanisms.

[53 FR 43370, Oct. 26, 1988; 53 FR 51274, Dec. 21, 1988]

§ 280.101 State fund or other state assurance.

(a) An owner or operator may satisfy the requirements of § 280.93 for underground storage tanks located in a state, where EPA is administering the requirements of this subpart, which assures that monies will be available from a state fund or state assurance program to cover costs up to the limits specified in § 280.93 or otherwise assures that such costs will be paid if the Regional Administrator determines that the state's assurance is at least equivalent to the financial mechanisms specified in this subpart.

(b) The Regional Administrator will evaluate the equivalency of a state fund or other state assurance principally in terms of: Certainty of the availability of funds for taking corrective action and/or for compensating third parties; the amount of funds that will be made available; and the types of costs covered. The Regional Administrator may also consider other factors as is necessary.

(c) The state must submit to the Regional Administrator a description of the state fund or other state assurance to be supplied as financial assurance, along with a list of the classes of underground storage tanks to which the funds may be applied. The Regional Administrator may require the state to submit additional information as is deemed necessary to make a determination regarding the acceptability of the state fund or other state assurance. Pending the determination by the Regional Administrator, the owner or operator of a covered class of USTs will be deemed to be in compliance with the requirements of § 280.93 for the amounts and types of costs covered by the state fund or other state assurance.

(d) The Regional Administrator will notify the state of his determination regarding the acceptability of the state's fund or other assurance in lieu of financial mechanisms specified in this subpart. Within 60 days after the Regional Administrator notifies a state that a state fund or other state assurance is acceptable, the state must provide to each owner or operator for which it is assuming financial responsibility a letter or certificate describing the nature of the state's assumption of responsibility. The letter or certificate from the state must include, or have attached to it, the following information: the facility's name and address and the amount of funds for corrective action and/or for compensating third parties that is assured by the state. The owner or operator must maintain this letter or certificate on file as proof of financial responsibility in accordance with § 280.107(b)(5).

§ 280.102 Trust fund.

(a) An owner or operator may satisfy the requirements of § 280.93 by establishing a trust fund that conforms to the requirements of this section. The trustee must be an entity that has the authority to act as a trustee and whose trust operations are regulated and examined by a federal agency or an agency of the state in which the fund is established.

(b) The wording of the trust agreement must be identical to the wording specified in § 280.103(b)(1), and must be accompanied by a formal certification of acknowledgement as specified in § 280.103(b)(2).

(c) The trust fund, when established, must be funded for the full required amount of coverage, or funded for part of the required amount of coverage and used in combination with other mechanism(s) that provide the remaining required coverage.

(d) If the value of the trust fund is greater than the required amount of coverage, the owner or operator may submit a written request to the Director of the implementing agency for release of the excess.

(e) If other financial assurance as specified in this subpart is substituted for all or part of the trust fund, the owner or operator may submit a written request to the Director of the implementing agency for release of the excess.

(f) Within 60 days after receiving a request from the owner or operator for release of funds as specified in paragraph (d) or (e) of this section, the Director of the implementing agency will instruct the trustee to release to the owner or operator such funds as the Director specifies in writing.

§280.103 Standby trust fund.

(a) An owner or operator using any one of the mechanisms authorized by §§280.96, 280.98, or 280.99 must establish a standby trust fund when the mechanism is acquired. The trustee of the standby trust fund must be an entity that has the authority to act as a trustee and whose trust operations are regulated and examined by a Federal agency or an agency of the state in which the fund is established.

(b)(1) The standby trust agreement, or trust agreement, must be worded as follows, except that instructions in brackets are to be replaced with the relevant information and the brackets deleted:

TRUST AGREEMENT

Trust agreement, the "Agreement," entered into as of [date] by and between [name of the owner or operator], a [name of state] [insert "corporation," "partnership," "association," or "proprietorship"], the "Grantor," and [name of corporate trustee], [insert "Incorporated in the state of _____" or "a national bank"], the "Trustee."

Whereas, the United States Environmental Protection Agency, "EPA," an agency of the United States Government, has established certain regulations applicable to the Grantor, requiring that an owner or operator of an underground storage tank shall provide assurance that funds will be available when needed for corrective action and third-party compensation for bodily injury and property damage caused by sudden and nonsudden accidental releases arising from the operation of the underground storage tank. The attached Schedule A lists the number of tanks at each facility and the name(s) and address(es) of the facility(ies) where the tanks are located that are covered by the standpoint trust agreement.

[Whereas, the Grantor has elected to establish [insert either "a guarantee," "surety bond," or "letter of credit"] to provide all or part of such financial assurance for the underground storage tanks identified herein and is required to establish a standby trust fund able to accept payments from the instrument (This paragraph is only applicable to the standby trust agreement.)];

Whereas, the Grantor, acting through its duly authorized officers, has selected the Trustee to be the trustee under this agreement, and the Trustee is willing to act as trustee;

Now, therefore, the Grantor and the Trustee agree as follows:

Section 1. Definitions

As used in this Agreement:

(a) The term "Grantor" means the owner or operator who enters into this Agreement and any successors or assigns of the Grantor.

(b) The term "Trustee" means the Trustee who enters into this Agreement and any successor Trustee.

Section 2. Identification of the Financial Assurance Mechanism

This Agreement pertains to the [identify the financial assurance mechanism, either a guarantee, surety bond, or letter of credit, from which the standby trust fund is established to receive payments (This paragraph is only applicable to the standby trust agreement.)].

Section 3. Establishment of Fund

The Grantor and the Trustee hereby establish a trust fund, the "Fund," for the benefit of [implementing agency]. The Grantor and the Trustee intend that no third party have access to the Fund except as herein provided. [The Fund is established initially as a standby to receive payments and shall not consist of any property.] Payments made by the provider of financial assurance pursuant to [the Director of the implementing agency's] instruction are transferred to the Trustee and are referred to as the Fund, together with all earnings and profits thereon, less any payments or distributions made by the Trustee pursuant to this Agreement. The Fund shall be held by the Trustee, IN TRUST, as hereinafter provided. The Trustee shall not be responsible nor shall it undertake any responsibility for the amount or adequacy of, nor any duty to collect from the Grantor as provider of financial assurance, any payments necessary to discharge any liability of the Grantor established by [the state implementing agency]

Section 4. Payment for ["Corrective Action" and/or Third-Party Liability Claims"]

The Trustee shall make payments from the Fund as [the Director of the implementing agency] shall direct, in writing, to provide for the payment of the costs of [insert: "taking corrective action" and/or compensating third parties for bodily injury and property damage caused by" either "sudden accidental releases" or "nonsudden accidental releases" or "accidental releases"] arising from operating the tanks covered by the financial assurance mechanism identified in this Agreement.

The Fund may not be drawn upon to cover any of the following:

(a) Any obligation of [insert owner or operator] under a workers' compensation, disability benefits, or unemployment compensation law or other similar law;

(b) Bodily injury to an employee of [insert owner or operator] arising from, and in the course of employment by [insert owner or operator];

(c) Bodily injury or property damage arising from the ownership, maintenance, use, or entrustment to others of any aircraft, motor vehicle, or watercraft;

(d) Property damage to any property owned, rented, loaned to, in the care, custody, or control of, or occupied by [insert owner or operator] that is not the direct result of a release from a petroleum underground storage tank;

(e) Bodily injury or property damage for which [insert owner or operator] is obligated to pay damages by reason of the assumption of liability in a contract or agreement other than a contract or agreement entered into to meet the requirements of 40 CFR 280.93.

The Trustee shall reimburse the Grantor, or other persons as specified by [the Director], from the Fund for corrective action expenditures and/or third-party liability claims in such amounts as [the Director] shall direct in writing. In addition, the Trustee shall refund to the Grantor such amounts as [the Director] specifies in writing. Upon refund, such funds shall no longer constitute part of the Fund as defined herein.

Section 5. Payments Comprising the Fund

Payments made to the Trustee for the Fund shall consist of cash and securities acceptable to the Trustee.

Section 6. Trustee Management

The Trustee shall invest and reinvest the principal and income of the Fund and keep the Fund invested as a single fund, without distinction between principal and income, in accordance with general investment policies and guidelines which the Grantor may communicate in writing to the Trustee from time to time, subject, however, to the provisions of this Section. In investing, reinvesting, exchanging, selling, and managing the Fund, the Trustee shall discharge his duties with respect to the trust fund solely in the interest of the beneficiaries and with the care, skill, prudence, and diligence under the circumstances then prevailing which persons of prudence, acting in a like capacity and familiar with such matters, would use in the conduct of an enterprise of a like character and with like aims; except that:

(i) Securities or other obligations of the Grantor, or any other owner or operator of the tanks, or any of their affiliates as defined in the Investment Company Act of 1940, as amended, 15 U.S.C. 80a-2(a), shall not be acquired or held, unless they are securities or other obligations of the federal or a state government;

(ii) The Trustee is authorized to invest the Fund in time or demand deposits of the Trustee, to the extent insured by an agency of the federal or state government; and

(iii) The Trustee is authorized to hold cash awaiting investment or distribution uninvested for a reasonable time and without liability for the payment of interest thereon.

Section 7. Commingling and Investment

The Trustee is expressly authorized in its discretion:

(a) To transfer from time to time any or all of the assets of the Fund to any common, commingled, or collective trust fund created by the Trustee in which the Fund is eligible to participate, subject to all of the provisions thereof, to be commingled with the assets of other trusts participating therein; and

(b) To purchase shares in any investment company registered under the Investment Company Act of 1940, 15 U.S.C. 80a-1 et seq., including one which may be created, managed, underwritten, or to which investment advice is rendered or the shares of which are sold by the Trustee. The Trustee may vote such shares in its discretion.

Section 8. Express Powers of Trustee

Without in any way limiting the powers and discretions conferred upon the Trustee by the other provisions of this Agreement or by law, the Trustee is expressly authorized and empowered:

(a) To sell, exchange, convey, transfer, or otherwise dispose of any property held by it, by public or private sale. No person dealing with the Trustee shall be bound to see to the application of the purchase money or to inquire into the validity or expediency of any such sale or other disposition;

(b) To make, execute, acknowledge, and deliver any and all documents of transfer and conveyance and any and all other instruments that may be necessary or appropriate to carry out the powers herein granted;

(c) To register any securities held in the Fund in its own name or in the name of a nominee and to hold any security in bearer form or in book entry, or to combine certificates representing such securities with certificates of the same issue held by the Trustee in other fiduciary capacities, or to deposit or arrange for the deposit of such securities in a qualified central depository even though, when so deposited, such securities may be merged and held in bulk in the name of the nominee of such depository with other securities deposited therein by another person, or to deposit or arrange for the deposit of any securities issued by the United States Government, or any agency or instrumentality thereof, with a Federal Reserve bank, but the books and records of the Trustee shall at all times show that all such securities are part of the Fund;

(d) To deposit any cash in the Fund in interest-bearing accounts maintained or savings certificates issued by the Trustee, in its separate corporate capacity, or in any other banking institution affiliated with the Trustee, to the extent insured by an agency of the federal or state government; and

(e) To compromise or otherwise adjust all claims in favor of or against the Fund.

Section 9. Taxes and Expenses

All taxes of any kind that may be assessed or levied against or in respect of the Fund and all brokerage commissions incurred by the Fund shall be paid from the Fund. All other expenses incurred by the Trustee in connection with the administration of this Trust, including fees for legal services rendered to the Trustee, the compensation of the Trustee to the extent not paid directly by the Grantor, and all other proper charges and disbursements of the Trustee shall be paid from the Fund.

Section 10. Advice of Counsel

The Trustee may from time to time consult with counsel, who may be counsel to the Grantor, with respect to any questions arising as to the construction of this Agreement or any action to be taken hereunder. The Trustee shall be fully protected, to the extent permitted by law, in acting upon the advice of counsel.

Section 11. Trustee Compensation

The Trustee shall be entitled to reasonable compensation for its services as agreed upon in writing from time to time with the Grantor.

Section 12. Successor Trustee

The Trustee may resign or the Grantor may replace the Trustee, but such resignation or replacement shall not be effective until the Grantor has appointed a successor trustee and this successor accepts the appointment. The successor trustee shall have the same powers and duties as those conferred upon the Trustee hereunder. Upon the successor trustee's acceptance of the appointment, the Trustee shall assign, transfer, and pay over to the successor trustee the funds and properties then constituting the Fund. If for any reason the Grantor cannot or does not act in the event of the resignation of the Trustee, the Trustee may apply to a court of competent jurisdiction for the appointment of a successor trustee or for instructions. The successor trustee shall specify the date on which it assumes administration of the trust in writing sent to the Grantor and the present Trustee by certified mail 10 days before such change becomes effective. Any expenses incurred by the Trustee as a result of any of the acts contemplated by this Section shall be paid as provided in Section 9.

Section 13. Instructions to the Trustee

All orders, requests, and instructions by the Grantor to the Trustee shall be in writing, signed by such persons as are designated in the attached Schedule B or such other designees as the Grantor may designate by amendment to Schedule B. The Trustee shall be fully protected in acting without inquiry in accordance with the Grantor's orders, requests, and instructions. All orders, requests, and instructions by [the Director of the implementing agency] to the Trustee shall be in writing, signed by [the Director], and the Trustee shall act and shall be fully protected in acting in accordance with such orders, requests, and instructions. The Trustee shall have the right to assume, in the absence of written notice to the contrary, that no event constituting a change or a termination of the authority of any person to act on behalf of the Grantor or [the director] hereunder has occurred. The Trustee shall have no duty to act in the absence of such orders, requests, and instructions from the Grantor and/or [the Director], except as provided for herein.

Section 14. Amendment of Agreement

This Agreement may be amended by an instrument in writing executed by the Grantor and the Trustee, or by the Trustee and [the Director of the implementing agency] if the Grantor ceases to exist.

Section 15. Irrevocability and Termination

Subject to the right of the parties to amend this Agreement as provided in Section 14, this Trust shall be irrevocable and shall continue until terminated at the written direction of the Grantor and the Trustee, or by the Trustee and [the Director of the implementing agency], if the Grantor ceases to exist. Upon termination of the Trust, all remaining trust property, less final trust administration expenses, shall be delivered to the Grantor.

Section 16. Immunity and Indemnification

The Trustee shall not incur personal liability of any nature in connection with any act or omission, made in good faith, in the administration of this Trust, or in carrying out any directions by the Grantor or [the Director of the implementing agency] issued in accordance with this Agreement. The Trustee shall be indemnified and saved harmless by the Grantor, from and against any personal liability to which the Trustee may be subjected by reason of any act or conduct in its official capacity, including all expenses reasonably incurred in its defense in the event the Grantor fails to provide such defense.

Section 17. Choice of Law

This Agreement shall be administered, construed, and enforced according to the laws of the state of [insert name of state], or the Comptroller of the Currency in the case of National Association banks.

Section 18. Interpretation

As used in this Agreement, words in the singular include the plural and words in the plural include the singular. The descriptive headings for each section of this Agreement shall not affect the interpretation or the legal efficacy of this Agreement.

In Witness whereof the parties have caused this Agreement to be executed by their respective officers duly authorized and their corporate seals (if applicable) to be hereunto affixed and attested as of the date first above written. The parties below certify that the wording of this Agreement is identical to the wording specified in 40 CFR 280.103(b)(1) as such regulations were constituted on the date written above.

[Signature of Grantor]
[Name of the Grantor]
[Title]

Attest:
[Signature of Trustee]
[Name of the Trustee]
[Title]
[Seal]
[Signature of Witness]
[Name of the Witness]
[Title]
[Seal]

(2) The standby trust agreement, or trust agreement must be accompanied by a formal certification of acknowledgement similar to the following. State requirements may differ on the proper content of this acknowledgment.

State of ——————————

County of ——————————

On this [date], before me personally came [owner or operator] to me known, who, being by me duly sworn, did depose and say that she/he resides at [address], that she/he is [title] of [corporation], the corporation described in and which executed the above instrument; that she/he knows the seal of said corporation; that the seal affixed to such instrument is such corporate seal; that it was so affixed by order of the Board of Directors of said corporation; and that she/he signed her/his name thereto by like order.

[Signature of Notary Public]
[Name of Notary Public]

(c) The Director of the implementing agency will instruct the trustee to refund the balance of the standby trust fund to the provider of financial assurance if the Director determines that no additional corrective action costs or third-party liability claims will occur as a result of a release covered by the financial assurance mechanism for which the standby trust fund was established.

(d) An owner or operator may establish one trust fund as the depository mechanism for all funds assured in compliance with this rule.

[53 FR 43370, Oct. 26, 1988; 53 FR 51274, Dec. 21, 1988]

§ 280.104 Local government bond rating test.

(a) A general purpose local government owner or operator and/or local government serving as a guarantor may satisfy the requirements of § 280.93 by having a currently outstanding issue or issues of general obligation bonds of $1 million or more, excluding refunded obligations, with a Moody's rating of Aaa, Aa, A, or Baa, or a Standard & Poor's rating of AAA, AA, A, or BBB. Where a local government has multiple outstanding issues, or where a local government's bonds are rated by both Moody's and Standard and Poor's, the lowest rating must be used to determine eligibility. Bonds that are backed by credit enhancement other than municipal bond insurance may not be considered in determining the amount of applicable bonds outstanding.

(b) A local government owner or operator or local government serving as a guarantor that is not a general-purpose local government and does not have the legal authority to issue general obligation bonds may satisfy the requirements of § 280.93 by having a currently outstanding issue or issues of revenue bonds of $1 million or more, excluding refunded issues and by also having a Moody's rating of Aaa, A, A, or Baa, or a Standard & Poor's rating of AAA, AA, A, or BBB as the lowest rating for any rated revenue bond issued by the local government. Where bonds are rated by both Moody's and Standard & Poor's, the lower rating for each bond must be used to determine eligibility. Bonds that are backed by credit enhancement may not be considered in determining the amount of applicable bonds outstanding.

(c) The local government owner or operator and/or guarantor must maintain a copy of its bond rating published within the last 12 months by Moody's or Standard & Poor's.

(d) To demonstrate that it meets the local government bond rating test, the chief financial officer of a general purpose local government owner or operator and/or guarantor must sign a letter worded exactly as follows, except that the instructions in brackets are to be replaced by the relevant information and the brackets deleted:

LETTER FROM CHIEF FINANCIAL OFFICER

I am the chief financial officer of [insert: name and address of local government owner or operator, or guarantor]. This letter is in support of the use of the bond rating test to demonstrate financial responsibility for [insert: "taking corrective action" and/or "compensating third parties for bodily injury and property damage"] caused by [insert: "sudden accidental releases" and/or "nonsudden accidental releases"] in the amount of at least [insert: dollar amount] per occurrence and [insert: dollar amount] annual aggregate arising from operating (an) underground storage tank(s).

Underground storage tanks at the following facilities are assured by this bond rating test: [List for each facility: the name and address of the facility where tanks are assured by the bond rating test].

The details of the issue date, maturity, outstanding amount, bond rating, and bond rating agency of all outstanding bond issues that are being used by [name of local government owner or operator, or guarantor] to demonstrate financial responsibility are as follows: [complete table]

Issue date	Maturity date	Outstanding amount	Bond rating	Rating agency
				[Moody's or Standard & Poor's]

The total outstanding obligation of [insert amount], excluding refunded bond issues, exceeds the minimum amount of $1 million. All outstanding general obligation bonds issued by this government that have been rated by Moody's or Standard & Poor's are rated as at least investment grade (Moody's Baa or Standard & Poor's BBB) based on the most recent ratings published within the last 12 months. Neither rating service has provided notification within the last 12 months of downgrading of bond ratings below investment grade or of withdrawal of bond rating other than for repayment of outstanding bond issues.

I hereby certify that the wording of this letter is identical to the wording specified in 40 CFR Part 280.104(d) as such regulations were constituted on the date shown immediately below.

[Date] ______

[Signature] ______

[Name] ______

[Title] ______

(e) To demonstrate that it meets the local government bond rating test, the chief financial officer of local government owner or operator and/or guarantor other than a general purpose government must sign a letter worded exactly as follows, except that the instructions in brackets are to be replaced by the relevant information and the brackets deleted:

LETTER FROM CHIEF FINANCIAL OFFICER

I am the chief financial officer of [insert: name and address of local government owner or operator, or guarantor]. This letter is in support of the use of the bond rating test to demonstrate financial responsibility for [insert: "taking corrective action" and/or "compensating third parties for bodily injury and property damage"] caused by [insert : "sudden accidental releases" and/or "nonsudden accidental releases"] in the amount of at least [insert: dollar amount] per occurrence and [insert: dollar amount] annual aggregate arising from operating (an) underground storage tank(s). This local government is not organized to provide general governmental services and does not have the legal authority under state law or constitutional provisions to issue general obligation debt.

Underground storage tanks at the following facilities are assured by this bond rating test: [List for each facility: the name and address of the facility where tanks are assured by the bond rating test].

The details of the issue date, maturity, outstanding amount, bond rating, and bond rating agency of all outstanding revenue bond issues that are being used by [name of local government owner or operator, or guarantor] to demonstrate financial responsibility are as follows: [complete table]

Issue date	Maturity date	Outstanding amount	Bond rating	Rating agency
				[Moody's or Standard & Poor's]

The total outstanding obligation of [insert amount], excluding refunded bond issues, exceeds the minimum amount of $1 million. All outstanding revenue bonds issued by this government that have been rated by Moody's or Standard & Poor's are rated as at least investment grade (Moody's Baa or Standard & Poor's BBB) based on the most recent ratings published within the last 12 months. The revenue bonds listed are not backed by third-party credit enhancement or are insured by a municipal bond insurance company. Neither rating service has provided notification within the last 12 months of downgrading of bond ratings below investment grade or of withdrawal of bond rating other than for repayment of outstanding bond issues.

I hereby certify that the wording of this letter is identical to the wording specified in 40 CFR part 280.104(e) as such regulations were constituted on the date shown immediately below.

[Date] ______

[Signature] ______

[Name] ______

[Title] ______

(f) The Director of the implementing agency may require reports of financial condition at any time from the local government owner or operator, and/or local government guarantor. If the Director finds, on the basis of such reports or other information, that the local government owner or operator, and/or guarantor, no longer meets the local government bond rating test requirements of § 280.104, the local government owner or operator must obtain alternative coverage within 30 days after notification of such a finding.

(g) If a local government owner or operator using the bond rating test to provide financial assurance finds that it no longer meets the bond rating test requirements, the local government owner or operator must obtain alternative coverage within 150 days of the change in status.

[58 FR 9053, Feb. 18, 1993]

§ 280.105 Local government financial test.

(a) A local government owner or operator may satisfy the requirements of § 280.93 by passing the financial test specified in this section. To be eligible to use the financial test, the local government owner or operator must have the ability and authority to assess and levy taxes or to freely establish fees and charges. To pass the local government financial test, the owner or operator must meet the criteria of paragraphs (b)(2) and (b)(3) of this section based on year-end financial statements for the latest completed fiscal year.

(b)(1) The local government owner or operator must have the following information available, as shown in the year-end financial statements for the latest completed fiscal year:

(i) *Total revenues:* Consists of the sum of general fund operating and non-operating revenues including net local taxes, licenses and permits, fines and forfeitures, revenues from use of money and property, charges for services, investment earnings, sales (property, publications, etc.), intergovernmental revenues (restricted and unrestricted), and total revenues from all other governmental funds including enterprise, debt service, capital projects, and special revenues, but excluding revenues to funds held in a trust or agency capacity. For purposes of this test, the calculation of total revenues shall exclude all transfers between funds under the direct control of the local government using the financial test (interfund transfers), liquidation of investments, and issuance of debt.

(ii) *Total expenditures:* Consists of the sum of general fund operating and non-operating expenditures including public safety, public utilities, transportation, public works, environmental protection, cultural and recreational, community development, revenue sharing, employee benefits and compensation, office management, planning and zoning, capital projects, interest payments on debt, payments for retirement of debt principal, and total expenditures from all other governmental funds including enterprise, debt service, capital projects, and special revenues. For purposes of this test, the calculation of total expenditures shall exclude all transfers between funds under the direct control of the local government using the financial test (interfund transfers).

(iii) *Local revenues:* Consists of total revenues (as defined in paragraph (b)(1)(i) of this section) minus the sum of all transfers from other governmental entities, including all monies received from Federal, state, or local government sources.

(iv) *Debt service:* Consists of the sum of all interest and principal payments on all long-term credit obligations and all interest-bearing short-term credit obligations. Includes interest and principal payments on general obligation bonds, revenue bonds, notes, mortgages, judgments, and interest bearing warrants. Excludes payments on non-interest-bearing short-term obligations, interfund obligations, amounts owed in a trust or agency capacity, and advances and contingent loans from other governments.

(v) *Total funds:* Consists of the sum of cash and investment securities from all funds, including general, enterprise, debt service, capital projects, and special revenue funds, but excluding employee retirement funds, at the end of the local government's financial reporting year. Includes Federal securities, Federal agency securities, state and local government securities, and other securities such as bonds, notes and mortgages. For purposes of this test, the calculation of total funds shall exclude agency funds, private trust funds, accounts receivable, value of real property, and other non-security assets.

(vi) *Population* consists of the number of people in the area served by the local government.

(2) The local government's year-end financial statements, if independently audited, cannot include an adverse auditor's opinion or a disclaimer of opinion. The local government cannot have outstanding issues of general obligation or revenue bonds that are rated as less than investment grade.

(3) The local government owner or operator must have a letter signed by the chief financial officer worded as specified in paragraph (c) of this section.

(c) To demonstrate that it meets the financial test under paragraph (b) of this section, the chief financial officer of the local government owner or operator, must sign, within 120 days of the close of each financial reporting year, as defined by the twelve-month period for which financial statements used to support the financial test are prepared, a letter worded exactly as follows, except that the instructions in brackets are to be replaced by the relevant information and the brackets deleted:

LETTER FROM CHIEF FINANCIAL OFFICER

I am the chief financial officer of [insert: name and address of the owner or operator]. This letter is in support of the use of the local government financial test to demonstrate financial responsibility for [insert: "taking corrective action" and/or "compensating third parties for bodily injury and property damage"] caused by [insert: "sudden accidental releases" and/or "nonsudden accidental releases"] in the amount of at least [insert: dollar amount] per occurrence and [insert: dollar amount] annual aggregate arising from operating [an] underground storage tank[s].

Underground storage tanks at the following facilities are assured by this financial test [List for each facility: the name and address of the facility where tanks assured by this financial test are located. If separate mechanisms or combinations of mechanisms are being used to assure any of the tanks at this facility, list each tank assured by this financial test by the tank identification number provided in the notification submitted pursuant to 40 CFR Part 280.22 or the corresponding state requirements.]

This owner or operator has not received an adverse opinion, or a disclaimer of opinion from an independent auditor on its financial statements for the latest completed fiscal year. Any outstanding issues of general obligation or revenue bonds, if rated, have a

Moody's rating of Aaa, Aa, A, or Baa or a Standard and Poor's rating of AAA, AA, A, or BBB; if rated by both firms, the bonds have a Moody's rating of Aaa, Aa, A, or Baa and a Standard and Poor's rating of AAA, AA, A, or BBB.

WORKSHEET FOR MUNICIPAL FINANCIAL TEST

Part I: Basic Information

1. Total Revenues

a. Revenues (dollars) ________

Value of revenues excludes liquidation of investments and issuance of debt. Value includes all general fund operating and non-operating revenues, as well as all revenues from all other governmental funds including enterprise, debt service, capital projects, and special revenues, but excluding revenues to funds held in a trust or agency capacity.

b. Subtract interfund transfers (dollars) ________

c. Total Revenues (dollars) ________

2. Total Expenditures

a. Expenditures (dollars) ________

Value consists of the sum of general fund operating and non-operating expenditures including interest payments on debt, payments for retirement of debt principal, and total expenditures from all other governmental funds including enterprise, debt service, capital projects, and special revenues.

b. Subtract interfund transfers (dollars) ________

c. Total Expenditures (dollars) ________

3. Local Revenues

a. Total Revenues (from 1c) (dollars) ________

b. Subtract total intergovernmental transfers (dollars) ________

c. Local Revenues (dollars) ________

4. Debt Service

a. Interest and fiscal charges (dollars) ________

b. Add debt retirement (dollars) ________

c. Total Debt Service (dollars) ________

5. Total Funds (Dollars) ________

(Sum of amounts held as cash and investment securities from all funds, excluding amounts held for employee retirement funds, agency funds, and trust funds)

6. Population (Persons) ________

Part II: Application of Test

7. Total Revenues to Population

a. Total Revenues (from 1c) ________
b. Population (from 6) ________
c. Divide 7a by 7b ________
d. Subtract 417 ________
e. Divide by 5,212 ________
f. Multiply by 4.095 ________

8. Total Expenses to Population

a. Total Expenses (from 2c) ________
b. Population (from 6) ________
c. Divide 8a by 8b ________
d. Subtract 524 ________
e. Divide by 5,401 ________
f. Multiply by 4.095 ________

9. Local Revenues to Total Revenues

a. Local Revenues (from 3c) ________
b. Total Revenues (from 1c) ________
c. Divide 9a by 9b ________
d. Subtract .695 ________
e. Divide by .205 ________
f. Multiply by 2.840 ________

10. Debt Service to Population

a. Debt Service (from 4d) ________
b. Population (from 6) ________
c. Divide 10a by 10b ________
d. Subtract 51 ________
e. Divide by 1,038 ________
f. Multiply by −1.866 ________

11. Debt Service to Total Revenues

a. Debt Service (from 4d) ________
b. Total Revenues (from 1c) ________
c. Divide 11a by 11b ________
d. Subtract .068 ________
e. Divide by .259 ________
f. Multiply by −3.533 ________

12. Total Revenues to Total Expenses

a. Total Revenues (from 1c) ________
b. Total Expenses (from 2c) ________
c. Divide 12a by 12b ________
d. Subtract .910 ________
e. Divide by .899 ________
f. Multiply by 3.458 ________

13. Funds Balance to Total Revenues

a. Total Funds (from 5) ________
b. Total Revenues (from 1c) ________
c. Divide 13a by 13b ________
d. Subtract .891 ________
e. Divide by 9.156 ________
f. Multiply by 3.270 ________

14. Funds Balance to Total Expenses

a. Total Funds (from 5) ________
b. Total Expenses (from 2c) ________
c. Divide 14a by 14b ________
d. Subtract .866 ________
e. Divide by 6.409 ________
f. Multiply by 3.270 ________

15. Total Funds to Population ________

a. Total Funds (from 5) ________
b. Population (from 6) ________
c. Divide 15a by 15b ________

d. Subtract 270 ________
e. Divide by 4,548 ________
f. Multiply by 1.866 ________

16. Add 7f + 8f + 9f + 10f + 11f + 12f + 13f + 14f + 15f + 4.937________

I hereby certify that the financial index shown on line 16 of the worksheet is greater than zero and that the wording of this letter is identical to the wording specified in 40 CFR part 280.105(c) as such regulations were constituted on the date shown immediately below.

[Date]
[Signature]
[Name]
[Title]

(d) If a local government owner or operator using the test to provide financial assurance finds that it no longer meets the requirements of the financial test based on the year-end financial statements, the owner or operator must obtain alternative coverage within 150 days of the end of the year for which financial statements have been prepared.

(e) The Director of the implementing agency may require reports of financial condition at any time from the local government owner or operator. If the Director finds, on the basis of such reports or other information, that the local government owner or operator no longer meets the financial test requirements of § 280.105 (b) and (c), the owner or operator must obtain alternate coverage within 30 days after notification of such a finding.

(f) If the local government owner or operator fails to obtain alternate assurance within 150 days of finding that it no longer meets the requirements of the financial test based on the year-end financial statements or within 30 days of notification by the Director of the implementing agency that it no longer meets the requirements of the financial test, the owner or operator must notify the Director of such failure within 10 days.

[58 FR 9054, Feb. 18, 1993]

§ 280.106 Local government guarantee.

(a) A local government owner or operator may satisfy the requirements of § 280.93 by obtaining a guarantee that conforms to the requirements of this section. The guarantor must be either the state in which the local government owner or operator is located or a local government having a "substantial governmental relationship" with the owner and operator and issuing the guarantee as an act incident to that relationship. A local government acting as the guarantor must:

(1) demonstrate that it meets the bond rating test requirement of § 280.104 and deliver a copy of the chief financial officer's letter as contained in § 280.104(c) to the local government owner or operator; or

(2) demonstrate that it meets the worksheet test requirements of § 280.105 and deliver a copy of the chief financial officer's letter as contained in § 280.105(c) to the local government owner or operator; or

(3) demonstrate that it meets the local government fund requirements of § 280.107(a), § 280.107(b), or § 280.107(c) and deliver a copy of the chief financial officer's letter as contained in § 280.107 to the local government owner or operator.

(b) If the local government guarantor is unable to demonstrate financial assurance under any of §§ 280.104, 280.105, 280.107(a), 280.107(b), or 280.107(c), at the end of the financial reporting year, the guarantor shall send by certified mail, before cancellation or non-renewal of the guarantee, notice to the owner or operator. The guarantee will terminate no less than 120 days after the date the owner or operator receives the notification, as evidenced by the return receipt. The owner or operator must obtain alternative coverage as specified in § 280.114(c).

(c) The guarantee agreement must be worded as specified in paragraph (d) or (e) of this section, depending on which of the following alternative guarantee arrangements is selected:

(1) If, in the default or incapacity of the owner or operator, the guarantor guarantees to fund a standby trust as directed by the Director of the implementing agency, the guarantee shall be worded as specified in paragraph (d) of this section.

(2) If, in the default or incapacity of the owner or operator, the guarantor guarantees to make payments as directed by the Director of the implementing agency for taking corrective action or compensating third parties

for bodily injury and property damage, the guarantee shall be worded as specified in paragraph (e) of this section.

(d) If the guarantor is a state, the local government guarantee with standby trust must be worded exactly as follows, except that instructions in brackets are to be replaced with relevant information and the brackets deleted:

Local Government Guarantee With Standby Trust Made by a State

Guarantee made this [date] by [name of state], herein referred to as guarantor, to [the state implementing agency] and to any and all third parties, and obliges, on behalf of [local government owner or operator].

Recitals

(1) Guarantor is a state.

(2) [Local government owner or operator] owns or operates the following underground storage tank(s) covered by this guarantee: [List the number of tanks at each facility and the name(s) and address(es) of the facility(ies) where the tanks are located. If more than one instrument is used to assure different tanks at any one facility, for each tank covered by this instrument, list the tank identification number provided in the notification submitted pursuant to 40 CFR part 280 or the corresponding state requirement, and the name and address of the facility.] This guarantee satisfies 40 CFR part 280, subpart H requirements for assuring funding for [insert: "taking corrective action" and/or "compensating third parties for bodily injury and property damage caused by" either "sudden accidental releases" or "nonsudden accidental releases" or "accidental releases"; if coverage is different for different tanks or locations, indicate the type of coverage applicable to each tank or location] arising from operating the above-identified underground storage tank(s) in the amount of [insert dollar amount] per occurrence and [insert dollar amount] annual aggregate.

(3) Guarantor guarantees to [implementing agency] and to any and all third parties that:

In the event that [local government owner or operator] fails to provide alternative coverage within 60 days after receipt of a notice of cancellation of this guarantee and the [Director of the implementing agency] has determined or suspects that a release has occurred at an underground storage tank covered by this guarantee, the guarantor, upon instructions from the [Director] shall fund a standby trust fund in accordance with the provisions of 40 CFR part 280.112, in an amount not to exceed the coverage limits specified above.

In the event that the [Director] determines that [local government owner or operator] has failed to perform corrective action for releases arising out of the operation of the above-identified tank(s) in accordance with 40 CFR part 280, subpart F, the guarantor upon written instructions from the [Director] shall fund a standby trust fund in accordance with the provisions of 40 CFR part 280.112, in an amount not to exceed the coverage limits specified above.

If [owner or operator] fails to satisfy a judgment or award based on a determination of liability for bodily injury or property damage to third parties caused by ["sudden" and/or "nonsudden"] accidental releases arising from the operation of the above-identified tank(s), or fails to pay an amount agreed to in settlement of a claim arising from or alleged to arise from such injury or damage, the guarantor, upon written instructions from the [Director], shall fund a standby trust in accordance with the provisions of 40 CFR part 280.112 to satisfy such judgment(s), award(s), or settlement agreement(s) up to the limits of coverage specified above.

(4) Guarantor agrees to notify [owner or operator] by certified mail of a voluntary or involuntary proceeding under Title 11 (Bankruptcy), U.S. Code naming guarantor as debtor, within 10 days after commencement of the proceeding.

(5) Guarantor agrees to remain bound under this guarantee notwithstanding any modification or alteration of any obligation of [owner or operator] pursuant to 40 CFR part 280.

(6) Guarantor agrees to remain bound under this guarantee for so long as [local government owner or operator] must comply with the applicable financial responsibility requirements of 40 CFR part 280, subpart H for the above identified tank(s), except that guarantor may cancel this guarantee by sending notice by certified mail to [owner or operator], such cancellation to become effective no earlier than 120 days after receipt of such notice by [owner or operator], as evidenced by the return receipt.

(7) The guarantor's obligation does not apply to any of the following:

(a) Any obligation of [local government owner or operator] under a workers' compensation, disability benefits, or unemployment compensation law or other similar law;

(b) Bodily injury to an employee of [insert: local government owner or operator] arising from, and in the course of, employment by [insert: local government owner or operator];

(c) Bodily injury or property damage arising from the ownership, maintenance, use, or entrustment to others of any aircraft, motor vehicle, or watercraft;

(d) Property damage to any property owned, rented, loaded to, in the care, custody, or control of, or occupied by [insert:

local government owner or operator] that is not the direct result of a release from a petroleum underground storage tank;

(e) Bodily damage or property damage for which [insert owner or operator] is obligated to pay damages by reason of the assumption of liability in a contract or agreement other than a contract or agreement entered into to meet the requirements of 40 CFR part 280.93.

(8) Guarantor expressly waives notice of acceptance of this guarantee by [the implementing agency], by any or all third parties, or by [local government owner or operator].

I hereby certify that the wording of this guarantee is identical to the wording specified in 40 CFR part 280.106(d) as such regulations were constituted on the effective date shown immediately below.

Effective date: ____________________

[Name of guarantor]

[Authorized signature for guarantor]

[Name of person signing]

[Title of person signing]

Signature of witness or notary:

If the guarantor is a local government, the local government guarantee with standby trust must be worded exactly as follows, except that instructions in brackets are to be replaced with relevant information and the brackets deleted:

LOCAL GOVERNMENT GUARANTEE WITH STANDBY TRUST MADE BY A LOCAL GOVERNMENT

Guarantee made this [date] by [name of guaranteeing entity], a local government organized under the laws of [name of state], herein referred to as guarantor, to [the state implementing agency] and to any and all third parties, and obliges, on behalf of [local government owner or operator].

Recitals

(1) Guarantor meets or exceeds [select one: the local government bond rating test requirements of 40 CFR part 280.104, the local government financial test requirements of 40 CFR part 280.105, or the local government fund under 40 CFR part 280.107(a), 280.107(b), or 280.107(c)].

(2) [Local government owner or operator] owns or operates the following underground storage tank(s) covered by this guarantee: [List the number of tanks at each facility and the name(s) and address(es) of the facility(ies) where the tanks are located. If more than one instrument is used to assure different tanks at any one facility, for each tank covered by this instrument, list the tank identification number provided in the notification submitted pursuant to 40 CFR part 280 or the corresponding state requirement, and the name and address of the facility.] This guarantee satisfies 40 CFR part 280, subpart H requirements for assuring funding for [insert: "taking corrective action" and/or "compensating third parties for bodily injury and property damage caused by" either "sudden accidental releases" or "nonsudden accidental releases" or "accidental releases"; if coverage is different for different tanks or locations, indicate the type of coverage applicable to each tank or location] arising from operating the above-identified underground storage tank(s) in the amount of [insert dollar amount] per occurrence and [insert: dollar amount] annual aggregate.

(3) Incident to our substantial governmental relationship with [local government owner or operator], guarantor guarantees to [implementing agency] and to any and all third parties that:

In the event that [local government owner or operator] fails to provide alternative coverage within 60 days after receipt of a notice of cancellation of this guarantee and the [Director of the implementing agency] has determined or suspects that a release has occurred at an underground storage tank covered by this guarantee, the guarantor, upon instructions from the [Director] shall fund a standby trust fund in accordance with the provisions of 40 CFR part 280.112, in an amount not to exceed the coverage limits specified above.

In the event that the [Director] determines that [local government owner or operator] has failed to perform corrective action for releases arising out of the operation of the above-identified tank(s) in accordance with 40 CFR part 280, subpart F, the guarantor upon written instructions from the [Director] shall fund a standby trust fund in accordance with the provisions of 40 CFR part 280.112, in an amount not to exceed the coverage limits specified above.

If [owner or operator] fails to satisfy a judgment or award based on a determination of liability for bodily injury or property damage to third parties caused by ["sudden" and/or "nonsudden"] accidental releases arising from the operation of the above-identified tank(s), or fails to pay an amount agreed to in settlement of a claim arising from or alleged to arise from such injury or damage, the guarantor, upon written instructions from the [Director], shall fund a standby trust in accordance with the provisions of 40 CFR part 280.112 to satisfy such judgment(s), award(s), or settlement agreement(s) up to the limits of coverage specified above.

(4) Guarantor agrees that, if at the end of any fiscal year before cancellation of this guarantee, the guarantor fails to meet or exceed the requirements of the financial responsibility mechanism specified in paragraph (1), guarantor shall send within 120 days of such failure, by certified mail, notice to [local government owner or operator], as evidenced by the return receipt.

(5) Guarantor agrees to notify [owner or operator] by certified mail of a voluntary or involuntary proceeding under Title 11 (Bankruptcy), U.S. Code naming guarantor as debtor, within 10 days after commencement of the proceeding.

(6) Guarantor agrees to remain bound under this guarantee notwithstanding any modification or alteration of any obligation of [owner or operator] pursuant to 40 CFR part 280.

(7) Guarantor agrees to remain bound under this guarantee for so long as [local government owner or operator] must comply with the applicable financial responsibility requirements of 40 CFR part 280, subpart H for the above identified tank(s), except that guarantor may cancel this guarantee by sending notice by certified mail to [owner or operator], such cancellation to become effective no earlier than 120 days after receipt of such notice by [owner or operator], as evidenced by the return receipt.

(8) The guarantor's obligation does not apply to any of the following:

(a) Any obligation of [local government owner or operator] under a workers' compensation, disability benefits, or unemployment compensation law or other similar law;

(b) Bodily injury to an employee of [insert: local government owner or operator] arising from, and in the course of, employment by [insert: local government owner or operator];

(c) Bodily injury or property damage arising from the ownership, maintenance, use, or entrustment to others of any aircraft, motor vehicle, or watercraft;

(d) Property damage to any property owned, rented, loaned to, in the care, custody, or control of, or occupied by [insert: local government owner or operator] that is not the direct result of a release from a petroleum underground storage tank;

(e) Bodily damage or property damage for which [insert: owner or operator] is obligated to pay damages by reason of the assumption of liability in a contract or agreement other than a contract or agreement entered into to meet the requirements of 40 CFR part 280.93.

(9) Guarantor expressly waives notice of acceptance of this guarantee by [the implementing agency], by any or all third parties, or by [local government owner or operator].

I hereby certify that the wording of this guarantee is identical to the wording specified in 40 CFR part 280.106(d) as such regulations were constituted on the effective date shown immediately below.

Effective date: ______

[Name of guarantor]

[Authorized signature for guarantor]

[Name of person signing]

[Title of person signing]

Signature of witness or notary: ______

(e) If the guarantor is a state, the local government guarantee without standby trust must be worded exactly as follows, except that instructions in brackets are to be replaced with relevant information and the brackets deleted:

Local Government Guarantee Without Standby Trust Made by a State

Guarantee made this [date] by [name of state], herein referred to as guarantor, to [the state implementing agency] and to any and all third parties, and obliges, on behalf of [local government owner or operator].

Recitals

(1) Guarantor is a state.

(2) [Local government owner or operator] owns or operates the following underground storage tank(s) covered by this guarantee: [List the number of tanks at each facility and the name(s) and address(es) of the facility(ies) where the tanks are located. If more than one instrument is used to assure different tanks at any one facility, for each tank covered by this instrument, list the tank identification number provided in the notification submitted pursuant to 40 CFR part 280 or the corresponding state requirement, and the name and address of the facility.] This guarantee satisfies 40 CFR part 280, subpart H requirements for assuring funding for [insert: "taking corrective action" and/or "compensating third parties for bodily injury and property damage caused by" either "sudden accidental releases" or "nonsudden accidental releases" or "accidental releases"; if coverage is different for different tanks or locations, indicate the type of coverage applicable to each tank or location] arising from operating the above-identified underground storage tank(s) in the amount of [insert: dollar amount] per occurrence and [insert: dollar amount] annual aggregate.

(3) Guarantor guarantees to [implementing agency] and to any and all third parties and obliges that:

In the event that [local government owner or operator] fails to provide alternative coverage within 60 days after receipt of a notice of cancellation of this guarantee and the [Director of the implementing agency] has determined or suspects that a release has occurred at an underground storage tank covered by this guarantee, the guarantor, upon written instructions from the [Director] shall make funds available to pay for corrective actions and compensate third parties for bodily injury and property damage in an amount not to exceed the coverage limits specified above.

In the event that the [Director] determines that [local government owner or operator] has failed to perform corrective action for releases arising out of the operation of the above-identified tank(s) in accordance with 40 CFR part 280, subpart F, the guarantor upon written instructions from the [Director] shall make funds available to pay for corrective actions in an amount not to exceed the coverage limits specified above.

If [owner or operator] fails to satisfy a judgment or award based on a determination of liability for bodily injury or property damage to third parties caused by ["sudden" and/or "nonsudden"] accidental releases arising from the operation of the above-identified tank(s), or fails to pay an amount agreed to in settlement of a claim arising from or alleged to arise from such injury or damage, the guarantor, upon written instructions from the [Director], shall make funds available to compensate third parties for bodily injury and property damage in an amount not to exceed the coverage limits specified above.

(4) Guarantor agrees to notify [owner or operator] by certified mail of a voluntary or involuntary proceeding under Title 11 (Bankruptcy), U.S. Code naming guarantor as debtor, within 10 days after commencement of the proceeding.

(5) Guarantor agrees to remain bound under this guarantee notwithstanding any modification or alteration of any obligation of [owner or operator] pursuant to 40 CFR part 280.

(6) Guarantor agrees to remain bound under this guarantee for so long as [local government owner or operator] must comply with the applicable financial responsibility requirements of 40 CFR part 280, subpart H for the above identified tank(s), except that guarantor may cancel this guarantee by sending notice by certified mail to [owner or operator], such cancellation to become effective no earlier than 120 days after receipt of such notice by [owner or operator], as evidenced by the return receipt. If notified of a probable release, the guarantor agrees to remain bound to the terms of this guarantee for all charges arising from the release, up to the coverage limits specified above, notwithstanding the cancellation of the guarantee with respect to future releases.

(7) The guarantor's obligation does not apply to any of the following:

(a) Any obligation of [local government owner or operator] under a workers' compensation disability benefits, or unemployment compensation law or other similar law;

(b) Bodily injury to an employee of [insert local government owner or operator] arising from, and in the course of, employment by [insert: local government owner or operator];

(c) Bodily injury or property damage arising from the ownership, maintenance, use, or entrustment to others of any aircraft, motor vehicle, or watercraft;

(d) Property damage to any property owned, rented, loaded to, in the care, custody, or control of, or occupied by [insert: local government owner or operator] that is not the direct result of a release from a petroleum underground storage tank;

(e) Bodily damage or property damage for which [insert: owner or operator] is obligated to pay damages by reason of the assumption of liability in a contract or agreement other than a contract or agreement entered into to meet the requirements of 40 CFR part 280.93.

(8) Guarantor expressly waives notice of acceptance of this guarantee by [the implementing agency], by any or all third parties, or by [local government owner or operator].

I hereby certify that the wording of this guarantee is identical to the wording specified in 40 CFR part 280.106(e) as such regulations were constituted on the effective date shown immediately below.

Effective date: ______________________

[Name of guarantor]

[Authorized signature for guarantor]

[Name of person signing]

[Title of person signing]

Signature of witness or notary:

If the guarantor is a local government, the local government guarantee without standby trust must be worded exactly as follows, except that instructions in brackets are to be replaced with relevant information and the brackets deleted:

LOCAL GOVERNMENT GUARANTEE WITHOUT STANDBY TRUST MADE BY A LOCAL GOVERNMENT

Guarantee made this [date] by [name of guaranteeing entity], a local government organized under the laws of [name of state], herein referred to as guarantor, to [the state implementing agency] and to any and all third parties, and obliges, on behalf of [local government owner or operator].

Recitals

(1) Guarantor meets or exceeds [select one: the local government bond rating test requirements of 40 CFR part 280.104, the local government financial test requirements of 40 part CFR 280.105, the local government fund under 40 CFR part 280.107(a), 280.107(b), or 280.107(c).

(2) [Local government owner or operator] owns or operates the following underground storage tank(s) covered by this guarantee: [List the number of tanks at each facility and the name(s) and address(es) of the facility(ies) where the tanks are located. If more than one instrument is used to assure different tanks at any one facility, for each tank covered by this instrument, list the tank identification number provided in the notification submitted pursuant to 40 CFR

part 280 or the corresponding state requirement, and the name and address of the facility.] This guarantee satisfies 40 CFR part 280, subpart H requirements for assuring funding for [insert: "taking corrective action" and/or "compensating third parties for bodily injury and property damage caused by" either "sudden accidental releases" or "nonsudden accidental releases" or "accidental releases"; if coverage is different for different tanks or locations, indicate the type of coverage applicable to each tank or location] arising from operating the above-identified underground storage tank(s) in the amount of [insert: dollar amount] per occurrence and [insert: dollar amount] annual aggregate.

(3) Incident to our substantial governmental relationship with [local government owner or operator], guarantor guarantees to [implementing agency] and to any and all third parties and obliges that:

In the event that [local government owner or operator] fails to provide alternative coverage within 60 days after receipt of a notice of cancellation of this guarantee and the [Director of the implementing agency] has determined or suspects that a release has occurred at an underground storage tank covered by this guarantee, the guarantor, upon written instructions from the [Director] shall make funds available to pay for corrective actions and compensate third parties for bodily injury and property damage in an amount not to exceed the coverage limits specified above.

In the event that the [Director] determines that [local government owner or operator] has failed to perform corrective action for releases arising out of the operation of the above-identified tank(s) in accordance with 40 CFR part 280, subpart F, the guarantor upon written instructions from the [Director] shall make funds available to pay for corrective actions in an amount not to exceed the coverage limits specified above.

If [owner or operator] fails to satisfy a judgment or award based on a determination of liability for bodily injury or property damage to third parties caused by ["sudden" and/or "nonsudden"] accidental releases arising from the operation of the above-identified tank(s), or fails to pay an amount agreed to in settlement of a claim arising from or alleged to arise from such injury or damage, the guarantor, upon written instructions from the [Director], shall make funds available to compensate third parties for bodily injury and property damage in an amount not to exceed the coverage limits specified above.

(4) Guarantor agrees that if at the end of any fiscal year before cancellation of this guarantee, the guarantor fails to meet or exceed the requirements of the financial responsibility mechanism specified in paragraph (1), guarantor shall send within 120 days of such failure, by certified mail, notice to [local government owner or operator], as evidenced by the return receipt.

(5) Guarantor agrees to notify [owner or operator] by certified mail of a voluntary or involuntary proceeding under Title 11 (Bankruptcy), U.S. Code naming guarantor as debtor, within 10 days after commencement of the proceeding.

(6) Guarantor agrees to remain bound under this guarantee notwithstanding any modification or alteration of any obligation of [owner or operator] pursuant to 40 CFR part 280.

(7) Guarantor agrees to remain bound under this guarantee for so long as [local government owner or operator] must comply with the applicable financial responsibility requirements of 40 CFR part 280, subpart H for the above identified tank(s), except that guarantor may cancel this guarantee by sending notice by certified mail to [owner or operator], such cancellation to become effective no earlier than 120 days after receipt of such notice by [owner or operator], as evidenced by the return receipt. If notified of a probable release, the guarantor agrees to remain bound to the terms of this guarantee for all charges arising from the release, up to the coverage limits specified above, notwithstanding the cancellation of the guarantee with respect to future releases.

(8) The guarantor's obligation does not apply to any of the following:

(a) Any obligation of [local government owner or operator] under a workers' compensation disability benefits, or unemployment compensation law or other similar law;

(b) Bodily injury to an employee of [insert: local government owner or operator] arising from, and in the course of, employment by [insert: local government owner or operator];

(c) Bodily injury or property damage arising from the ownership, maintenance, use, or entrustment to others of any aircraft, motor vehicle, or watercraft;

(d) Property damage to any property owned, rented, loaded to, in the care, custody, or control of, or occupied by [insert: local government owner or operator] that is not the direct result of a release from a petroleum underground storage tank;

(e) Bodily damage or property damage for which [insert: owner or operator] is obligated to pay damages by reason of the assumption of liability in a contract or agreement other than a contract or agreement entered into to meet the requirements of 40 CFR part 280.93.

(9) Guarantor expressly waives notice of acceptance of this guarantee by [the implementing agency], by any or all third parties, or by [local government owner or operator].

I hereby certify that the wording of this guarantee is identical to the wording specified in 40 CFR part 280.106(e) as such regulations were constituted on the effective date shown immediately below.

Effective date: ______________________
[Name of guarantor]
[Authorized signature for guarantor]
[Name of person signing]
[Title of person signing]
Signature of witness or notary:

[58 FR 9056, Feb. 18, 1993]

§280.107 Local government fund.

A local government owner or operator may satisfy the requirements of §280.93 by establishing a dedicated fund account that conforms to the requirements of this section. Except as specified in paragraph (b), a dedicated fund may not be commingled with other funds or otherwise used in normal operations. A dedicated fund will be considered eligible if it meets one of the following requirements:

(a) The fund is dedicated by state constitutional provision, or local government statute, charter, ordinance, or order to pay for taking corrective action and for compensating third parties for bodily injury and property damage caused by accidental releases arising from the operation of petroleum underground storage tanks and is funded for the full amount of coverage required under §280.93, or funded for part of the required amount of coverage and used in combination with other mechanism(s) that provide the remaining coverage; or

(b) The fund is dedicated by state constitutional provision, or local government statute, charter, ordinance, or order as a contingency fund for general emergencies, including taking corrective action and compensating third parties for bodily injury and property damage caused by accidental releases arising from the operation of petroleum underground storage tanks, and is funded for five times the full amount of coverage required under §280.93, or funded for part of the required amount of coverage and used in combination with other mechanism(s) that provide the remaining coverage. If the fund is funded for less than five times the amount of coverage required under §280.93, the amount of financial responsibility demonstrated by the fund may not exceed one-fifth the amount in the fund; or

(c) The fund is dedicated by state constitutional provision, or local government statute, charter, ordinance or order to pay for taking corrective action and for compensating third parties for bodily injury and property damage caused by accidental releases arising from the operation of petroleum underground storage tanks. A payment is made to the fund once every year for seven years until the fund is fully-funded. This seven year period is hereafter referred to as the "pay-in-period." The amount of each payment must be determined by this formula:

$$\frac{TF - CF}{Y}$$

Where TF is the total required financial assurance for the owner or operator, CF is the current amount in the fund, and Y is the number of years remaining in the pay-in-period, and;

(1) The local government owner or operator has available bonding authority, approved through voter referendum (if such approval is necessary prior to the issuance of bonds), for an amount equal to the difference between the required amount of coverage and the amount held in the dedicated fund. This bonding authority shall be available for taking corrective action and for compensating third parties for bodily injury and property damage caused by accidental releases arising from the operation of petroleum underground storage tanks, or

(2) The local government owner or operator has a letter signed by the appropriate state attorney general stating that the use of the bonding authority will not increase the local government's debt beyond the legal debt ceilings established by the relevant state laws. The letter must also state that prior voter approval is not necessary before use of the bonding authority.

(d) To demonstrate that it meets the requirements of the local government fund, the chief financial officer of the local government owner or operator and/or guarantor must sign a letter worded exactly as follows, except that the instructions in brackets are to be replaced by the relevant information and the brackets deleted:

LETTER FROM CHIEF FINANCIAL OFFICER

I am the chief financial officer of [insert: name and address of local government owner or operator, or guarantor]. This letter is in support of the use of the local government fund mechanism to demonstrate financial responsibility for [insert: "taking corrective action" and/or "compensating third parties for bodily injury and property damage"] caused by [insert: "sudden accidental releases" and/or "nonsudden accidental releases"] in the amount of at least [insert: dollar amount] per occurrence and [insert: dollar amount] annual aggregate arising from operating (an) underground storage tank(s).

Underground storage tanks at the following facilities are assured by this local government fund mechanism: [List for each facility: the name and address of the facility where tanks are assured by the local government fund].

[Insert: "The local government fund is funded for the full amount of coverage required under § 280.93, or funded for part of the required amount of coverage and used in combination with other mechanism(s) that provide the remaining coverage." or "The local government fund is funded for ten times the full amount of coverage required under § 280.93, or funded for part of the required amount of coverage and used in combination with other mechanisms(s) that provide the remaining coverage," or "A payment is made to the fund once every year for seven years until the fund is fully-funded and [name of local government owner or operator] has available bonding authority, approved through voter referendum, of an amount equal to the difference between the required amount of coverage and the amount held in the dedicated fund" or "A payment is made to the fund once every year for seven years until the fund is fully-funded and I have attached a letter signed by the State Attorney General stating that (1) the use of the bonding authority will not increase the local government's debt beyond the legal debt ceilings established by the relevant state laws and (2) that prior voter approval is not necessary before use of the bonding authority"].

The details of the local government fund are as follows:

Amount in Fund (market value of fund at close of last fiscal year): ——————

[If fund balance is incrementally funded as specified in § 280.107(c), insert:

Amount added to fund in the most recently completed fiscal year: ——————

Number of years remaining in the pay-in period: _______]

A copy of the state constitutional provision, or local government statute, charter, ordinance or order dedicating the fund is attached.

I hereby certify that the wording of this letter is identical to the wording specified in 40 CFR 280.107(d) as such regulations were constituted on the date shown immediately below.

[Date]
[Signature]
[Name]
[Title]

[58 FR 9059, Feb. 18, 1993]

§ 280.108 Substitution of financial assurance mechanisms by owner or operator.

(a) An owner or operator may substitute any alternate financial assurance mechanisms as specified in this subpart, provided that at all times he maintains an effective financial assurance mechanism or combination of mechanisms that satisfies the requirements of § 280.93.

(b) After obtaining alternate financial assurance as specified in this subpart, an owner or operator may cancel a financial assurance mechanism by providing notice to the provider of financial assurance.

[53 FR 43370, Oct. 26, 1988. Redesignated at 58 FR 9051, Feb. 18, 1993]

§ 280.109 Cancellation or nonrenewal by a provider of financial assurance.

(a) Except as otherwise provided, a provider of financial assurance may cancel or fail to renew an assurance mechanism by sending a notice of termination by certified mail to the owner or operator.

(1) Termination of a local government guarantee, a guarantee, a surety bond, or a letter of credit may not occur until 120 days after the date on which the owner or operator receives the notice of termination, as evidenced by the return receipt.

(2) Termination of insurance or risk retention coverage, except for non-payment or misrepresentation by the insured, or state-funded assurance may not occur until 60 days after the date on which the owner or operator receives the notice of termination, as evidenced by the return receipt. Termination for non-payment of premium or misrepresentation by the insured may not occur until a minimum of 10 days after the date on which the owner or operator receives the notice of termi-

nation, as evidenced by the return receipt.

(b) If a provider of financial responsibility cancels or fails to renew for reasons other than incapacity of the provider as specified in §280.114, the owner or operator must obtain alternate coverage as specified in this section within 60 days after receipt of the notice of termination. If the owner or operator fails to obtain alternate coverage within 60 days after receipt of the notice of termination, the owner or operator must notify the Director of the implementing agency of such failure and submit:

(1) The name and address of the provider of financial assurance;

(2) The effective date of termination; and

(3) The evidence of the financial assistance mechanism subject to the termination maintained in accordance with §280.107(b).

[58 FR 9051, Feb. 18, 1993]

§280.110 Reporting by owner or operator.

(a) An owner or operator must submit the appropriate forms listed in §280.111(b) documenting current evidence of financial responsibility to the Director of the implementing agency:

(1) Within 30 days after the owner or operator identifies a release from an underground storage tank required to be reported under §280.53 or §280.61;

(2) If the owner or operator fails to obtain alternate coverage as required by this subpart, within 30 days after the owner or operator receives notice of:

(i) Commencement of a voluntary or involuntary proceeding under Title 11 (Bankruptcy), U.S. Code, naming a provider of financial assurance as a debtor,

(ii) Suspension or revocation of the authority of a provider of financial assurance to issue a financial assurance mechanism,

(iii) Failure of a guarantor to meet the requirements of the financial test,

(iv) Other incapacity of a provider of financial assurance; or

(3) As required by §280.95(g) and §280.109(b).

(b) An owner or operator must certify compliance with the financial responsibility requirements of this part as specified in the new tank notification form when notifying the appropriate state or local agency of the installation of a new underground storage tank under §280.22.

(c) The Director of the Implementing Agency may require an owner or operator to submit evidence of financial assurance as described in §280.111(b) or other information relevant to compliance with this subpart at any time.

[58 FR 9051, Feb. 18, 1993]

§280.111 Recordkeeping.

(a) Owners or operators must maintain evidence of all financial assurance mechanisms used to demonstrate financial responsibility under this subpart for an underground storage tank until released from the requirements of this subpart under §208.113. An owner or operator must maintain such evidence at the underground storage tank site or the owner's or operator's place of work. Records maintained off-site must be made available upon request of the implementing agency.

(b) An owner or operator must maintain the following types of evidence of financial responsibility:

(1) An owner or operator using an assurance mechanism specified in §§280.95 through 280.100 or §280.102 or §§280.104 through 280.107 must maintain a copy of the instrument worded as specified.

(2) An owner or operator using a financial test or guarantee, or a local government financial test or a local government guarantee supported by the local government financial test must maintain a copy of the chief financial officer's letter based on year-end financial statements for the most recent completed financial reporting year. Such evidence must be on file no later than 120 days after the close of the financial reporting year.

(3) An owner or operator using a guarantee, surety bond, or letter of credit must maintain a copy of the signed standby trust fund agreement and copies of any amendments to the agreement.

(4) A local government owner or operator using a local government guarantee under §280.106(d) must maintain a copy of the signed standby trust fund

agreement and copies of any amendments to the agreement.

(5) A local government owner or operator using the local government bond rating test under § 280.104 must maintain a copy of its bond rating published within the last twelve months by Moody's or Standard & Poor's.

(6) A local government owner or operator using the local government guarantee under § 280.106, where the guarantor's demonstration of financial responsibility relies on the bond rating test under § 280.104 must maintain a copy of the guarantor's bond rating published within the last twelve months by Moody's or Standard & Poor's.

(7) An owner or operator using an insurance policy or risk retention group coverage must maintain a copy of the signed insurance policy or risk retention group coverage policy, with the endorsement or certificate of insurance and any amendments to the agreements.

(8) An owner or operator covered by a state fund or other state assurance must maintain on file a copy of any evidence of coverage supplied by or required by the state under § 280.101(d).

(9) An owner or operator using a local government fund under § 280.107 must maintain the following documents:

(i) A copy of the state constitutional provision or local government statute, charter, ordinance, or order dedicating the fund, and

(ii) Year-end financial statements for the most recent completed financial reporting year showing the amount in the fund. If the fund is established under § 280.107(a)(3) using incremental funding backed by bonding authority, the financial statements must show the previous year's balance, the amount of funding during the year, and the closing balance in the fund.

(iii) If the fund is established under § 280.107(a)(3) using incremental funding backed by bonding authority, the owner or operator must also maintain documentation of the required bonding authority, including either the results of a voter referendum (under § 280.107(a)(3)(i)), or attestation by the State Attorney General as specified under § 280.107(a)(3)(ii).

(10) A local government owner or operator using the local government guarantee supported by the local government fund must maintain a copy of the guarantor's year-end financial statements for the most recent completed financial reporting year showing the amount of the fund.

(11)(i) An owner or operator using an assurance mechanism specified in §§ 280.95 through 280.107 must maintain an updated copy of a certification of financial responsibility worded as follows, except that instructions in brackets are to be replaced with the relevant information and the brackets deleted:

Certification of Financial Responsibility

[Owner or operator] hereby certifies that it is in compliance with the requirements of subpart H of 40 CFR part 280.

The financial assurance mechanism(s) used to demonstrate financial responsibility under subpart H of 40 CFR part 280 is (are) as follows:

[For each mechanism, list the type of mechanism, name of issuer, mechanism number (if applicable), amount of coverage, effective period of coverage and whether the mechanism covers "taking corrective action" and/or "compensating third parties for bodily injury and property damage caused by" either "sudden accidental releases" or "nonsudden accidental releases" or "accidental releases."]

[Signature of owner or operator]
[Name of owner or operator]
[Title]
[Date]
[Signature of witness or notary]
[Name of witness or notary]
[Date]

(ii) The owner or operator must update this certification whenever the financial assurance mechanism(s) used to demonstrate financial responsibility change(s).

[58 FR 9051, Feb. 18, 1993]

§ 280.112 Drawing on financial assurance mechanisms.

(a) Except as specified in paragraph (d) of this section, the Director of the implementing agency shall require the guarantor, surety, or institution issuing a letter of credit to place the amount of funds stipulated by the Director, up to the limit of funds provided by the financial assurance mechanism, into the standby trust if:

(1)(i) The owner or operator fails to establish alternate financial assurance within 60 days after receiving notice of cancellation of the guarantee, surety bond, letter of credit, or, as applicable, other financial assurance mechanism; and

(ii) The Director determines or suspects that a release from an underground storage tank covered by the mechanism has occurred and so notifies the owner or operator or the owner or operator has notified the Director pursuant to subparts E or F of a release from an underground storage tank covered by the mechanism; or

(2) The conditions of paragraph (b)(1) or (b)(2) (i) or (ii) of this section are satisfied.

(b) The Director of the implementing agency may draw on a standby trust fund when:

(1) The Director makes a final determination that a release has occurred and immediate or long-term corrective action for the release is needed, and the owner or operator, after appropriate notice and opportunity to comply, has not conducted corrective action as required under 40 CFR part 280, subpart F; or

(2) The Director has received either:

(i) Certification from the owner or operator and the third-party liability claimant(s) and from attorneys representing the owner or operator and the third-party liability claimant(s) that a third-party liability claim should be paid. The certification must be worded as follows, except that instructions in brackets are to be replaced with the relevant information and the brackets deleted:

Certification of Valid Claim

The undersigned, as principals and as legal representatives of [insert: owner or operator] and [insert: name and address of third-party claimant], hereby certify that the claim of bodily injury [and/or] property damage caused by an accidental release arising from operating [owner's or operator's] underground storage tank should be paid in the amount of $[__________].

[Signatures]
Owner or Operator
Attorney for Owner or Operator
(Notary)
Date

[Signatures]
Claimant(s)
Attorney(s) for Claimant(s)
(Notary)
Date

or (ii) A valid final court order establishing a judgment against the owner or operator for bodily injury or property damage caused by an accidental release from an underground storage tank covered by financial assurance under this subpart and the Director determines that the owner or operator has not satisfied the judgment.

(c) If the Director of the implementing agency determines that the amount of corrective action costs and third-party liability claims eligible for payment under paragraph (b) of this section may exceed the balance of the standby trust fund and the obligation of the provider of financial assurance, the first priority for payment shall be corrective action costs necessary to protect human health and the environment. The Director shall pay third-party liability claims in the order in which the Director receives certifications under paragraph (b)(2)(i) of this section, and valid court orders under paragraph (b)(2)(ii) of this section.

(d) A governmental entity acting as guarantor under § 280.106(e), the local government guarantee without standby trust, shall make payments as directed by the Director under the circumstances described in § 280.112 (a), (b), and (c).

[58 FR 9052, Feb. 18, 1993]

§ 280.113 Release from the requirements.

An owner or operator is no longer required to maintain financial responsibility under this subpart for an underground storage tank after the tank has been properly closed or, if corrective action is required, after corrective action has been completed and the tank has been properly closed as required by 40 CFR part 280, subpart G.

[53 FR 43370, Oct. 26, 1988. Redesignated at 58 FR 9051, Feb. 18, 1993]

§ 280.114 Bankruptcy or other incapacity of owner or operator or provider of financial assurance.

(a) Within 10 days after commencement of a voluntary or involuntary proceeding under Title 11 (Bankruptcy), U.S. Code, naming an owner or operator as debtor, the owner or operator must notify the Director of the implementing agency by certified mail of such commencement and submit the appropriate forms listed in § 280.111(b) documenting current financial responsibility.

(b) Within 10 days after commencement of a voluntary or involuntary proceeding under Title 11 (Bankruptcy), U.S. Code, naming a guarantor providing financial assurance as debtor, such guarantor must notify the owner or operator by certified mail of such commencement as required under the terms of the guarantee specified in § 280.96.

(c) Within 10 days after commencement of a voluntary or involuntary proceeding under Title 11 (Bankruptcy), U.S. Code, naming a local government owner or operator as debtor, the local government owner or operator must notify the Director of the implementing agency by certified mail of such commencement and submit the appropriate forms listed in § 280.111(b) documenting current financial responsibility.

(d) Within 10 days after commencement of a voluntary or involuntary proceeding under Title 11 (Bankruptcy), U.S. Code, naming a guarantor providing a local government financial assurance as debtor, such guarantor must notify the local government owner or operator by certified mail of such commencement as required under the terms of the guarantee specified in § 280.106.

(e) An owner or operator who obtains financial assurance by a mechanism other than the financial test of self-insurance will be deemed to be without the required financial assurance in the event of a bankruptcy or incapacity of its provider of financial assurance, or a suspension or revocation of the authority of the provider of financial assurance to issue a guarantee, insurance policy, risk retention group coverage policy, surety bond, letter of credit, or state-required mechanism. The owner or operator must obtain alternate financial assurance as specified in this subpart within 30 days after receiving notice of such an event. If the owner or operator does not obtain alternate coverage within 30 days after such notification, he must notify the Director of the implementing agency.

(f) Within 30 days after receipt of notification that a state fund or other state assurance has become incapable of paying for assured corrective action or third-party compensation costs, the owner or operator must obtain alternate financial assurance.

[58 FR 9053, Feb. 18, 1993]

§ 280.115 Replenishment of guarantees, letters of credit, or surety bonds.

(a) If at any time after a standby trust is funded upon the instruction of the Director of the implementing agency with funds drawn from a guarantee, local government guarantee with standby trust, letter of credit, or surety bond, and the amount in the standby trust is reduced below the full amount of coverage required, the owner or operator shall by the anniversary date of the financial mechanism from which the funds were drawn:

(1) Replenish the value of financial assurance to equal the full amount of coverage required, or

(2) Acquire another financial assurance mechanism for the amount by which funds in the standby trust have been reduced.

(b) For purposes of this section, the full amount of coverage required is the amount of coverage to be provided by § 280.93 of this subpart. If a combination of mechanisms was used to provide the assurance funds which were drawn upon, replenishment shall occur by the earliest anniversary date among the mechanisms.

[58 FR 9053, Feb. 18, 1993]

§ 280.116 Suspension of enforcement. [Reserved]

APPENDIX C

EPA-APPROVED Notification FORM for Underground Storage Tanks

(40 C.F.R. Part 280, Appendix I (1992))

APPENDIX I TO PART 280—NOTIFICATION FOR UNDERGROUND STORAGE TANKS (FORM)

Notification for Underground Storage Tanks

FORM APPROVED
OMB NO. 2050-0068
APPROVAL EXPIRES 9-30-91

EPA estimates public reporting burden for this form to average 30 minutes per response, including time for reviewing instructions, gathering and maintaining the data needed, and completing and reviewing the form. Send comments regarding this burden estimate to Chief, Information Policy Branch, PM-223, U.S. Environmental Protection Agency, 401 M St., S.W., Washington, D.C. 20460; and to the Office of Information and Regulatory Affairs, Office of Management and Budget, Washington, D.C. 20503, marked "Attention: Desk Officer for EPA."

STATE USE ONLY

I D Number

Date Received

GENERAL INFORMATION

Notification is required by Federal law for all underground tanks that have been used to store regulated substances since January 1, 1974, that are in the ground as of May 8, 1986, or that are brought into use after May 8, 1986. The information requested is required by Section 9002 of the Resource Conservation and Recovery Act, (RCRA), as amended.

The primary purpose of this notification program is to locate and evaluate underground tanks that store or have stored petroleum or hazardous substances. It is expected that the information you provide will be based on reasonably available records, or, in the absence of such records, your knowledge, belief, or recollection.

Who Must Notify? Section 9002 of RCRA, as amended, requires that, unless exempted, owners of underground tanks that store regulated substances must notify designated State or local agencies of the existence of their tanks. Owner means—

(a) in the case of an underground storage tank in use on November 8, 1984, or brought into use after that date, any person who owns an underground storage tank used for the storage, use, or dispensing of regulated substances, and

(b) in the case of any underground storage tank in use before November 8, 1984, but no longer in use on that date, any person who owned such tank immediately before the discontinuation of its use.

What Tanks Are Included? Underground storage tank is defined as any one or combination of tanks that (1) is used to contain an accumulation of "regulated substances," and (2) whose volume (including connected underground piping) is 10% or more beneath the ground. Some examples are underground tanks storing: 1. gasoline, used oil, or diesel fuel, and 2. industrial solvents, pesticides, herbicides or fumigants.

What Tanks Are Excluded? Tanks removed from the ground are not subject to notification. Other tanks excluded from notification are:

1. farm or residential tanks of 1,100 gallons or less capacity used for storing motor fuel for noncommercial purposes.
2. tanks used for storing heating oil for consumptive use on the premises where stored.
3. septic tanks.
4. pipeline facilities (including gathering lines) regulated under the Natural Gas Pipeline Safety Act of 1968, or the Hazardous Liquid Pipeline Safety Act of 1979, or which is an intrastate pipeline facility regulated under State laws.
5. surface impoundments, pits, ponds, or lagoons.
6. storm water or waste water collection systems.
7. flow-through process tanks.
8. liquid traps or associated gathering lines directly related to oil or gas production and gathering operations.
9. storage tanks situated in an underground area (such as a basement, cellar, mineworking, drift, shaft, or tunnel) if the storage tank is situated upon or above the surface of the floor.

What Substances Are Covered? The notification requirements apply to underground storage tanks that contain regulated substances. This includes any substance defined as hazardous in section 101 (14) of the Comprehensive Environmental Response, Compensation and Liability Act of 1980 (CERCLA), with the exception of those substances regulated as hazardous waste under Subtitle C of RCRA. It also includes petroleum, e.g., crude oil or any fraction thereof which is liquid at standard conditions of temperature and pressure (60 degrees Fahrenheit and 14.7 pounds per square inch absolute).

Where To Notify? Completed notification forms should be sent to the address given at the top of this page.

When To Notify? 1. Owners of underground storage tanks in use or that have been taken out of operation after January 1, 1974, but still in the ground, must notify by May 8, 1986. 2. Owners who bring underground storage tanks into use after May 8, 1986, must notify within 30 days of bringing the tanks into use.

Penalties: Any owner who knowingly fails to notify or submits false information shall be subject to a civil penalty not to exceed $10,000 for each tank for which notification is not given or for which false information is submitted.

INSTRUCTIONS

Please type or print in ink all items except "signature" in Section V. This form must be completed for each location containing underground storage tanks. If more than 5 tanks are owned at this location, photocopy the reverse side, and staple continuation sheets to this form.

Indicate number of continuation sheets attached ☐

I. OWNERSHIP OF TANK(S)

Owner Name (Corporation, Individual, Public Agency, or Other Entity)

Street Address

County

City State ZIP Code

Area Code Phone Number

Type of Owner ***(Mark all that apply ☒)***

☐ Current ☐ State or Local Gov't ☐ Private or Corporate

☐ Former ☐ Federal Gov't (GSA facility I.D. no. ____________) ☐ Ownership uncertain

II. LOCATION OF TANK(S)

(If same as Section 1, mark box here ☐)

Facility Name or Company Site Identifier, as applicable

Street Address or State Road, as applicable

County

City (nearest) State ZIP Code

Indicate number of tanks at this location ☐

Mark box here if tank(s) are located on land within an Indian reservation or on other Indian trust lands ☐

III. CONTACT PERSON AT TANK LOCATION

Name (If same as Section I, mark box here ☐) Job Title Area Code Phone Number

IV. TYPE OF NOTIFICATION

☐ Mark box here only if this is an amended or subsequent notification for this location

V. CERTIFICATION (Read and sign after completing Section VI.)

I certify under penalty of law that I have personally examined and am familiar with the information submitted in this and all attached documents, and that based on my inquiry of those individuals immediately responsible for obtaining the information, I believe that the submitted information is true, accurate, and complete.

Name and official title of owner or owner's authorized representative	Signature	Date Signed

CONTINUE ON REVERSE SIDE

EPA Form 7530-1 (Revised 9-88) Page 1

Owner Name (from Section I) ______ Location (from Section II) ______ Page No. ____ of ____ Pages

VI. DESCRIPTION OF UNDERGROUND STORAGE TANKS *(Complete for each tank at this location.)*

Tank Identification No. (e.g., ABC-123), or Arbitrarily Assigned Sequential Number (e.g., 1,2,3...)	Tank No.	Tank No.	Tank No.	Tank No.	Tank No.
1. Status of Tank *(Mark all that apply ☒)*					
Currently in Use	☐	☐	☐	☐	☐
Temporarily Out of Use	☐	☐	☐	☐	☐
Permanently Out of Use	☐	☐	☐	☐	☐
Brought into Use after 5/8/86	☐	☐	☐	☐	☐
2. Estimated Age (Years)					
3. Estimated Total Capacity (Gallons)					
4. Material of Construction *(Mark one ☒)*					
Steel	☐	☐	☐	☐	☐
Concrete	☐	☐	☐	☐	☐
Fiberglass Reinforced Plastic	☐	☐	☐	☐	☐
Unknown	☐	☐	☐	☐	☐
Other, Please Specify					
5. Internal Protection *(Mark all that apply ☒)*					
Cathodic Protection	☐	☐	☐	☐	☐
Interior Lining (e.g., epoxy resins)	☐	☐	☐	☐	☐
None	☐	☐	☐	☐	☐
Unknown	☐	☐	☐	☐	☐
Other, Please Specify					
6. External Protection *(Mark all that apply ☒)*					
Cathodic Protection	☐	☐	☐	☐	☐
Painted (e.g., asphaltic)	☐	☐	☐	☐	☐
Fiberglass Reinforced Plastic Coated	☐	☐	☐	☐	☐
None	☐	☐	☐	☐	☐
Unknown	☐	☐	☐	☐	☐
Other, Please Specify					
7. Piping *(Mark all that apply ☒)*					
Bare Steel	☐	☐	☐	☐	☐
Galvanized Steel	☐	☐	☐	☐	☐
Fiberglass Reinforced Plastic	☐	☐	☐	☐	☐
Cathodically Protected	☐	☐	☐	☐	☐
Unknown	☐	☐	☐	☐	☐
Other, Please Specify					
8. Substance Currently or Last Stored in Greatest Quantity by Volume *(Mark all that apply ☒)*					
a. Empty	☐	☐	☐	☐	☐
b. Petroleum					
Diesel	☐	☐	☐	☐	☐
Kerosene	☐	☐	☐	☐	☐
Gasoline (including alcohol blends)	☐	☐	☐	☐	☐
Used Oil	☐	☐	☐	☐	☐
Other, Please Specify					
c. Hazardous Substance	☐	☐	☐	☐	☐
Please Indicate Name of Principal CERCLA Substance					
OR					
Chemical Abstract Service (CAS) No.					
Mark box ☒ if tank stores a mixture of substances	☐	☐	☐	☐	☐
d. Unknown	☐	☐	☐	☐	☐
9. Additional Information (for tanks permanently taken out of service)					
a. Estimated date last used (mo/yr)	/	/	/	/	/
b. Estimated quantity of substance remaining (gal.)					
c. Mark box ☒ if tank was filled with inert material (e.g., sand, concrete)	☐	☐	☐	☐	☐

EPA Form 7530-1 (Revised 9-88) Reverse

Page 2

Owner Name (from Section I) ____________ Location (from Section II) ____________ Page No. _____ of _____ Pages

VII CERTIFICATION OF COMPLIANCE (COMPLETE FOR ALL NEW TANKS AT THIS LOCATION)

10 Installation (mark all that apply)

- ☐ The installer has been certified by the tank and piping manufacturers
- ☐ The installer has been certified or licensed by the implementing agency
- ☐ The installation has been inspected and certified by a registered professional engineer
- ☐ The installation has been inspected and approved by the implementing agency
- ☐ All work listed on the manufacturer's installation checklists has been completed
- ☐ Another method was used as allowed by the implementing agency Please specify

____

11 Release Detection (mark all that apply)

- ☐ Manual tank gauging
- ☐ Tank tightness testing with inventory controls
- ☐ Automatic tank gauging
- ☐ Vapor monitoring
- ☐ Ground water monitoring
- ☐ Interstitial monitoring within a secondary barrier
- ☐ Interstitial monitoring within secondary containment
- ☐ Automatic line leak detectors
- ☐ Line tightness testing
- ☐ Another method allowed by the implementing agency Please specify

____

12 Corrosion Protection (if applicable)

- ☐ As specified for coated steel tanks with cathodic protection
- ☐ As specified for coated steel piping with cathodic protection
- ☐ Another method allowed by the implementing agency Please specify

____

13 I have financial responsibility in accordance with Subpart I Please specify

Method ____

Insurer ____

Policy Number ____

14 OATH I certify that the information concerning installation provided in Item 10 is true to the best of my belief and knowledge

Installer ______________________________ ______________________
Name Date

____
Position

____
Company

EPA Form 7530 1 (9 88) Page 3

APPENDIX D

Agencies Designated to Receive Notifications

(40 C.F.R. Part 280, Appendix II, 1993)

APPENDIX II TO PART 280—LIST OF AGENCIES DESIGNATED TO RECEIVE NOTIFICATIONS

Alabama (EPA Form), Alabama Department of Environmental Management, Ground Water Section/Water Division, 1751 Congressman W.L. Dickinson Drive, Montgomery, Alabama 36130, 205/271-7823

Alaska (EPA Form), Department of Environmental Conservation, Box 0, Juneau, Alaska 99811-1800, 970/465-2653

American Samoa (EPA Form), Executive Secretary, Environmental Quality Commission, Office of the Governor, American Samoan Government, Pago Pago, American Samoa 96799; Attention: UST Notification

Arizona (EPA Form), Attention: UST Coordinator, Arizona Department of Environmental Quality, Environmental Health Services, 2005 N. Central, Phoenix, Arizona 85004

Arkansas (EPA Form), Arkansas Department of Pollution Control and Ecology, P.O. Box 9583, Little Rock, Arkansas 72219, 501/562-7444

California (State Form), Executive Director, State Water Resources Control Board, P.O. Box 100, Sacramento, California 95801, 916/445-1533

Colorado (EPA Form), Section Chief, Colorado Department of Health, Waste Management Division, Underground Tank Program, 4210 East 11th Avenue, Denver, Colorado 80220, 303/320-8333

Connecticut (State Form), Hazardous Materials Management Unit, Department of Environmental Protection, State Office Building, 165 Capitol Avenue, Hartford, Connecticut 06106

Delaware (State Form), Division of Air and Waste Management, Department of Natural Resources and Environmental Control, P.O. Box 1401, 89 Kings Highway, Dover, Delaware 19903, 302/726-5409

District of Columbia (EPA Form), Attention: UST Notification Form, Department of Consumer and Regulatory Affairs, Pesticides and Hazardous Waste Management Branch, Room 114, 5010 Overlook Avenue SW., Washington, DC 20032

Florida (State Form), Florida Department of Environmental Regulation, Solid Waste Section, Twin Towers Office Building, 2600 Blair Stone Road, Tallahassee, Florida 32399, 904/487-4398

Georgia (EPA Form), Georgia Department of Natural Resources, Environmental Protection Division, Underground Storage Tank Program, 3420 Norman Berry Drive, 7th Floor, Hapeville, Georgia 30354, 404/656-7404

Guam (State Form), Administrator, Guam Environmental Protection Agency, P.O. Box 2999, Agana, Guam 96910, Overseas Operator (Commercial call 646-8863)

Hawaii (EPA Form), Administrator, Hazardous Waste Program, 645 Halekauwila Street, Honolulu, Hawaii 96813, 808/548-2270

Idaho (EPA Form), Underground Storage Tank Coordinator, Water Quality Bureau, Division of Environmental Quality, Idaho Department of Health and Welfare, 450 W. State Street, Boise, Idaho 83720, 208/334-4251

Illinois (EPA Form), Underground Storage Tank Coordinator, Division of Fire Prevention, Office of State Fire Marshal, 3150 Executive Park Drive, Springfield, Illinois 62703-4599

Indiana (EPA Form), Underground Storage Tank Program, Office of Environmental Response, Indiana Department of Environmental Management, 105 South Meridian Street, Indianapolis, Indiana 46225

Iowa (State Form), UST Coordinator, Iowa Department of Natural Resources, Henry A. Wallace Building, 900 East Grand, Des Moines, Iowa 50219, 512/281-8135

Kansas (EPA Form), Kansas Department of Health and Environment, Forbes Field, Building 740, Topeka, Kansas 66620, 913/296-1594

Kentucky (State Form), Department of Environmental Protection, Hazardous Waste Branch, Fort Boone Plaza, Building #2, 18 Reilly Road, Frankfort, Kentucky 40601, 501/564-6716

Louisiana (State Form), Secretary, Louisiana Department of Environmental Quality, P.O. Box 44066, Baton Rouge, Louisiana 70804, 501/342-1265

Maine (State Form), Attention: Underground Tanks Program, Bureau of Oil and Hazardous Material Control, Department of Environmental Protection, State House—Station 17, Augusta, Maine 04333

Maryland (EPA Form), Science and Health Advisory Group, Office of Environmental Programs, 201 West Preston Street, Baltimore, Maryland 21201

Massachusetts (EPA Form), UST Registry, Department of Public Safety, 1010 Commonwealth Avenue, Boston, Massachusetts 02215, 617/566-4500

Michigan (EPA Form), Michigan Department of State Police, Fire Marshal Division, General Office Building, 7150 Harris Drive, Lansing, Michigan 48913

Minnesota (State Form), Underground Storage Tank Program, Division of Solid and Hazardous Wastes, Minnesota Pollution Control Agency, 520 West Lafayette Road, St. Paul, Minnesota 55155

Mississippi (State Form), Department of Natural Resources, Bureau of Pollution Control, Underground Storage Tank Section, P.O. Box 10385, Jackson, Mississippi 39209, 601/961-5171

Missouri (EPA Form), UST Coordinator, Missouri Department of Natural Resources, P.O. Box 176, Jefferson City, Missouri 65102, 314/751-7428

Montana (EPA Form), Solid and Hazardous Waste Bureau, Department of Health and Environmental Science, Cogswell Bldg., Room B-201, Helena, Montana 59620

Nebraska (EPA Form), Nebraska State Fire Marshal, P.O. Box 94677, Lincoln, Nebraska 68509-4677, 402/471-9465

Nevada (EPA Form), Attention: UST Coordinator, Division of Environmental Protection, Department of Conservation and Natural Resources, Capitol Complex 201 S. Fall Street, Carson City, Nevada 89710, 800/992-0900, Ext. 4670, 702/885-4670

New Hampshire (EPA Form), NH Dept. of Environmental Services, Water Supply and Pollution Control Division, Hazen Drive, P.O. Box 95, Concord, New Hampshire 03301, Attention: UST Registration

New Jersey (State Form), Underground Storage Tank Coordinator, Department of Environmental Protection, Division of Water Resources (CN-029), Trenton, New Jersey 08625, 609/292-0424

New Mexico (EPA Form), New Mexico Environmental Improvement Division, Groundwater/Hazardous Waste Bureau, P.O. Box 968, Santa Fe, New Mexico 37504, 505/827-2933

New York (EPA Form), Bulk Storage Section, Division of Water, Department of Environmental Conservation, 50 Wolf Road, Room 326, Albany, New York 12233-0001, 518/457-4351

North Carolina (EPA Form), Division of Environmental Management, Ground-Water Operations Branch, Department of Natural Resources and Community Development, P.O. Box 27687, Raleigh, North Carolina 27611, 919/733-3221

North Dakota (State Form), Division of Hazardous Management and Special Studies, North Dakota Department of Health, Box 5520, Bismarck, North Dakota 58502-5520

Northern Mariana Islands (EPA Form), Chief, Division of Environmental Quality, P.O. Box 1304, Commonwealth of Northern Mariana Islands, Saipan, CM 96950, Cable Address: Gov. NMI Saipan, Overseas Operator: 6984

Ohio (State Form), State Fire Marshal's Office, Department of Commerce, 8895 E. Main Street, Reynoldsburg, Ohio 43068, State Hotline: 800/282-1927

Oklahoma (EPA Form), Underground Storage Tank Program, Oklahoma Corporation Comm., Jim Thorpe Building, Oklahoma City, Oklahoma 73105

Oregon (State Form), Underground Storage Tank Program, Hazardous and Solid Waste Division, Department of Environmental Quality, 811 S.W. Sixth Avenue, Portland, Oregon 98204, 503/229-5788

Pennsylvania (EPA Form), PA Department of Environmental Resources, Bureau of Water Quality Management, Ground Water Unit, 9th Floor Fulton Building, P.O. Box 2063, Harrisburg, Pennsylvania 17120

Puerto Rico (EPA Form), Director, Water Quality Control Area, Environmental Quality Board, Commonwealth of Puerto Rico, Santurce, Puerto Rico, 809/725-0717

Rhode Island (EPA Form), UST Registration, Department of Environmental Management, 83 Park Street, Providence, Rhode Island 02903, 401/277-2234

South Carolina (State Form), Ground-Water Protection Division, South Carolina Department of Health and Environmental Control, 2600 Bull Street, Columbia, South Carolina 29201, 803/758-5213

South Dakota (EPA Form), Office of Water Quality, Department of Water and Natural Resources, Joe Foss Building, Pierre, South Dakota 57501,

Tennessee (EPA Form), Tennessee Department of Health and Environment, Division of Superfund Underground Storage Tank Section, 150 Ninth Avenue, North, Nashville, Tennessee 37219-5404, 615/741-0690

Texas (EPA Form), Underground Storage Tank Program, Texas Water Commission, P.O. Box 13087, Austin, Texas 78711

Utah (EPA Form), Division of Envirormental Health, P.O. Box 45500, Salt Lake City, Utah 84145-0500

Vermont (State Form), Underground Storage Tank Program, Vermont AEC/Waste Management Division, State Office Building, Montpelier, Vermont 05602, 802/828-3395

Virginia (EPA Form), Virginia Water Control Board, P.O. Box 11143, Richmond, Virginia 23230-1143, 804/257-6685

Virgin Islands (EPA Form), 205(J) Coordinator, Division of Natural Resources Management, 14 F Building 111, Watergut Homes, Christianstead, St. Croix, Virgin Islands 00820

Washington (State Form), Underground Storage Tank Notification, Solid and Hazardous Waste Program, Department of Ecology, M/S PV-11, Olympia, Washington 98504-8711, 206/459-6316

West Virginia (EPA Form), Attention: UST Notification, Solid and Hazardous Waste, Ground Water Branch, West Virginia Department of Natural Resources, 1201 Greenbriar Street, Charleston, West Virginia 25311

Wisconsin (State Form), Bureau of Petroleum Inspection, P.O. Box 7969, Madison, Wisconsin 53707, 608/266-7605

Wyoming (EPA Form), Water Quality Division, Department of Environmental Quality, Herschler Building, 4th Floor West, 122 West 25th Street, Cheyenne, Wyoming 82002, 307/777-7781.

APPENDIX E

Statement for Shipping Tickets and Notices

(40 C.F.R. Part 280, Appendix III (1992))

(**Note**: This statement published by the EPA updates 40 C.F.R. section 380.3 Appendix III (1988)).

Note—A federal law (the Resource Conservation and Recovery Act (RCRA), as amended (Pub. L. No. 98-616)) requires owners of certain underground storage tanks to notify designated state or local agencies by May 8, 1986, of the existence of their tanks. Notifications for tanks brought into use after May 8, 1986, must be made within thirty days. Consult EPA's regulations, issued on November 8, 1985 (40 C.F.R. Part 280) to determine if you are affected by this law.

APPENDIX F

Compliance Timetable for Existing Underground Storage Tanks to Meet Leak Detection Requirements

(40 C.F.R. § 280.40)

SCHEDULE FOR PHASE-IN OF RELEASE DETECTION

Year system was installed	Year when release detection is required (by Dec. 22 of year indicated)				
	1989	1990	1991	1992	1993
Before 1965 or date unknown	RD	P			
1965-69		P/RD			
1970-74		P	RD		
1975-79		P		RD	
1980-1989		P			RD
New tanks (after December 22) immediately upon installation					

P=Must begin release detection for all pressurized piping as defined in § 280.41(b)(1).

RD=Must begin release detection for tanks and suction piping in accordance with § 280.41(a), § 280.41(b)(2), and § 280.42.

APPENDIX G

Summary of Reporting and Recordkeeping Requirements

(40 C.F.R. §§ 280.20–280.107)

I. OWNER'S REPORTS TO THE AGENCY

A. Notification of

- New tanks (in use after 5/8/86)
  -Within 30 days of certification of installation-putting in use
  § 280.20(e), §280.22

II. OWNER'S AND OPERATOR'S REPORTS TO THE AGENCY

A. Release Reports

- Suspected releases
  -Within 24 hours
  § 280.50

- Spills and overfills
  -Within 24 hours
  § 280.53

- Confirmed releases
  -Within 24 hours
  § 280.61

B. Corrective Actions

- Initial abatement measure
  -Within 20 days of release
  § 280.62(b)

- Initial site characterization
  -Within 45 days of release confirmation
  § 280.63(b)

- Free product removal
  -Within 45 days of release confirmation
  § 280.64(d)

- Investigation of soil and groundwater clean up
  -As soon as practicable or as required by agency
  § 280.65(b)

Corrective action plan
-As required by agency
§ 280.66(a)

C. Closure or Change-in-Service

-At least 30 days before beginning
§ 280.71(a)

III. RECORDKEEPING BY OWNERS AND OPERATORS

The following documents must be maintained by either the owner or operator for each underground storage tank system. Any records which are maintained off site must be made available upon request by the implementing agency.

- Corrosion expert's report if no protection equipment is used
  § 280.20(a)(4); § 280.20(b)(3)

- Documentation of operation of corrosion protection equipment
  § 280.31(d)

- Documentation of UST system repairs
  § 280.33(f)

- Recent compliance with release detection requirements
  § 280.45

- Permanent closure site investigation results
  § 280.74

- Evidence of all financial assurance mechanisms used to demonstrate financial responsibility
  § 280.107 (renumbered as § 280.111 by 58 Fed. Reg.9026 at 9051)

APPENDIX H

Compliance Dates for Financial Assurance by Owners of Underground Storage Tanks

(40 C.F.R. § 280.91)

DATE	CLASS OF OWNER
January 24, 1989	Petroleum marketing firms owning 1,000 or more USTs, and all other UST owners with a tangible net worth of $20 million or more (except that compliance with § 280.94(b) is required by July 24, 1989)
October 26, 1989	All petroleum marketing firms owning 100–999 USTs
April 26, 1991	All petroleum marketing firms owning 13–99 USTs at more than one facility
December 31, 1993	All other petroleum UST owners, excluding local government entities
One year from the date of promulgation of additional mechanisms for use by local government entities to comply with financial responsibility requirements for USTs containing petroleum (i.e., Feb. 18, 1994)*	All local government entities

* 58 Fed. Reg. 9026 at 9027 (Feb. 18, 1993).

APPENDIX I

EPA Final Rules Regarding Financial Responsibility Requirements

58 Fed. Reg. 9026 (Feb. 18, 1993)

federal register

Thursday
February 18, 1993

Part II

Environmental Protection Agency

40 CFR Part 280

Underground Storage Tanks Containing Petroleum; Financial Responsibility Requirements; Final Rule

ENVIRONMENTAL PROTECTION AGENCY

40 CFR Part 280

[FRL-4128-9]

RIN 2050-AC67

Underground Storage Tanks Containing Petroleum; Financial Responsibility Requirements

AGENCY: Environmental Protection Agency.

ACTION: Final rule.

SUMMARY: The Environmental Protection Agency (EPA, or the Agency) is promulgating financial responsibility requirements applicable to local governmental owners and operators of underground storage tanks containing petroleum. EPA promulgates these requirements under the authority of section 9003 (c) and (d) of the Resource Conservation and Recovery Act as amended by the Hazardous and Solid Waste Amendments of 1984 (HSWA) and the Superfund Amendments and Reauthorization Act of 1986 (SARA). This rule establishes four alternative mechanisms for use by local governments to demonstrate financial responsibility for taking corrective action and compensating third parties for bodily injury and property damage caused by sudden and nonsudden accidental underground storage tank releases. The Agency is adding these local governmental financial assurance mechanisms to the existing mechanisms contained in the financial responsibility rule promulgated October 26, 1988. These additional mechanisms will allow a greater number of local governmental entities to comply with the financial assurance requirements and will result in a net cost savings to local governments estimated at approximately $32 million over a ten year period.

EFFECTIVE DATE: This rule becomes effective on March 22, 1993.

FOR FURTHER INFORMATION CONTACT: The RCRA/Superfund Hotline at (800) 424-9346 (toll free) or (703) 412-9810 in Virginia, or Sammy Ng in EPA's Office of Underground Storage Tanks at (703) 308-8882.

SUPPLEMENTARY INFORMATION: The contents of today's preamble are listed in the following outline:

I. Authority

These regulations are issued under the authority of sections 2002, 9001, 9002, 9003, 9004, 9005, 9006, 9007, and 9009 of the Solid Waste Disposal Act, as amended. The principal amendments to this Act have been under the Resource Conservation and Recovery Act of 1976, the Hazardous and Solid Waste Amendments of 1984 (Pub. L. 98-616) and the Superfund Amendments and Reauthorization Act of 1986 (Pub. L. 99-499) (42 U.S.C. 6912, 6991, 6991a, 6991b, 6991c, 6991d, 6991e, 6991f, and 6991h).

II. Background

This section provides the legislative and regulatory background for this rule and summarizes today's additional mechanisms for financial responsibility for local government entities.

A. Legislative and Regulatory Overview

This section discusses the statutory authority for financial responsibility regulations for UST owners and operators, the provisions of the financial responsibility regulations promulgated on October 26, 1988 and the scope of the financial responsibility regulations being promulgated today.

1. RCRA Subtitle I

The Hazardous and Solid Waste Amendments of 1984 (HSWA) extended and strengthened the provisions of the Resource Conservation and Recovery Act (RCRA). HSWA added Subtitle I to RCRA, establishing provisions for the development and implementation of a regulatory program for underground storage tanks (USTs) containing certain substances, including petroleum and other regulated substances (such non-petroleum regulated substances are hereinafter referred to as "hazardous substances"). Section 9003(a) of Subtitle I requires the EPA Administrator to promulgate requirements for release detection, prevention, and correction as necessary to protect human health and the environment. These technical standards were promulgated at 53 FR 37082 (September 23, 1988).

The Superfund Amendments and Reauthorization Act of 1986 (SARA) amended sections 9003 (c) and (d) of Subtitle I to mandate that the Agency establish financial responsibility requirements for UST owners and operators to assure the costs of corrective action and third-party liability caused by sudden and nonsudden accidental releases from USTs. SARA also modified Subtitle I by specifying the minimum statutory levels of financial responsibility for petroleum marketers and the factors that EPA may consider in setting minimum levels for non-marketers. The objective of the financial responsibility requirements is to ensure that owners and operators can respond promptly to clean up releases and to compensate third parties for any injuries or damages associated with UST releases.

2. October 26, 1988 Rule

The final financial responsibility rule, promulgated on October 26, 1988 applies to owners or operators of "petroleum UST systems" with the following exceptions:

(1) Federal or State entities that own or operate USTs containing petroleum; and

(2) Owners and operators of tank systems excluded from the technical standards.

To cover the potential costs of corrective action and third-party liability claims from sudden and nonsudden accidental releases from USTs, the rule requires the following parties to obtain financial assurance of at least $1 million per occurrence:

(1) All owners or operators of petroleum USTs at facilities engaged in petroleum production, refining, or marketing; and

(2) Owners or operators of USTs with an average monthly throughput of more than 10,000 gallons.

Owners or operators of USTs at facilities not engaged in petroleum production, refining, or marketing with an average monthly throughput of 10,000 gallons or less must maintain financial assurance of at least $500,000 per occurrence. All owners or operators must maintain an annual aggregate of $1 million or $2 million, depending on the number of USTs assured. The responsibility for cleanup and third-party compensation in the event of UST releases was established under the technical standards published in September 1988. The October 1988 financial responsibility rule made owners and operators responsible for complying with the financial responsibility requirements, but otherwise imposed no new liability; rather, the rule was intended to verify that local government owners or operators of USTs would be able to meet their liabilities in the event of an UST release. It is important to note that exemption from the financial responsibility requirements would not exempt an owner or operator from their liabilities in the event of an UST release.

UST owners or operators may use the following mechanisms to satisfy the requirements: Insurance or risk retention group coverage, surety bond, guarantee, letter of credit, financial test of self-insurance, trust fund, a State-required mechanism, or a State fund or other State assurance. (Under the October 26, 1988 rule, only private companies reporting to credit reporting agencies, publicly-held companies reporting to the Securities and Exchange Commission, and public utilities reporting to specified agencies are eligible to use the financial test of self-insurance.) Mechanisms can be used alone or in combination to cover the costs of taking corrective action and compensating third parties as long as a mechanism or a combination of mechanisms provides the full amount of required assurance. The only combination of mechanisms that is not allowed is the financial test of self-insurance and a guarantee where the financial statements of the owner or operator and the guarantor are consolidated.

The October 26, 1988 final rule requires owners or operators to submit documentation of financial responsibility to the implementing agency for three occurrences: (1) After a known or suspected release occurs, (2) when a provider becomes incapable of providing assurance, and (3) when a provider revokes a mechanism and the owner or operator is unable to obtain alternate coverage. Owners or operators must also submit documentation of financial responsibility if requested by the implementing agency. In addition, UST owners or operators must notify the implementing agency of their methods of demonstrating financial responsibility upon installation of new tanks. Owners or operators must also maintain records of the financial assurance mechanisms used to satisfy these requirements on-site or at their place of business.

The October 26, 1988 rule also contains provisions that require third-party providers of financial assurance (i.e., sureties, insurance companies, risk retention groups, guarantors, and providers of letters of credit) to provide notice of cancellation with an adequate time period for the UST owners and operators to seek alternative coverage and to determine whether there has been a release that would trigger the third-party mechanism. On November 9, 1989, EPA published an interim final rule that modified the required language of endorsements required for insurance policies as they relate to cancellation (54 FR 47077).

The State program approval objective for financial responsibility of owners and operators of petroleum UST systems was also promulgated October 26, 1988. This objective outlines two general provisions: (1) The considerations used to determine whether States' financial responsibility requirements will be considered "no less stringent" than the corresponding Federal requirements standard, and (2) the standards that must be met to demonstrate adequate enforcement of compliance.

3. Discussion of the Financial Responsibility Requirements for Governments in the October 26, 1988 Rule

Although the final financial responsibility rule (53 FR 43322, October 26, 1988) exempts those government entities whose debts and liabilities are the debts and liabilities of Federal or State governments, local government entities are required to provide financial assurance for USTs that they own or operate. Under the Agency's schedule for phased compliance with the final rule, local government entities have been given until February 18, 1994, one year from the promulgation of today's rule, to comply. In the October 1988 final rule, the Agency stated its intention to develop a financial test in the interim that would allow local governments to demonstrate that they have the requisite financial strength and stability to pay the costs associated with UST releases. After passing this financial self-test, local government entities will be allowed to demonstrate financial responsibility in a manner similar to private companies that meet the criteria of the corporate financial test of self-insurance.

Under the compliance schedule, Indian tribes are required to comply with financial responsibility requirements under the same schedule as local governments; that is, within one year from the promulgation of today's rule (i.e., before February 18, 1994).

4. The Proposed Rule

The proposed rule was published on June 18, 1990. The Agency received comments from 23 commenters. Most supported the development of the new financial responsibility mechanisms, stating that these additional mechanisms allow more local governments to comply with the financial assurance requirements, and that they would be able to do so at lower cost. Some commenters suggested changes or additions to the mechanisms proposed. Where appropriate, the Agency has adopted these suggestions. The specific issues raised and the Agency's responses are addressed in "Summary of Comments and Responses on Proposed Additional Financial Responsibility Mechanisms for Local Governments Subject to Subtitle I of the Resource Conservation and Recovery Act."

One commenter proposed as a new alternative mechanism that EPA issue regulations allowing implementing agencies to redirect funds from Federal- or State-funded programs to pay for the expenses associated with corrective actions. The Agency rejected this suggestion because it has no statutory authority to redirect funds from other State or Federal programs.

B. Key Provisions in Rule

In today's rule, the Agency is providing additional mechanisms that will allow local governments to comply with the financial responsibility requirements. These mechanisms do not replace the existing methods; rather, they supplement them. These mechanisms are similar in intent to the corporate guarantee and the financial test of self-insurance now allowed as mechanisms for corporations. Local governments eligible to use the mechanisms may use them alone or in combination with other mechanisms, as described below.

One commenter questioned the language indicating that all local governments "may use" the new financial assurance mechanisms, since the criteria associated with using the mechanisms by definition restricts their use by certain entities. The Agency emphasizes that all local governments may seek to use all mechanisms, but only those that meet all qualifying criteria may use a specific mechanism to demonstrate financial responsibility.

EPA is promulgating four additional mechanisms for use by local government entities to demonstrate financial responsibility:

(1) Bond rating test. Local government entities with $1 million or more of total outstanding issues of general obligation bonds (excluding refunded obligations) and having investment-grade ratings would be eligible to demonstrate financial responsibility. Non-general purpose local governments (e.g., special districts and school districts) with $1 million or more of investment-grade revenue bonds may also use this mechanism if they do not have the authority to issue general obligation bonds. General obligation bonds that are backed by credit enhancement mechanisms other than bond insurance may not be included in the bond rating test. Revenue bonds that are backed by any type of credit enhancement mechanism may not be included in the bond rating test. Bonds with investment grade ratings are defined as those having a Moody's bond rating of Baa or higher (i.e., Aaa, Aa or A), or a Standard and Poor's bond rating of BBB or higher (i.e., AAA, AA, or A). Passing the bond rating test will be considered a sufficient demonstration of financial responsibility.

(2) Worksheet test. A worksheet test has been developed for use by local government entities that do not have general obligation or revenue bond ratings or that have less than $1 million in outstanding issues of investment-grade-rated general obligation or revenue bonds. (Governments meeting the requirements of both the bond rating test and the worksheet test may use either mechanism but are assumed to use the bond rating test as a matter of administrative convenience.) Local governmental entities having outstanding issues of general obligation or revenue bonds that are rated as less than investment grade are not eligible to use the worksheet test. The worksheet incorporates several financial criteria designed to measure a local government entity's financial stability. Passing the worksheet test will be a sufficient demonstration of financial responsibility.

(3) Guarantee. A local government entity can demonstrate financial responsibility by obtaining a binding guarantee from another governmental entity able to demonstrate financial responsibility assurance through the alternative mechanisms. The guarantor must have the authority to provide a guarantee to the local government entity seeking financial assurance. For example, a town may serve as the guarantor for a special district, a county may serve as the guarantor for a school district, a State may serve as the guarantor for a county, or a city may act as a guarantor to a special district (e.g., a transportation authority or a government utility). A guarantee for the entire aggregate limit for which a local government must demonstrate financial responsibility will be a sufficient demonstration of financial responsibility. A guarantee for a lesser amount may be used in combination with one or more other allowable mechanisms to demonstrate financial responsibility.

(4) Maintenance of a funded balance. Local government entities may satisfy the financial responsibility regulations by developing a self-administered emergency response fund to finance an UST corrective action and pay for third-party damages. A fund balance established for the entire aggregate limit for which a local government must demonstrate financial responsibility will be a sufficient demonstration of financial responsibility. A fund balance established for a lesser amount may be used in combination with one or more other allowable mechanisms to demonstrate financial responsibility.

The October 1988 rule allows the use of combinations of financial responsibility mechanisms. This feature is extended to include the financial self-test mechanisms being promulgated today. For example, a local government entity may use the guarantee or funded balance mechanisms to satisfy the deductible amounts of insurance policies. Local governmental entities may use the mechanisms being promulgated today in addition to the mechanisms allowed by the October 1988 rule: insurance, risk retention group (RRG) coverage, surety bond, letter of credit, State-required mechanisms, or a State fund or other State assumption of responsibility.

In contrast to the specifications for the corporate self-test, EPA does not believe that local governments will use consolidated financial statements to support both the worksheet and the guarantee mechanisms. Local governments are separate legal and financial entities from States and from each other. The situation wherein a local government will consolidate its financial statements with a State, or vice versa, and use the consolidated statements to support both the worksheet and the guarantee, cannot occur. In addition, most local governments are independently chartered. By the nature of the local government charters, local government operations that are consolidated, such as utility operations accounted for as enterprise funds, never issue stand-alone financial statements, because they have no independent standing. Thus, there is no potential that the consolidated entities could first use their own financial statements for the worksheet, and then rely on the consolidated financial statements for a guarantee, because they have no independent financial statements. Independent authorities (e.g., independent school districts) are independent because they have separate charters and/or articles of incorporation; they operate independently and their financial statements are never consolidated with the statements of the nearby general purpose governments.

To support this rule, the Agency has prepared a Background Document, "Background Document in Support of Financial Self-Test for Local Governments Subject to the Financial Responsibility Requirements of Subtitle I of the Resource Conservation and Recovery Act," that describes in detail the methodology and analyses used to evaluate potential financial responsibility mechanisms.

C. Rationale for Agency's Approach

The Agency had four main goals in developing the additional alternatives being promulgated today for local governments to demonstrate financial responsibility under Subtitle I. First, the Agency wanted to recognize fundamental differences between governmental entities and private entities. Second, the Agency wanted to keep the rule as flexible as possible to allow local governments a variety of choices in demonstrating financial responsibility. Thus, the Agency is promulgating several financial assurance mechanisms for local governments. Third, the Agency wanted to keep the mechanisms as simple as possible to minimize the administrative burden on local governments as well as the implementing agency. Thus, the

Agency is promulgating options that use data believed to be readily available to local governmental entities or that are consistent with governmental practices and is maintaining the same approach to reporting requirements adopted in the regulations published in the October 1988 rule. Fourth, the Agency wanted financial responsibility mechanisms that could realistically be used by local governments.

In the October 1988 rule, the Agency provided a mechanism whereby financially secure corporations can self-insure. The rule provided two alternatives for corporations. Under Alternative I, a firm can self-insure if it meets four criteria: (1) Tangible net worth equal to 10 times the sum of its financial responsibility amounts for underground storage tanks, its closure, post-closure care, liability coverage, and/or corrective action costs for Subtitle C facilities, and its plugging and abandonment costs for Class I Hazardous Waste Injection Wells, (2) tangible net worth equal to at least $10 million, (3) annual filing of its financial statements with the Securities and Exchange Commission (SEC), the Rural Electrification Administration (REA), the Energy Information Administration (EIA), or Dun & Bradstreet (which must have assigned a financial strength rating of 4A or 5A), and (4) annual reports which, if independently audited, did not include an adverse auditor's opinion or a disclaimer of opinion. Under Alternative II, a corporation can self-insure if it meets four criteria: (1) Tangible net worth of at least $10 million, (2) tangible net worth at least six times its UST obligation, (3) U.S. assets equal to at least 90 percent of total assets, or at least six times its UST obligations, and (4) net working capital equal to at least six times the required amount of UST aggregate coverage, or a current Standard and Poor's bond rating of AAA, AA, A, or BBB, or a current Moody's bond rating of Aaa, Aa, A, or Baa. In addition, a firm using Alternative II must either report its financial information to the SEC, the EIA, or the REA or obtain a special auditor's report.

Local government entities, however, differ in several important characteristics from corporations, which makes the application of the corporate self-test mechanism in the October 1988 rule impractical for local governments. For example, "general purpose" local governments (counties, municipalities, and townships) generally use accounting systems that do not recognize assets in a manner similar to private companies. For example, municipal buildings and infrastructure (e.g., streets and utility lines) are not generally carried as assets on the local government financial statements. Thus, a test based on "tangible net worth" is, by definition, unworkable for many local governments. (It should be noted, however, that government-owned utilities that provide financial data to the Rural Electrification Administration or the Energy Information Administration are allowed to use the corporate financial test under the October 1988 rule.) Also, the accounting standards used by most local governmental entities are not the same as the Generally Accepted Accounting Principals ("GAAP") used by private entities. Most local governments use either cash basis accounting (often mandated by State law) or "modified" accrual accounting, where the recognition of revenues may be delayed. Consequently, a test based on "net working capital" may be unworkable for most local governmental entities. In addition, local governments are not generally required to report financial information to a regulatory agency similar to the Securities and Exchange Commission. Thus, it is impossible to incorporate mandatory reporting to an independent organization into a self-test.

Nevertheless, the Agency believes that a mechanism parallel to self-insurance is particularly appropriate for local government entities. The Agency has determined that local government entities are, in general, more financially stable than private companies. Most local governments, unlike private entities, have the authority to levy taxes or to independently set rates, which provide a consistent, reliable source of income. In contrast to corporations, they are less likely to dissolve or merge with other entities which means that they are less likely to have abrupt changes in financial structure. They are, by definition, geographically fixed, eliminating potential concerns that they may move and abandon their USTs. They rarely go bankrupt, suggesting that they are, as a class, more financially stable. As discussed in the background document, the available literature suggests that even bankruptcy does not allow local government entities to void their legal obligations. Additionally, unlike some private companies, local governments are generally required to make their financial data publicly available.

These factors suggest that a self-test for municipalities does not necessarily require the same level of built-in safeguards as required of private entities. Assurance that local government owners and operators will be financially responsible for their UST-related obligations, therefore, can be demonstrated more easily than assurance for private entities. Consequently, the primary concern of the Agency in developing this rule is that local governments show evidence of financial stability and prudent financial management.

D. Description of the Regulated Community

This section describes the nature of the local governmental entities that would be regulated under today's rule, including a description of their UST ownership characteristics, a brief description of their operation, and an overview of the considerations the Agency has used in developing today's rule.

The Agency estimates that about 62,000 petroleum USTs that are subject to Subtitle I jurisdiction are owned or operated by approximately 25,000 local government entities. Most of these USTs store petroleum products for purposes other than retail motor fuel sales. A local government entity may, for example, own USTs that store gasoline to fill police and fire vehicle tanks.

Local government entities include both general purpose local governments and special purpose local government entities. General purpose local government entities include municipalities, counties, townships, towns, villages, parishes, and New England towns. Special purpose local governments include entities that perform a single function or a limited range of functions. Special purpose local governments are generally designated as either public authorities or special districts such as school districts, water and sewer authorities, transit authorities, redevelopment authorities, irrigation districts, or power authorities. All local governments, both general and special purpose, are subject to this rule and are eligible to use the new financial assurance mechanisms described in today's rule. Several commenters requested an expansion or clarification of the definition of local government entities to include local public transit systems and redevelopment authorities. The Agency originally intended these types of local government entities to be included in the definition, and has clarified the definition as requested by the commenters.

The Agency's research has shown an extremely low rate of fiscal emergencies among governmental entities through the 1970s and 1980s. A 1983 study by the Advisory Council on Intergovernmental Relations (ACIR) found only three incidents of bankruptcy among general purpose governments, only one of which caused a general purpose governmental body to void a legally binding agreement. In all other cases, even local government entities that entered bankruptcy were forced to make full restitution, although sometimes over a stretched-out payment term. Since 1983, only five additional general purpose governments are known to have declared bankruptcy. There has been a similarly low rate of bankruptcy among special purpose districts. Between 1972 and 1989, 29 utility special districts, two school districts, and six other special purpose districts and hospitals filed for bankruptcy (out of a total of more than 40,000 school districts and special purpose districts).

Although bankruptcy is an extreme condition, the Agency believes this very low incidence (0.003 percent per year) reflects general stability of local government entities. In contrast, 56,423 (1.3 percent) of the 4,256,243 private companies in operation filed bankruptcy petitions in 1982.[1] This number increased to 88,278 in 1987. Combined with the relatively low costs of UST financial responsibility obligations (relative to other environmental obligations and most governmental activities in general), the relative stability of local governments is interpreted by EPA to indicate a general ability to meet financial obligations under Subtitle I.

In addition, the Agency's research has shown relatively few cases where releases were known to have come from local government-owned USTs. For releases that did occur, local government entities were generally able to clean up and to pay for the costs of corrective actions associated with the releases. Because of the limited data regarding local government responses to UST releases, however, the Agency has relied primarily on data and analyses regarding the overall financial health of local governments. One commenter indicated that cleanups of UST releases at airports are generally funded from operations or funds for construction projects. The Agency interprets this statement as additional support for allowing local governments to demonstrate financial responsibility based on their internal financial condition, rather than requiring the use of third-party mechanisms.

[1] "Statistical Abstract of the United States," 109th Edition. United States Department of Commerce, Washington, D.C., 1989; and "General Report on Industrial Organization," 1982 Enterprise Statistics. Issued October 1986.

III. Section-by-Section Analysis

A. Applicability

Today's rule would apply to all non-exempt governmental owners and operators of underground storage tanks containing petroleum. 40 CFR § 280.90(c) exempted from financial responsibility requirements State and Federal government entities whose debts and liabilities are the debts and liabilities of a State or the United States. Although the October 1988 rule excluded State and Federal governments, it required local government entities to demonstrate financial assurance for USTs that are owned or operated by the government.

Data available to the Agency in preparing the Regulatory Impact Analysis for the October 1988 rule suggest that local government entities collectively own approximately 62,000 USTs. Additional analysis of the New York State tank notification data base suggests that larger local government entities are more likely to own USTs and are more likely to own multiple USTs, but a specific breakdown of how many of each type of local government own USTs is not available from the data available to EPA. Overall, EPA estimates that about approximately 25,000 local governments own USTs.

Local government entities are created under State law, and consequently vary significantly from State to State. All local government entities recognized under State law may seek to use the financial assurance mechanisms being promulgated today. As recognized by the Bureau of the Census, local government entities generally fall into the following categories:

County Governments: Organized county governments are found throughout the nation except for Connecticut, Rhode Island, the District of Columbia, and limited portions of other States. In Louisiana, the county governments are officially designated as "parish" governments, and the "borough" governments of Alaska resemble county governments in other States. In general, county governments are defined in terms of a geographical area served, rather than a specific population.

Municipal Governments: Municipal governments include active government units officially designated as cities, boroughs (except in Alaska), towns (except in the six New England States and Minnesota, New York, and Wisconsin), and villages. This concept corresponds to the "incorporated places" that are recognized in Census Bureau reporting of population and housing statistics.

Township Governments: Township governments exist to serve inhabitants of areas without regard to population concentrations. This category includes governments officially designated as "towns" in the six New England States, New York, and Wisconsin, some "plantations" in Maine, and "locations" in New Hampshire, as well as governments called townships in other areas. In Minnesota, the terms "town" and "township" are used interchangeably.

School Districts Governments: Forty-five States have established public school systems with sufficient autonomy and fiscal authority that they can be classified as independent local government entities.

Special Purpose Districts: Special purpose districts are governmental entities created to perform a single or limited range of functions (e.g., school districts, park and recreation districts, libraries, fire protection districts, cemeteries, transit districts, redevelopment authorities, etc.). These districts may be subdivided into any of the following distinct categories: (1) Local or metropolitan districts; (2) districts dependent on or independent of a municipality for their creation or operation; and (3) districts created by State enactment or by municipal resolution. They have sufficient administrative and fiscal autonomy to qualify as separate governments.

Indian Tribes: Indian Tribes are included in the statutory definition of municipality in RCRA Section 1004(13) and are, therefore, required to comply with the financial responsibility requirements by the same compliance date as other local government entities. This rule treats Indian Lands as local government entities and allows them to use the self-test mechanisms to demonstrate financial responsibility.

Several commenters requested exemptions from the UST financial responsibility requirements for local governments. Commenters gave the following reasons for such an exemption: (1) Local governments, as a class, have sufficient financial strength and stability to pay for corrective actions without the need to demonstrate financial responsibility; and (2) the adverse effects on the ability of local governments to fund emergency services if required to divert funds to pay for assurance mechanisms. One commenter, a small rural town, indicated that it cannot qualify to self-insure and added that the financial responsibility regulations impose financial burdens with which the town, and presumably other towns, could not possibly comply.

EPA believes that commenters may have failed to distinguish between: (1) The need for local governments to pay for costs associated with UST releases, as required under the technical standards; and (2) the financial responsibility regulations, which merely require that UST owners be able to demonstrate that they will be able to meet such costs if they occur. Even if EPA were to exempt local governments from the requirement to demonstrate financial responsibility, such an exemption would not, under Subtitle I, relieve them from the legal liability to pay for the costs of UST releases and to compensate third parties for damages caused by releases.

The Agency agrees that most local government entities do have the resources and the will to meet financial responsibilities. This belief underlies the effort to develop mechanisms by which local governments can demonstrate compliance with the financial responsibility requirements without the need to obtain insurance or the use of other third-party mechanisms.

The Agency also agrees with commenters who noted that some local governments may not have the resources to meet their UST-related financial obligations. Consequently, it would not be appropriate to exempt all local governments from the need to demonstrate financial responsibility. Further, EPA believes that exempting all local governments from the requirement to demonstrate financial responsibility would not be consistent with statutory intent as discussed in 9003(d)(5).

The Agency notes the concern about the potential impact on local governmental services. The Agency believes, however, that the mechanisms provided will allow any fiscally solvent local government to demonstrate financial responsibility and continue to operate its USTs, and will do so at minimum cost to the affected local governments. EPA encourages governments unable to demonstrate financial responsibility using the worksheet, bond rating, or fund balance mechanisms to seek guarantees from neighboring jurisdictions or from county governments. EPA believes that such entities are better able to determine the strengths of the government seeking the guarantee, and to measure how essential are the services offered, than the Agency would be in developing a uniform national standard.

B. Definition of Terms

1. Bond Ratings

A bond rating is an "evaluation of the credit quality of notes and bonds usually made by independent rating services . . . Ratings generally measure the probability of the timely repayment of principal and interest of municipal bonds." [2] In this rule, only ratings made by Moody's Investors Service and Standard & Poor's will be considered eligible for use in demonstrating financial responsibility.

2. Investment Grade Bonds

As defined by the Comptroller of the Currency, investment grade bonds are generally regarded as eligible for bank investment. In addition, the legal investment laws of various States may impose certain ratings or other standards for obligations eligible for investment by savings banks, trust companies, and fiduciaries generally. For purposes of this rule, investment grade bonds are considered to include bonds rated Aaa, Aa, A, and Baa by Moody's, or AAA, AA, A, and BBB by Standard and Poor's.[3]

3. General Obligation Bonds

General obligation (G.O.) bonds, also known as "full faith and credit" bonds, are secured by their issuers' ability to levy ad valorem taxes or to draw from other unrestricted revenue sources, such as sales or income taxes. These bonds are important mechanisms for financing municipal capital improvements such as schools, streets, and municipal buildings. The bond issuer's ability to generate revenues is evaluated by analyzing factors in four categories: socioeconomic, finance, debt, and administration.[4]

4. Revenue Bonds

A revenue bond is a long-term debt instrument that is issued to finance a specific public enterprise and that is payable solely from enterprise earnings or from a dedicated tax.[5] The Agency has determined that most revenue bonds issued by general purpose governments (i.e., counties, municipalities, and townships) are issued to fund specific projects with dedicated revenue streams not necessarily central to the operations of that government, and that the evaluation criteria associated with these revenue bonds may not fully reflect the socioeconomic, financial, and administrative condition of a general purpose government. Instead, the ratings reflect a more limited set of criteria pertaining to the specific project financed. In contrast, the Agency has determined that revenue bonds issued by special districts are generally used to finance projects central to the operations of the special districts, so that the ratings encompass a broader view of the overall financial condition of the issuing entities. In this rule, the Agency allows only special districts and school districts that do not have the authority to issue general obligation debt to use investment-grade ratings on revenue bonds to demonstrate financial responsibility.

5. Substantial Governmental Relationship

The October 26, 1988 rule authorized owners and operators to obtain a corporate guarantee to meet their financial responsibility requirements. The corporate guarantor must: (a) Have a controlling interest in the owner or operator or in a specified related firm; or (b) issue the guarantee as an act incident to a "substantial business relationship" with the owner or operator (§ 280.96). The object of the corporate guarantee is a valid and enforceable contract. Additionally, to insure that State insurance laws will not impair the enforceability or validity of the mechanism, a corporate guarantee may be used only if it is certified for use by the Attorney General of the State in which the USTs are located.

Local governments, however, do not have "controlling interests" in one another, and their interactions may not be of an economic nature constituting a "substantial business relationship." As with the corporate guarantee, the Agency is concerned that local governmental guarantees be valid and enforceable, and that they do not conflict with State insurance laws. Thus, a municipality using a local governmental guarantee must certify that there is a "substantial governmental relationship" underlying the guarantee. Such a relationship must include a clear commonality of interests, such as common constituencies served, overlapping geographical jurisdiction, or mutual impact in the event of an UST release. In addition, a local government acting as a guarantor must have the authority to enter into such agreements.

Examples of governmental guarantees could include: (1) A guarantee offered

[2] Moody's Investors Service, Inc., "Moody's on Municipals: An Introduction to Issuing Debt," 1989, p. 75.

[3] Both Standard and Poor's and Moody's recognize groupings within the major bond rating classes. Moody's signifies higher ranking bonds within a class with a "1" (e.g., Baa1), while Standard and Poor's uses a +/− system to designate higher and lower ranking bonds. This proposed rule does not consider these groupings. Thus, a Baa1 rating is classified as a Baa rating for the purposes of the test, while an AA+ or AA− rating is classified as an AA rating.

[4] Standard & Poor's Corporation, "Standard & Poor's Debt Ratings Criteria: Municipal Overview," 1986.

[5] Standard & Poor's Corporation, "Standard & Poor's Municipal Finance Criteria," 1989.

by a county to an incorporated city located partially or entirely within the limits of the county; (2) a guarantee offered by one county to another if both counties cover a common aquifer subject to contamination by UST releases; (3) a guarantee offered by the State to a local government within the State; or (4) a guarantee offered by a general purpose local government to an independent school district, water district, utility district, or other special district serving the guarantor in whole or in part. One commenter questioned what types of publicly owned utilities would be eligible to receive a guarantee. Any special district is eligible to receive a guarantee if it has its own governing body and an independent accounting system.

Additional examples of appropriate intergovernmental relationships for a governmental guarantee would be joint operating agreements for emergency responses across jurisdictional boundaries, or purchase of non-UST-related services such as water or education.

One commenter asked three questions pertaining to activities that constitute a "substantial governmental relationship": (1) Whether a governmental entity may act as a guarantor for more than one entity; (2) whether a contractual relationship (under an intergovernmental pooling arrangement) of a pool to provide safety and risk management services in addition to risk pooling will be recognized as a "substantial governmental relationship"; and (3) what criteria determine that a relationship is "sufficiently non-monetary."

The Agency concludes that a local government may act as guarantor for multiple entities. A guarantee from a risk pool, however, is not considered a governmental guarantee for the purposes of establishing financial responsibility. The role of a risk pool is almost exclusively monetary, similar to that of insurance. Issuance of a guarantee would not change the nature of that relationship. The Agency recognizes that participation in a risk pool provides a means for local governments to reduce their liability for large unforeseen events. However, risk pools have not been approved as a Federal financial responsibility mechanism because no comprehensive yet manageable set of Federal guidelines could be developed to ensure that all risk pools would have adequate oversight to make them comparable to the other financial responsibility mechanisms allowed.

The Agency notes that, under § 280.100, risk pools can be adopted as Federal financial responsibility mechanisms by individual States as State-required mechanisms. That is, a State may allow or require local governments to demonstrate financial responsibility through participation in a risk pool if the State can demonstrate to the Agency that the risk pool would be at least equivalent to the other financial responsibility mechanisms allowed.

C. Amount and Scope

The amount and scope of financial responsibility is not being changed from the requirements established in the October 1988 rule. Governmental entities owning or operating USTs at facilities with a monthly throughput of less than 10,000 gallons must demonstrate financial responsibility in the amount of $500,000 per occurrence. Governmental owners and operators owning or operating one or more USTs at facilities with a monthly throughput of 10,000 gallons or more must demonstrate financial responsibility in the amount of $1 million. In addition, owners and operators of USTs must demonstrate financial responsibility in the amount of an appropriate annual aggregate. Owners and operators of 100 or fewer USTs must demonstrate financial responsibility in the annual aggregate amount of $1 million, and owners and operators of more than 100 USTs must demonstrate financial responsibility in the annual aggregate amount of $2 million.

One commenter suggested incorporating a mechanism in the rule that would allow for reductions in the required level of assurance when tanks are replaced with intrinsically safe tanks or upgraded to be intrinsically safe. The commenter believed that this proposal would result in more equitable and less burdensome requirements for assurance. The Agency disagrees with the commenter's suggestion for the reasons cited in the October 1988 final rule and the June 1990 proposed rule.

Another commenter indicated that disclosing the amount of money that will be paid per release by an assurance mechanism may adversely affect a local government's position in litigation or settlement negotiations. The commenter recommended deleting this provision from the financial officer's letter. EPA believes that the commenter may have misinterpreted the intent of the financial officer's letter. The amount assured, as cited in the financial officer's letter, is not meant to be a minimum amount that must be paid in the event of a release, but rather the minimum amount that a local government must be able to pay if required to meet corrective action costs and third-party liabilities. EPA assumes that governments will use all defenses and mechanisms to ensure that payments for third-party liabilities are fair and equitable. Conversely, the amount of financial assurance demonstrated does not limit a local government's potential liability in the event of a release. Local governments are liable for all costs resulting from a release, regardless of the amount for which they demonstrate financial responsibility. EPA requires that an amount be specified in the financial officer's letter to ensure that senior officials of the government are aware of their potential obligations as UST owners.

IV. New Mechanisms for Demonstrating Financial Responsibility

A. Description of New Mechanisms

Today's rule promulgates four additional financial assurance mechanisms for use by local government entities that own or operate USTs containing petroleum: A bond rating test, a worksheet test, a governmental guarantee, and maintenance of a funded balance. These additional mechanisms are described below. In addition to these mechanisms, local governments that are owners and operators of USTs may use any of the financial responsibility mechanisms authorized under 40 CFR § 280.94 (i.e., insurance, Risk Retention Group (RRG) coverage, surety bonds, letters of credit, fully-funded trust funds, State-required mechanisms, a State fund, or other State assumption of responsibility). The Background Document prepared in conjunction with this rule explains in more detail the data and methodology used to develop the new mechanisms now being finalized.

1. Bond Rating Test (§ 280.104)

In order to pass the bond rating test, local government entities must have outstanding issues of general obligation bonds that are currently rated at least "investment grade" by Moody's or Standard & Poor's. Special districts, such as school districts or airport authorities, that do not have the authority to issue general obligation bonds may substitute investment grade revenue bonds for general obligation debt to satisfy the bond rating test. In both cases, the municipality's total outstanding obligation must be $1 million or more, excluding refunded obligations. Investment grade bonds are those with a current Standard and Poor's bond rating of AAA, AA, A, or

BBB, or a current Moody's bond rating of Aaa, Aa, A, or Baa. If a local government has multiple outstanding issues of general obligation or revenue bonds with different ratings, or if the ratings assigned to a single class or issue of bonds by different ratings agencies differ, the *lowest* rating must satisfy the criterion of the test.

If a local government owner or operator using the bond rating test to provide financial assurance finds that it no longer meets the bond rating test requirements, the local government owner or operator must obtain alternative coverage within 150 days of the change in status.

The Agency is aware that municipal bonds are often insured by third-party insurance companies, and that the rating assigned to such insured bonds is established primarily by the credit-worthiness of the insurer. After examining the criteria used by the rating companies to evaluate bond insurance companies, however, the Agency has concluded that the provisions for on-going review and intervention granted to the bond insurance companies under the insurance agreements provides a level of third-party oversight comparable to that provided directly by the bond rating companies. For purposes of this rule, therefore, the Agency is not distinguishing between general obligation bonds that are uninsured or insured by a bond insurance company.

EPA has not found evidence that other providers of other methods of credit enhancement, such as letters of credit, provide a degree of oversight equivalent to that provided by bond insurers. Consequently, ratings that are supported by means of credit enhancement other than bond insurance may not be used to demonstrate financial responsibility.

The Agency has selected the existence of investment-grade bond ratings on general obligation debt as an option for demonstrating financial responsibility for several reasons. First, EPA took into consideration the use of bond ratings as a standard measure of risk by banks and other fiduciary entities. As a result of a 1938 agreement issued jointly by the Comptroller of the Currency, the Federal Deposit Insurance Corporation, the Board of Governors of the Federal Reserve System and the Executive Committee of the National Association of Supervisors of State Banks, these agencies have given municipal bonds in the first four rating categories (Aaa through Baa or AAA through BBB) privileged status as investment securities. Banks are permitted to hold only a certain number of low or unrated bonds, and they must balance such holdings with higher rated or more credit-worthy securities. Second, bond ratings serve as one of the only independent evaluations of local government entities' financial health. To perform their evaluations, the bond rating companies must consider a variety of factors that affect both local government entities' current ability to pay and the likelihood of continued ability to pay in the future. In particular, the costs of environmental obligations are included in the evaluations. Thus, the costs of underground storage tanks, solid waste landfills, hazardous waste landfills, sewage treatment plants, and associated environmental liabilities are factored into the rating analysis.[6] Third, general obligation bonds are secured by the full faith and credit of the borrower, and backed by the issuers' ability to levy taxes or make legislative appropriations. The Agency considers this underlying security equivalent to the requisite level of financial responsibility intended under Subtitle I. Fourth, bonds are re-rated on a periodic basis. Local governments are required to provide current financial data annually; failure to do so can result in removal of the bond rating. Also, the rating agencies receive local newspapers from around the country to monitor local conditions.[7]

Today's rule allows the use of insured issues of general obligation bonds. Information from bond rating companies indicates that local governments do not purchase insurance as a means of earning an investment grade rating, but rather to increase the rating from a lower investment grade (e.g., Baa, Baa1, or A) to the very highest (Aaa). In exchange for the cost of the insurance, the local governments obtain a lower interest rate for the life of the bond. Analysis undertaken by Moody's of four major bond insurers shows that virtually all of the insured debt would have earned an investment grade rating without the insurance, and so would qualify under the bond rating test.[8] In addition, bond insurers, unlike bond rating agencies, have a strong financial interest in the soundness of the local governments. If a local government defaults on a payment, the bond insurers must meet the payment. Consequently, bond insurers track the financial obligations of insured local governments closely and often have covenants that allow them to intervene in local government operations. Insurers, for example, may insist on more conservative fiscal policies to preserve the financial strength of a community, which in turn, lowers the risk and cost associated with the bond insurance. Although the bond rating of insured bonds does not directly indicate a local government's financial condition, it does demonstrate both that the government has assured the insurance company of its ability to meet its debts, and that a third party has a strong confidence in the financial health of the local government.

Two commenters agreed with and endorsed the methodology of the bond rating test, stating that the test will serve as a simple method for demonstrating financial responsibility and will provide the Agency with the assurance it seeks without imposing too great a burden on the regulated community.

Several commenters suggested that the Agency expand the bond rating test to include revenue bonds and other sorts of debt instruments as well as general obligation bonds. The Agency has researched the criteria used to assign credit ratings on short-term notes, certificates of participation, lease rental debt, and revenue bonds, and examined how well the credit rating addresses the financial health and fiscal management practices of local governments. The Agency also reviewed the default rates of these types of securities.

EPA is expanding the bond rating mechanism to allow non-general purpose governments (i.e., special districts and school districts) that do not have the authority to issue general obligation bonds to demonstrate financial responsibility if they have earned an investment-grade rating on at least $1 million in outstanding revenue bond issues not backed by any form of credit enhancement.

EPA has determined that revenue bond financing is central to the operation of most special districts and that the ratings on revenue bonds issued by special districts therefore provide an adequate representation of their financial strength. Special districts are created for a specific purpose, such as to provide airport services to a community. The revenue stream underlying the strength of a special district is the same as the base underlying its associated revenue

[6] Linda Reidt Critchfield, EPA Office of Underground Storage Tanks, memorandum to the record, "Conversation with Al Medioli, Moody's Investor Services on August 29, 1989," September 15, 1989.

[7] Ibid.

[8] Memorandum from Kate Donaldson, James Dickson, and Tony Bansal, ICF Incorporated, to Stephanie Bergman, EPA Office of Underground Storage Tanks, "Municipal Bond Insurance," May 31, 1989; memorandum from Kate Donaldson, James Dickson, and Tony Bansal, ICF Incorporated, to Stephanie Bergman, EPA Office of Underground Storage Tanks, "Municipal Bond Insurance Companies," June 22, 1989.

bonds. Ratings on revenue bonds are, therefore, appropriate measures of special districts' financial capabilities. (This is not the case for a general purpose government that issues revenue bonds, such as a city, because the revenue stream supporting a specific revenue bond is not equivalent to the overall tax base supporting the local government.) In addition, EPA determined that there has been a low incidence of default of investment-rated revenue bonds not enhanced by third-party support—e.g., bond insurance or a letter of credit. EPA examined information on revenue bond defaults between January 1989 and May 1991. Over that time period, approximately 150 issues defaulted. EPA estimates that no more than five of these issues had unenhanced investment-grade bond ratings from Moody's at the time of default, representing a default rate of less than 0.1 percent per year of rated bonds. Eight of the defaulted issues were backed by letters of credit, and two were insured by bond insurance companies.

Because the credit rating for revenue bonds issued by general purpose governments (e.g., townships, cities, and counties) would not measure the financial health and fiscal management practices of that type of government as a whole, and because revenue bonds are not usually used to finance projects central to the operation of a general purpose government, the Agency has determined that general purpose governments with the authority to issue general obligation debt may not use revenue bonds to demonstrate financial responsibility.

Similarly, because the credit rating for short-term notes, lease rental debt, and certificates of participation does not provide sufficient information on the financial strength of local governments, local governments may not use these instruments to demonstrate financial responsibility.

Two commenters asserted that the bond rating test is unavailable to many local governments simply because the amount of outstanding debt is less than one million dollars and suggested that the required amount of outstanding debt should be decreased. EPA intends the bond rating mechanism to be used by local governments that have shown their capability to sustain debts comparable in size to the minimum level of financial assurance as determined by statute. Governments that are not able to demonstrate such capability may use another mechanism to demonstrate financial responsibility. Based on the analysis conducted for the proposed rule, the Agency estimates that approximately 87 percent of general obligation bonds are issued for aggregate amounts greater than $1 million.

One commenter endorsed the bond rating test, but noted that a governmental entity will no longer qualify for the bond rating test if it reduces its total debt below $1 million. The commenter suggested that the amount of unused debt capacity may be more important than the amount of debt. Another commenter stated that the essential factor in the test should not be the dollar limit outstanding, but rather the statutory right of the authority to issue bonds and the credit ratings which have been established for that particular government entity on debt which has or could be issued. Because a local government entity does not have a credit rating from Standard and Poor's or Moody's unless it has outstanding debt, the commenter urged the Agency to devise some test, presumably a worksheet test, to measure credit worthiness if bond ratings have not been issued.

The Agency has determined that it is appropriate to require $1 million in outstanding debt as part of the bond rating mechanism. The requirement ensures that the bond rating used to demonstrate financial responsibility is based on a level of outstanding debt consistent with the amount of financial responsibility being demonstrated. Although there may be merit in the argument that the level of debt capacity is an indicator of potential financial abilities, EPA does not believe that incorporating available debt capacity would be feasible. First, calculating levels of available capacity is more difficult than applying the bond rating test as written, and is subject to greater uncertainties. Second, the fact that the local government has available debt capacity does not ensure that it will be able to issue the debt and maintain its bond rating, particularly if the amount of outstanding debt is substantially less than the amount of required financial assurance. The Agency notes that excess bond authority may be used as one part of one alternative of the fund balance mechanism.

Because bond rating information is easily obtainable, the use of bond ratings as a self-test mechanism will impose minimal administrative burden in determining a local government entity's eligibility. Many local government entities, however, do not currently have general obligation bond ratings. As of July, 1991 Moody's had ratings for a total of 7,653 investment-rated general obligation bonds issued by local government entities that were "investment grade" and were not insured.[9] (Because some local government entities may have multiple issues of general obligation bonds, the number of local governments with rated bonds may be lower.) Although Standard & Poor's rates additional entities, there is a substantial overlap—one study found that 94 percent of cities of 2,500 or more residents with a rating from Standard & Poor's also had a rating from Moody's.[10] In contrast, there are more than 80,000 local government entities in the United States, of which an estimated 25,000 own USTs. To provide local governments with as many compliance choices as possible to meet the requirement, the Agency has also developed additional self-test mechanisms to demonstrate financial responsibility.

2. Local Government Financial Test (§ 280.105)

As part of the underground storage tank requirements proposed on June 18, 1990, EPA included a local government financial test that could be used by local government owners and operators of USTs to satisfy the financial responsibility requirements of § 280.93. The local government financial test, or "worksheet test", was designed for local governments that cannot use the bond rating test (§ 280.104) because they have less than $1 million in outstanding investment grade bonds. As described in the preamble to the proposed rule, however, local government entities that have applicable outstanding debt rated lower than investment grade, even if this amount is less than $1 million, cannot use the worksheet test. This limitation on the use of the worksheet test applies, therefore, to the general obligation debt of general purpose local governments and to outstanding revenue bonds of those local government entities that are legally restricted from issuing general obligation bonds.

As described in the preamble to the proposed rule, the Agency designed and developed the worksheet test to capture local government variation using an index of financial strength. The index assigns a rank to each of the general purpose governments in the Census of Governments. After arraying the governments according to their rank on the index, the test establishes a cut-off

[9] Brenda Ramos, Moody's Investors Service, Public Finance Department, letter to Linda Critchfield, EPA, July, 1991.

[10] Cliff, George S., and Farnham, Paul G., "Standard & Poor's vs. Moody's: Which City Characteristics Influence Bond Ratings?", Quarterly Review of Economics and Business, Board of Trustees of the University of Illinois, Volume 24, No. 3, 1984.

point that, in the Agency's opinion, excludes that bottom fraction of local governmental entities that might not be able to meet their financial obligations in the event of an UST release. The procedures used to develop the index and establish the threshold cut-off are discussed in the preamble to the proposed rule, the background documents to this rulemaking, and in subsequent sections.

The test was designed to isolate the fraction of governmental entities that are in poor financial condition from those other governments that, in general, have sufficient resources and flexibility to respond to an UST release. Consequently, the Agency is not establishing the worksheet test as a precedent for other Agency regulations affecting local governments, because other regulations may require either larger required levels of funds or more certain cash flows.

Features of the Proposed Local Government Financial Test

The proposed worksheet test had the following features:

• Using a worksheet, an eligible local governmental entity would calculate nine financial ratios using easily available financial data. The nine ratios were:

—Debt service to total revenues,
—Total funds to total expenses,
—Total revenues to total expenses,
—Debt service to population,
—Total revenues to population,
—Total expenses to population,
—Total funds to total revenues,
—Total funds to population, and
—Local revenues to current expenditures.

• Each of the nine ratios was compared to the national distribution of that ratio to calculate a z-score, which is a measure of how far above or below the national average the municipality's ratio lies.

• The individual z-scores for the nine ratios were then weighted and added to calculate a total score, or index.

• Governments with a total score that passed the specified threshold could use the test as a mechanism for demonstrating financial responsibility for UST corrective action and third-party liability claims. To simplify the use of the worksheet test, the threshold value was incorporated into the calculation of the final score, so that governments achieving a final score greater than zero passed the worksheet test.

Comments on the Proposed Local Government Financial Test

EPA received several comments on its proposed financial test for local governments. The comments focused on (1) the exclusion of local governments with less than investment grade debt; (2) use of the term "self-insurance"; (3) updating the worksheet test using 1987 Census of Governments data; (4) deleting specific ratios from the test; (5) lowering the threshold level; and (6) the appropriateness of the worksheet test for non-general purpose local governments. The substance of the major comments received is briefly summarized below, followed by the Agency's rationale for accepting or rejecting the commenters' recommendations in the final worksheet test requirements.

(1) *Exclusion of Local Governments with Less than Investment Grade Debt.* One commenter believed the worksheet test should be available to all local governments, even those with outstanding debt rated below investment grade. The commenter reasoned that bond rating entities are not always accurate and, moreover, provide ratings that allow investors to assess a potential investment, a different purpose than assessing financial responsibility to respond to an UST release. The commenter stated that allowing use of the worksheet test would recognize these realities without undercutting the purpose of the test.

For reasons cited in the preamble to the proposed rule, however, EPA does not agree that local governments with bond ratings of less than investment grade should be eligible to use the worksheet test. The Agency notes that (1) failure to earn an investment rating is costly to local governments, (2) local governments have the incentive and ability to work with bond rating agencies to establish policies and procedures that would raise the bond ratings, and (3) the bond rating process involves a more detailed examination of local government financial condition than can be accomplished through a simple worksheet test.

(2) *Use of the Term "Self-insurance."* One commenter stated that State law might prohibit certain otherwise eligible government entities from using the worksheet test. The commenter noted that New York State law authorizes specific programs for self-insurance and that, because they have not been specially authorized for this purpose, component school districts cannot use the worksheet test (or, indeed, the bond rating test) to demonstrate the ability to self-insure.

EPA understands that the term "self-insurance" has specific legal meanings that may be limiting and has, therefore, modified the rule to delete references to "self-insurance." The modifications clarify that the use of the worksheet test mechanism is to demonstrate compliance with the financial responsibility regulations, and not to "self-insure."

(3) *Updating the Worksheet Test Using 1987 Census of Governments.* Although not proposing specific amendments to the worksheet, two commenters criticized the use of data from the 1982 Census of Governments in developing the worksheet test. One commenter believed that use of decade-old data could introduce inaccuracies in the results of the worksheet test. As an example, the commenter pointed out that changes in the financial practices of local governments, such as an increase in the size of new debt issues, could mean that the reality of what makes a local government financially strong is different now than it was in 1982.

EPA agrees with the commenter and, in response, has updated the analyses used to develop the worksheet test using data from the 1987 Census of Governments, which was not available when this rule was proposed. As further described below and in the Background Document, the new analyses show that the ratios included in the proposed worksheet test were highly correlated with similar factors in the analyses of both the 1982 and 1987 Census of Governments data, and that incorporation of the 1987 data did not significantly alter the structure of the worksheet test. In particular, EPA confirmed that ratios incorporating population (for example, total revenues per capita) and fiscal autonomy (e.g., local revenues to current expenditures) are important indicators of the relative financial strengths of governments. In addition, EPA has updated the worksheet to reflect changes in the means, standard deviations, and weights associated with each of the ratios.

(4) *Deleting Specific Ratios from the Worksheet Test.* One commenter urged the Agency to delete Factor 5, "local coverage" (local revenue to current expenses), from the worksheet test as inappropriate for use in assessing a local government's level of financial responsibility. The commenter argued that Factor 5 disadvantages those local governments that rely more heavily on State funding than others. While this factor is designed to measure local autonomy and the ability of local governments to redirect funds to meet the cost of UST releases, the commenter argued that a significant proportion of

the funds that local governments receive from States is not tied to specific purposes and may be used as the local government deems appropriate.

EPA believes that, because local governments do not control and cannot assure the continuance of State or Federal aid, local governments with a high dependence on non-local sources are less assured of the ability to respond to UST releases, whether the funds are dedicated to specific programs or not. The Agency notes that a local government may be weak in a particular variable but still pass the worksheet test. For example, a government with a high reliance on intergovernmental aid may still pass the worksheet test if its overall financial situation is predominantly sound as measured by the remaining variables. The selection of factors was developed through extensive statistical analysis of local government financial conditions. For reasons described below, however, the Agency has modified the proposed worksheet test to replace the ratio of local revenues to current expenditures with the ratio of local revenues to total revenues, an alternate ratio representative of "local coverage".

(5) *Lowering the Threshold Level.* One commenter recommended that the threshold value should be reduced from 15 to a maximum of 10 percent. The commenter argued that EPA's own statements in the preamble that local governments rarely go bankrupt, are not permitted to void obligations through bankruptcy, and generally possess the ability to meet financial obligations through taxation were inconsistent with the finding that 15 percent of local governments should be disqualified from using the worksheet test to demonstrate financial responsibility. In addition, the commenter believed that the worksheet analysis exaggerated the actual impacts likely to occur by not including consideration of incidence of UST ownership. The commenter reasoned that small local governments, the ones that are most likely to rely on a worksheet test, are much less likely to own USTs than larger local governments. Thus, the average impacts assumed exaggerate actual impacts likely to occur. The commenter concluded that these factors suggested that a 15 percent failure rate was too stringent, but that a maximum cutoff of 10 percent would recognize the reality of the financial strength of local governments.

The Agency notes that costs associated with clean-ups can range widely and that different standards cannot be applied to different owners. In fact, if standards were based on size of the local government, proportionately fewer smaller governments would be able to demonstrate financial responsibility because of the more limited total resources of small local governments. As shown in the background document, however, EPA believes that smaller governments are more likely to pass the worksheet test than are their larger counterparts. Consequently, the Agency believes its overall approach used to set the threshold is appropriate.

Two commenters pointed out that the difference between the 10 and 15 percent cutoffs in the Agency's analysis was not great. Another commenter stated that the threshold should be reconsidered or justified because the commenter did not believe the preamble or supporting documents contained evidence that 15 percent of local government entities are, in fact, financially unstable and, even if they are generally unstable, that they will be incapable of meeting their UST obligations.

As described below, the Agency has updated the worksheet test using the 1987 Census of Governments, including updated means, standard deviations, and weights for each ratio, as well as a reevaluation of the threshold level. Based on its review of the updated information, the Agency has determined that a threshold level that allows 90 percent of local governments to demonstrate financial responsibility based on the worksheet test represents a reasonable balance between the statutory requirement that UST owners demonstrate financial responsibility and the demonstrated stability of most local governments. Consequently, the Agency agrees with the commenters that a 10 percent threshold offers adequate safeguards.

(6) *Appropriateness of the Worksheet Test for Non-general Purpose Local Governments.* Two commenters stated that a financial test, such as the worksheet, designed to measure the financial strength of general purpose governments, is unsuitable for special purpose organizations such as airports, bridge and toll road authorities, and publicly-owned utilities. Unlike general purpose governments, one commenter argued, these so-called "proprietary" government entities conform to generally accepted accounting procedures similar to accounting systems employed in the private sector, rather than the modified accrual accounting terms and criteria appropriate to measure the success of a traditional government. Because the corporate test is similarly inappropriate for these special-purpose entities, the commenter requested that the Agency develop an alternative financial test for government entities required to use accrual accounting. The commenter suggested that the corporate test in 40 CFR 280.95, based on the accrual method, might be modified to take into account the substantially greater financial stability of publicly-owned utilities.

The Agency recognizes that specific data requirements preclude most special districts from using the worksheet test. In limited cases, however, some special districts (e.g., school districts that serve a specific population) may have the information necessary to complete the worksheet test (e.g., they can measure population). EPA believes that the new mechanisms, particularly with the inclusion of revenue bonds issued by special districts using the bond rating test, will allow most UST-owning governments to demonstrate financial responsibility without the need for an additional financial test targeted specifically at special districts.

Update of Worksheet Test Using 1987 Census of Governments

Although its basic features have not been modified, the Agency has updated the worksheet test using the 1987 Census of Governments. The procedures used to conduct the new analyses were the same as for the proposed rule, as documented in the preamble to the proposed rule and the background documents to this rulemaking, and as summarized below.

Starting with 78 different financial ratios and variables commonly used in financial analysis, the Agency used a statistical technique called "factor analysis" to group the variables. Factor analysis serves two purposes. First, it identifies underlying characteristics, or factors, that differentiate between the members of a population (in this case, between different counties, municipalities, and townships). Second, it tells how much of the difference (the "percent of variance explained") between the members of the population is accounted for by each factor. The Background Document contains a more detailed explanation of the statistical analyses performed, including the factor analysis.

The factor analysis identified a total of 15 factors that distinguish between local government entities. Based on its review of the results of the factor analysis, the Agency identified six factors that (1) captured the variation in financial performance of local governments and (2) appeared appropriate for the UST financial test. As with the proposed worksheet test, the final worksheet test includes the following six factors: (1) Debt burden, (2) funds coverage, (3) outlays per capita, (4) funds per capita, (5) local coverage, and (6) revenues to expenses. In selecting the factors and variables to be included in the worksheet test, however, the Agency rejected size, because the Agency did not wish to exclude financially strong smaller local government entities simply because of size.

After selecting the factors to be represented in the worksheet, it was necessary to select the specific ratios to represent the factors. In choosing ratios, the Agency wished to (1) keep the total number of ratios to a manageable level, while (2) retaining as large a number of specific indicators as feasible. The final worksheet uses nine ratios, which include the variables (1) debt service, (2) total revenues, (3) total expenditures, (4) population, (5) total funds, and (6) local revenues. The ratios selected and the factors that they represent are presented below.

Factor 1—debt burden: debt service to total revenues.

Factor 2—funds coverage: total funds to total revenues; total funds to total expenses.

Factor 3—outlays per capita: total revenues per capita; total expenses per capita.

Factor 4—funds per capita: total funds per capita; debt service per capita.

Factor 5—local coverage: local revenues to total revenues.

Factor 6—revenues to expenses: total revenues to total expenses.

EPA found that, in general, the same ratios included in the proposed worksheet test were important in the factor analyses of both the 1982 and 1987 Census of Governments data. There is, however, one change to the worksheet test ratios as a result of the updated factor analysis. Factor 5, "local coverage", is now represented by the ratio of local revenues to total revenues rather than the ratio of local revenues to current expenses. The factor analysis of 1987 Census of Governments data found that the ratio of local revenues to total revenues was very highly correlated with Factor 5, while the ratio of local revenues to current expenses was correlated less highly and was also correlated with several different factors. The preamble to the proposed rule provides a detailed description of the importance of each of these factors. One other minor addition to the final test is inclusion of payments for retirement of debt principal (not just interest payments) in the calculation of total expenses. This was inadvertently omitted from the proposed rule. (EPA has modified the parameters of the test to reflect the revised definition of total expenses.)

Together, these factors provide a balanced view of the stability and financial strength of a local government entity. The Agency does not believe that any single factor or variable can provide a sufficient indication of overall financial stability. Specifically, EPA does not believe a focus on funds alone, without adequate safeguards, would provide as good an indication of the ability of local government entities to provide financial assurance for an UST release.

These factors serve to achieve the Agency's goal of identifying and eliminating those local government entities that have overall financial characteristics that are in the bottom fraction of all local government entities, and that may, therefore, be at sufficient risk of experiencing financial distress that would prevent them from meeting their UST obligations.

As described at proposal, in developing the worksheet the Agency determined that performance on the specific ratios selected to represent the six factors should be standardized so that all ratios are placed on an equal basis. This is done by calculating the "z-score" for each of the ratios in the test. The z-score of an individual ratio is calculated by first subtracting the mean, and then dividing by the standard deviation:

$$z = \frac{\text{(ratio—mean)}}{\text{standard deviation}}$$

The distribution of the z-scores will always have a mean of 0 and a standard deviation of 1, thereby placing each variable in the index on a common level. To calculate a single index value, the z-scores are then weighted and added together; the weights are based on the percentage of variance explained by the underlying factors.

Selection of the Final Threshold Value

Having updated the financial index, the Agency then examined different threshold levels to determine a cut-off for selecting those local governments that have fiscal characteristics adequate to demonstrate financial responsibility to meet UST obligations. As described in the preamble to the proposed rule, EPA evaluated the impacts of a $1 million release to determine an appropriate threshold for allowing local governments to demonstrate financial responsibility through the worksheet test. In selecting a threshold, the Agency was guided by two important considerations: (1) most local governmental entities are expected to be able to meet their financial obligations under Subtitle I, so a cut-off threshold in the lower range (i.e., 1 to 30 percent) is appropriate, and (2) local governmental entities on the margin of the selected threshold should clearly be able to pay the emergency response and corrective action costs of an average UST release.

For purposes of the evaluation, EPA assumed that the release costs would be financed by a mortgage-type loan over a 20 year period at an interest rate of 10 percent. This interest rate is meant to be illustrative; local governments may be able to borrow at rates lower than 10 percent. Under a mortgage-type loan, repayment is made in equal annual installments consisting of both interest payments and principal repayment. The annual payment of a $1 million loan over 20 years at an interest rate of 10 percent is $117,459; the first year's payment consists of $100,000 interest and $17,459 principal repayment.

To evaluate whether a debt of $1 million would be too burdensome, the Agency considered the post-release performance on the nine ratios used to develop the index. The Agency paid specific attention to two financial parameters that financial institutions regularly use to evaluate prospective debtors: Debt service capability and accumulated funds. The Agency felt that it is important to consider the potential debtor's debt servicing capability because excessive debt would require excessive funds for debt servicing, which could result in a negative cash flow (expenditures greater than revenues) for weak debtors. Continuous negative cash flows increase the risk of financial instability in the short run and financial insolvency in the long run. It is important to consider the amount of accumulated funds available to a prospective debtor because a reserve of accumulated funds provides an extra "cushion" for those emergencies when routine cash flows are disrupted as a result of unforeseen circumstances. As long as a local government that is on the margin of the cut-off threshold being evaluated can demonstrate that it can service its debts and has a "cushion" of accumulated funds for emergencies, the Agency feels comfortable that it will be able to perform its routine business when faced with an UST release.

In its evaluation, however, the Agency did not use a precise yardstick for evaluating the impacts of a $1 million release. It is the Agency's belief that proposing a cut-off threshold that is applicable to the majority of local governments with diverse size, demographic, and financial characteristics is more a matter of informed judgment than one of precise measurement.

Impacts were evaluated on the bottom 30 percent of all general purpose local governments in the 1987 Census of Governments with data sufficient to calculate the index score (11,487 governments). For each government, the following adjustments were made to 1987 financial performance in accordance with the definitions of terms used in calculating the index:

- Total expenses were increased by $117,459 (total incremental debt service);
- Current expenses were increased by $117,459 (total incremental debt service);
- Total debt was increased by $982,451 (loan amount of $1 million minus first-year principal repayment);
- Total funds were reduced by $117,459 (total incremental debt service); and
- Debt service was increased by $117,459 (total incremental debt service).

In essence, the evaluation was made as if the release had been incurred in 1987 and reflected in end-of-year financial data, with no adjustments made by the local government to redirect funds or to increase revenues.

After adjusting the financial values, each of the nine ratios in the index test was recalculated. Impacts were examined by looking at the "marginal" local governments at each threshold in one percent increments. That is, to evaluate the effects of selecting a threshold of −6.425 (the index value exceeded by 95 percent of all general purpose local governments), EPA examined the 383 local governments scoring between −6.425 and −6.043 (the index value exceeded by 94 percent of all local governments). The remainder of this discussion presents results of the "post-release" ratios for each of five different threshold levels: −6.425, −4.937, −3.990, −3.242, and −2.586. Details of the results are provided in the Background Document supporting this rule.

It should be noted that no attempt was made to weight the potential impacts in terms of the likelihood of UST ownership. That is, although only about 2,764 of the 26,189 general purpose local governments serving fewer than 2,500 residents are believed to own USTs, the release costs were imposed on all local governments.[11] Consequently, the average impacts shown exaggerate the actual impacts likely to occur. (Nevertheless, an individual government experiencing an UST release may experience the full effects assumed in estimating the average impacts.) Also, the results assume that the local governments make no efforts to mitigate the financial impacts, either through increasing taxes and fees or reducing other expenditures.

Because the index ranks local governments in terms of a smooth array, there is unlikely to be a single value at which clear differences in performance appear. Instead, an evaluation of impacts is likely to show increasing performance and ability to accommodate the costs of an UST release with increasing threshold value.

Evaluation of Threshold of −6.425. The marginal local governments meeting a threshold of −6.425 (those between the fifth and sixth percentiles on the index test) have an average post-release fund balance of about $3,052,000 and a median post-release fund balance of about −$49,000.[12,13] About 62 percent of the marginal local governments have a negative fund balance, with the median of total funds per capita equal to −$46. The median debt service per capita is $167. The median ratio of local revenues to total revenues equals 32 percent. For the median local government, total revenues are about 47 percent of total expenditures.

Evaluation of Threshold of −4.937. With an increase in threshold to −4.937 (corresponding to the 10 percentile value), the average post-release fund balance of the marginal local governments is $1,673,000 and the median post-release fund balance increases to −$34,400. The percentage of local governments with negative cash balances improves to about 56 percent. The median ratio of total funds per capita improves marginally to −$25. The median annual debt service per capita decreases to $131. The median ratio of locally derived revenues to total revenues increases to 40 percent, whereas the median ratio of total revenues to total expenses increases slightly to about 49 percent.

Evaluation of Threshold of −3.990. When the minimum score is changed to −3.990 (corresponding to the 15 percentile value), the average post-release fund balance is $4,180,000 and the median fund balance decreases slightly to about −$35,000. The percentage of local governments with negative fund balances increases slightly, to about 58 percent, while the median ratio of funds per capita improves slightly to −$24. The median ratio of debt service per capita decreases to $127. The median ratio of local revenues to revenues increases to 47 percent, whereas the median of total revenues to total expenses decreases slightly to 48 percent.

Evaluation of Threshold of −3.242. At a threshold of −3.242 (corresponding to the 20 percentile value), the average post-release fund balance is $5,693,000 and the median post-release fund balance increases to about −$20,000. The percentage of local governments with negative fund balances decreases to 55 percent, while the median fund balance per capita increases to −$14. The median ratio of debt service per capita decreases to $119. The median ratio of local revenues to total revenues increases to 52 percent, whereas the ratio of total revenues to total expenses remains steady at 48 percent.

Evaluation of Threshold at −2.586. At a threshold of −2.586 (corresponding to the 25 percentile value), the average post-release fund balance is $6,651,000 and the median post-release fund balance improves to about −$15,000. The percentage of local governments with negative fund balances decreases to 52 percent. The median ratio of fund balance to population improves to $10. The median value of debt service per capita increases slightly to $123. Local governments show increasing coverage of their expenses, including an increase in the median ratio of local revenues to total revenues to 57 percent and in the median ratio of total revenues to total expenses to about 54 percent.

Analysis of Impacts on Households. EPA has considered the impacts of tank closures that may be caused by the inability of local governments to demonstrate financial responsibility. EPA estimated the impacts on households of compliance with the financial responsibility requirements for the median size, marginal government at each threshold examined. EPA estimates that the median "marginal" general purpose government (by population) at the 5 percentile threshold serves

[11] EPA, "Economic Impact Analysis of Additional Mechanisms for Local Government Entities to Demonstrate Financial Responsibility for Underground Storage Tanks," EPA Office of Underground Storage Tanks, November 1992.

[12] A threshold value set at the 5 percentile would exclude the local governments with index values in the lowest five percent and would include the remaining 95 percent. A threshold value set at the 10 percentile would be more stringent—it would exclude the local governments with index values in the lowest 10 percent, and allow only those local governments with index values in the upper 90 percent to pass the worksheet test.

[13] The median value is the value for which half of the local governments have a higher value, and half have a lower value.

approximately 1,011 residents, or 389 households. EPA's analysis of UST ownership patterns suggests that governments of this size own an average of 1.1 USTs. Based on an average present value cost per UST closure of $7,000, the residents would incur an estimated present value cost of $19.80 per household.[14] The present value of closure costs are estimated to range from $18.00 to $54.00 per household for residents served by median governments owning one to three USTs, respectively.

Based on the average number of USTs owned by the median "marginal" general purpose local government at the 10 percentile, the costs to governments required to close their USTs are estimated to be $15.81 per household. Costs may range from $14.37 to $43.11 per household for residents served by median governments owning one to three USTs, respectively.

Based on the average number of USTs owned by the median "marginal" general purpose local government at the 15 percentile, the costs to governments required to close their USTs are estimated to be $16.81 per household. Costs may range from $15.28 to $45.86 per household for residents served by median governments owning one to three USTs, respectively.

Based on the average number of USTs owned by the median "marginal" general purpose local government at the 20 percentile, the costs to governments required to close their USTs are estimated to be $14.95 per household. Cost may range from $13.59 to $40.78 per household for residents served by median governments owning one to three USTs, respectively.

Based on the average number of USTs owned by the median "marginal" general purpose local government at the 25 percentile, the costs to governments required to close their USTs are estimated to be $13.90 per household. Costs may range from $12.64 to $37.91 per household for residents served by median governments owning one to three USTs, respectively.

These estimates tend to exaggerate the costs per household, because they use the total estimated aggregate cost over a ten-year period. Consequently, they represent the cost to households if the entire cost associated with closing USTs were incurred and levied in a single year, rather than paid out over time.

Summary. Based on its review, the Agency has concluded that there are significant improvements in the "post-release" financial conditions of governments as the threshold is increased to about the 10 percentile, modest improvements as the threshold is increased from about the 10 percentile to the 20 percentile, and further increases beyond the 20 percentile. Because the extent of the increases from the 10 to the 15 percentile is minor, the Agency has determined that a threshold level of 10 percent provides adequate safeguards, and is consistent with statutory intent.

3. Governmental Guarantee (§ 280.106)

In today's rule, EPA is providing for the use of a guarantee mechanism for governmental entities. This mechanism, although not strictly a "self-test" mechanism, provides local government entities with a financial assurance mechanism comparable to the corporate guarantee allowed for private owners and operators of USTs. To be eligible to act as a guarantor, a local government entity must pass the bond rating or worksheet test.

The governmental guarantee differs in several respects from the current corporate guarantee. Under the governmental guarantee, local governments would be allowed to choose between a guarantee with or without a standby trust requirement. Under the corporate guarantee, firms are required to use a standby trust. If a local government chooses the governmental guarantee without the standby trust option, it is required to pay for corrective actions as needed and as directed by the implementing agency. Under the standby trust option, local governments will be required to fund a separate trust fund to the full amount of coverage upon discovery of a release. Again, the Agency's decision to allow local governments the option of a guarantee without the standby trust fund is based on local governments' history of meeting obligations and on their ability to consistently raise revenue through taxation. In addition, the governmental guarantee requires that the local governments entering into the agreement demonstrate a "substantial governmental relationship." This parallels the requirement in the corporate guarantee for a "substantial business relationship," while recognizing that the types of relationships between governments is fundamentally different than business relationships and that they are primarily based on common or overlapping constituencies.

The requirement of a "substantial governmental relationship" reflects two concerns of the Agency. First, EPA wishes to ensure that the guarantee contract is founded on a sufficient basis to be held valid and enforceable. Second, EPA seeks to avoid conflict with State insurance laws and regulations. The existence of a "substantial governmental relationship" should provide sufficient nonmonetary consideration to address these concerns.

One commenter supported the requirement for a substantial governmental relationship, stating that the governmental guarantee mechanism needs to be based on a substantial governmental relationship, and that the relationship should incorporate the full faith and credit of the guaranteeing agency.

One commenter asked whether, in States that allow intergovernmental risk pooling, the contractual relationship of a pool to provide safety and risk management services in addition to risk pooling would be recognized as a "substantial governmental relationship," thereby allowing existing pools to act as guarantors to their members. EPA does not believe that a risk pool should be allowed to operate as a guarantor, because the nature of the relationship is strictly monetary and does not necessarily involve a substantial governmental relationship. It should also be noted that risk pools can be included as compliance mechanisms on a state-by-state basis as state-required mechanisms.

Another commenter claimed that EPA should explicitly recognize the relationship between a "joint action agency" and its member publicly-owned utilities as a "substantial governmental relationship", thus allowing these entities to qualify for use of the governmental guarantee mechanism. The commenter reasoned that these agencies, not-for-profit entities created by State law to allow publicly-owned utilities to combine resources for various purposes, could include the provision of a guarantee of UST financial responsibility within their operation. EPA has concluded that because joint action agencies are non-profit organizations and not governmental entities, they are not eligible to act as guarantors.

A guarantee is a promise by one party (the guarantor) to pay specified debts or satisfy the specified obligations of another party (the principal) in the event that the principal fails to satisfy its debts or obligations. In the corporate guarantee, if the owner or operator fails

[14] As discussed in the EIA, the present value cost of closure includes the costs of closure associated with the technical standards (e.g., tank excavation and removal, product removal, and site assessment), plus the present value of the incremental cost of fuel purchased at retail service stations, minus the present value of the expected cost of corrective action for UST releases if the USTs were not closed.

to perform corrective action or satisfy third-party claims, the guarantor agrees to fund a standby trust from which the implementing agency will direct the payment of corrective action costs or third-party claims.

EPA believes that the guarantee mechanism would work well for governments, and is establishing two possible constructions for such a mechanism (discussed below). Using this mechanism, a municipality, for example, might obtain a guarantee from the State, a town might obtain a guarantee from the surrounding county or parish, or a special district might obtain the guarantee of the sponsoring local government entity. Guarantors must demonstrate that they are qualified to provide financial assurance by satisfying the bond rating test under 40 CFR 280.104, the worksheet test under 40 CFR 280.105, or the fully-funded fund balance test under 40 CFR 280.107.

Several commenters supported the inclusion of the governmental guarantee mechanism, although some also noted specific cases where the mechanism might not be applicable. Two commenters did not believe that the mechanism would be used by certain classes of government entities, arguing that special districts would be unable to obtain guarantees from local governments and that local governments would be unable to obtain guarantees from their State governments.

EPA believes that the guarantee is likely to be used primarily by governments with close and long-standing ties. The Agency emphasizes that the guarantee mechanism was developed to allow governments with common interests to cooperate to keep necessary USTs in operation. The mechanism is not intended to require any government to act as a guarantor. Nevertheless, if even a small number of governments are able to qualify using this mechanism, it will serve the purpose intended.

Commenters agreed that the guarantor should not be required to fund a standby trust, arguing that (1) a standby trust is not appropriate for local governments, given their strong history of meeting their financial obligations and their ability to raise revenue consistently, (2) a standby trust was unnecessary for guarantees among governmental entities, and would add unnecessary paperwork and administrative costs that were contrary to the Agency's goal of reducing the burden on local government, and (3) the governmental guarantee would not necessarily be similar to a corporate guarantee because of State-by-State differences in statutory restrictions. EPA agrees with the commenters and has allowed for use of a governmental guarantee with or without a stand-by trust fund.

One commenter stated that certification by a State Attorney General was necessary because some States could presumably prohibit or restrict the ability of a municipal government to make such a guarantee. Another commenter supported the Agency's decision not to require a State Attorney General's certificate attesting to the legality of the governmental guarantee. The commenter agreed with the Agency that the added degree of certainty provided by this requirement was appropriate in the case of a corporate guarantee, but was unnecessary for guarantees among governmental entities, and would burden local governments with unnecessary paperwork and costs.

The Agency is not requiring certification by the State Attorney General prior to offering the guarantee. Local governments have strictly defined and enforced limitations on their abilities to enter into contracts. These limitations are codified in State law and constitution and vary by State. The Agency believes that these restrictions imposed on local government entities should, in general, act as a sufficient check to prevent local governments from entering into invalid guarantees, and that the nature and purpose of local governments will prevent the issuance of guarantees unless there is a clear governmental interest.

Because the Agency wants to avoid unnecessary paperwork and burden on the part of local governments, EPA intends to keep the rule as proposed. EPA encourages governments wishing to use the guarantee to seek clarification of their authority if they are unsure of whether they may issue guarantees.

EPA solicited comments on whether passing the fund balance test should qualify governmental entities to act as guarantors. The sole commenter on this issue stated that a government passing the fund balance test should qualify to act as a guarantor, assuming that State statute permitted a governmental entity to be a guarantor. After further review, EPA has decided that allowing governments using the fund balance mechanism to act as guarantors would be consistent with the overall approach taken in the development of the new mechanisms. The Agency has, therefore, modified the proposed rule to allow the fully-funded fund balance mechanism to serve as the basis for a governmental guarantee.

Government Guarantee With Standby Trust

The first alternative governmental guarantee parallels the corporate guarantee, in that it must include a pledge to fund a standby trust in the event of failure by the UST owner or operator to pay corrective action or third-party liability claims. In today's rule, the guarantor must have legal authority to issue the guarantee. The Agency anticipates that most guarantees will be based on a clear and significant governmental relationship such as overlapping geographical boundaries, taxing or service constituencies, or shared impact from an UST release

Government Guarantee Without Standby Trust Requirements

In a governmental guarantee without a standby trust requirement, the guarantor agrees to provide funds for corrective action and third-party compensation as directed by the implementing agency on an on-going basis, up to the limits of the guarantee. Rather than fully funding a standby trust, the guarantor would make the payments directly as funds are required.

The current corporate guarantee requires the establishment of a standby trust, and requires a guarantor to fund the trust (1) after notification that a guarantee will be cancelled if a release has been detected and no alternate coverage has been obtained, or (2) when a release has occurred and the owner or operator has failed to perform corrective action or payment of a settlement or judgment for third-party liability. The corporate guarantee requires funding of a standby trust for several reasons. First, the issuance of a guarantee is founded on the existence of a substantial business relationship; such relationships are subject to change over time. Second, the underlying mechanism used by the guarantor depends primarily on the existence of readily liquidated assets, rather than on-going financial strength. Consequently, the Agency wishes to insure that the funds are made available before adverse events can occur. Third, the Agency wishes to reduce the potential delay involved in enforcing first against the UST owner or operator, and then against the guarantor for payment.

These concerns are mitigated under the governmental guarantee. First, the Agency believes that the governmental relationships that are likely to lead to the issuance of guarantees will be founded on geographical proximity and service to a common constituency. These relationships are not subject to rapid change. Second, the Agency recognizes in this rule that local government entities, as a class, have greater financial stability than private corporations. It is, therefore, less critical to obtain funds immediately to pay for contingent liabilities (such as payment of third-party claims) that may not occur. Third, the Agency recognizes the difference in purpose between governmental and private organizations, specifically the role of local governments to serve the public. This service orientation may make local governments more likely to fulfill their financial obligations. Consequently, the Agency has less concern that the absence of a standby trust will result in a delay in securing cleanup actions by local government owners or operators. With its modified structure, the mechanism permits a "pay-as-you-go" approach. These provisions allow a guarantor to fund corrective action costs as they are incurred, instead of requiring the guarantor to fund the standby trust fully in advance of anticipated expenditures.

Commenters on this issue agree that the governmental guarantee provides adequate safeguards without the need for creation of a standby trust fund.

4. Maintenance of a Fund Balance (§ 280.107)

Under this option, the UST owner or operator would create a dedicated fund specifically for UST releases or general catastrophic events. The dedicated fund must meet the local government's aggregate financial responsibility requirements (or such amount needed to fulfill gaps in financial responsibility from other mechanisms used in combination with the funded balance). Use of the fund balance mechanism requires local governmental entities to establish irrevocable trusts pledged to use for UST response or use in responding to catastrophic events, including UST releases.

Control of the fund would continue to rest with the local government entity. Control and accounting for these funds would be administered following the standards appropriate for other insurance trusts already maintained by local government entities, including pension trusts and worker's compensation funds.

The fund balances must be held as cash or investment securities that will be available in the event of an UST release and must be irrevocably dedicated to use for UST response or for responding to catastrophic events, including UST releases. As discussed below, the Agency is providing three alternatives that may be used in establishing the fund.

Based on an analysis of Census data and data on Minnesota cities, the Agency believes that the fund balance mechanism is unlikely to be used widely by general purpose governments, because few who require an alternative mechanism to the bond rating and worksheet tests have adequate funds.[15] The fund balance mechanism may prove more useful for special districts and school districts that may not be able to use the worksheet test. The inclusion of a fund balance mechanism as a financial assurance option should increase the flexibility provided owners and operators in demonstrating financial assurance. Today EPA is providing the following three sub-options, any one of which may be used to demonstrate financial responsibility.

Fully-Funded Dedicated Fund

Under this alternative, the local government would establish a separate fund, dedicated to payment of UST corrective actions and third-party liability claims, in the amount of its aggregate financial responsibility requirements. The fund balance must be established as an irrevocable fiduciary or trust account, with proceeds invested in cash or readily marketable securities. This mechanism would be the most similar to the corporate trust fund option (§ 280.102 of subpart H) and is intended to be similar to "trust accounts" and "insurance accounts" held by local governments for pensions and insurance. Although there are currently no restrictions to local governments using the trust fund option under § 280.102, the fully-funded dedicated fund option would not require the local government to establish a third-party trustee for the fund. Instead, the fund would be administered by the treasurer or chief financial officer of the local government entity as a separate trust account.

Catastrophic Events Contingency Fund

Under this option, a municipality would be able to use a dedicated fund used for general emergency response and third-party liability (e.g., flood relief, hurricane relief, or other environmental cleanups) as evidence of UST financial responsibility. In the proposed rule, EPA required that a fund used to cover both UST costs and other emergency costs incurred by local governments be funded in the amount of $10 million to qualify as a mechanism for demonstrating financial responsibility. Numerous commenters requested that EPA reduce the required size of a combined emergency response fund. EPA conducted a limited survey of nine governments to determine the prevalence and typical size of emergency response funds.[16] The Agency found that most funds are relatively small (less than $5 million). Based on those findings, EPA has reduced the required size of a combined emergency response fund to $5 million. In making this determination, the Agency considered that, when a local government draws upon its emergency response fund, it must replenish the fund in order to be prepared to meet the costs of the next emergency that may arise. It should be noted that local governments may establish a dedicated fund equal to their aggregate annual UST liability if doing so requires sequestering less money.

A combined fund balance of $5 million will equal or exceed five times the aggregate financial assurance level for most local government entities, based on the number of USTs owned and operated. This requirement is analogous to the requirement in the corporate self-test that firms must have tangible net worth equal to at least ten times their aggregate financial assurance level. The fund balance must be established as an irrevocable fiduciary or trust account, with proceeds invested in cash or readily marketable securities. The fund may be administered by the treasurer or chief financial officer of the local government entity as a separate trust account.

In establishing this option, the Agency recognizes that States often permit local governments to administer fiduciary and trust accounts, such as pension funds and workers' compensation funds, while requiring private companies to establish third-party trustees or to subscribe to State-maintained funds. EPA believes the distinction between local government entities and private companies reflects differences in State oversight (e.g., State requirements that local government entities submit budgets or financial statements), differences in purpose (i.e., companies exist to make profits, whereas local government entities are created to provide a public service), and differences in financial stability.

The Agency is including this option to allow municipalities flexibility in establishing emergency response funds while ensuring that adequate funds are

[15] State Auditor of Minnesota, "Report of the State Auditor of Minnesota on the Revenues, Expenditures, and Debt of the Cities in Minnesota for the Fiscal Year Ended December 1987," November 1988.

[16] Memorandum from James Dickson, Rebecca Holmes, and William Driscoll, ICF Incorporated, to Andrea Osborne, EPA Office of Underground Storage Tanks, "Local Government Emergency Response Funds," October 13, 1992.

available to respond to an UST release. Although the Agency lacks data on municipal expenditures for general emergency response and third-party liability, it believes the $5 million requirement will assure the availability of funds for any number of UST releases should other catastrophic events occur in the same year. Thus, although the fund would not be solely dedicated to responding to UST releases, the required fund balance of five times the requirement for a dedicated UST fund will assure adequate resources to respond to an UST emergency. (UST clean-up costs currently average between $100,000–$150,000.)

Three commenters criticized the $10 million fund balance that would have been required in the proposed rule, arguing that (1) the funding level required by the catastrophic events contingency fund was too inflexible for local governments to use the option, (2) only a handful of existing utility funds that are or can be used to respond to releases currently meet the $10 million requirement, and (3) a fund balance level of $10 million would be infeasible for most transit systems. Some commenters argued that a lesser amount (e.g., three times the requirement for a dedicated UST fund) would be adequate for a general catastrophic events contingency fund.

The Agency emphasizes that the combined fund balance alternative was developed as an administrative convenience for those governments that already maintain large contingency funds. The Agency's primary concern is to ensure that funds will be available to meet the costs associated with UST releases. The Agency notes that a government able to reserve only three times the required amount (i.e., $3 million), as suggested by one commenter, could establish a $1 million UST emergency response fund and a separate $2 million catastrophic events fund. If commenters are correct that a combined $3 million fund would be adequate to meet all requirements of the catastrophic events fund while reserving the $1 million necessary to demonstrate financial responsibility, then a $2 million fund reserved for catastrophic events other than UST releases would be adequate to meet those costs. As discussed below, the Agency has determined that local governments have minimal financial incentives to commingle funds designated for UST-related costs with other funds.

Incrementally Funded Trust Fund Combined With Unused Bonding Authority

Under this option, a municipality would be required to fund a dedicated fund for UST releases incrementally, making payments equal to at least one-seventh of the aggregate liability each year. A municipality using this alternative must fully fund the trust fund by the beginning of the seventh year. The fund balance must be established as an irrevocable fiduciary or trust account, with proceeds invested in cash or readily marketable securities. The fund may be administered by the treasurer or chief financial officer of the local government entity as a separate trust account.

In addition, until the dedicated fund is fully funded, the municipality is required to demonstrate the authority to issue a specified amount of general obligation bonds to respond to an UST release. The authority may consist of either a voter-approved bond referendum specifically targeted for payment of the costs associated with an UST release, or certification from the State Attorney General that the government has the authority to issue the bonds without voter approval and that the proceeds of these bonds can be used to respond to an UST release. The Agency is requiring the unused bonding authority to ensure that municipalities have resources to respond to UST releases while allowing them to develop a dedicated fund over time.

The Agency believes this mechanism is appropriate for local government entities, but not private companies, for several reasons. First, local governments operate under statutory and constitutional limitations on debt issuance. By requiring prior voter approval or certification that such approval is not necessary, the Agency is relying on safeguards that do not exist for private companies. Second, local government entities exist to provide public services, whereas private companies do not. Third, local government entities have historically been much more stable than private companies. Fourth, local government entities have an ability to levy taxes or raise fees and charges that is not available to private companies.

In developing this option, the Agency learned that local government entities will frequently obtain a bond referendum before incurring costs related to specific projects, such as construction projects, and will delay the issuance of the bonds until the funds are needed. The Agency was also informed that New York law allows local governments to issue debt to pay certain obligations without passing a bond referendum. The Agency is also considering that the Tax Reform Act of 1986 penalizes local government entities for investing the proceeds of tax exempt bond issues. Thus, the Agency recognizes that there is both a precedent for having unused bonding authority and an incentive not to issue bonds unless necessary for actual payment of debts.

Commenters support the option of allowing use of a funded balance. One commenter noted that the availability of this option would give public entities greater flexibility in meeting their responsibilities under the financial responsibility requirements. Another commenter favored an incrementally funded trust fund as a possible financial assurance mechanism for local governments.

Two commenters identified particular situations in which local governments could be prohibited by State law from using a funded balance as a financial assurance mechanism. The Agency reiterates that it has provided multiple financial responsibility mechanisms to increase the options available to local governments so that each governmental entity can choose the mechanism most appropriate to its needs. EPA believes that State laws prohibiting the use of any of the mechanisms is consistent with the State Program Approval process, which allows States to set more stringent standards. EPA recognizes that limitations imposed by some States may act to disallow some or all of the mechanisms provided in the rule.

One commenter requested the Agency to clarify that an order or resolution of the governing body of a publicly-owned electric utility satisfies the requirement for an "order dedicating the fund." The commenter noted that this governing body would be the same entity which, for the vast majority of publicly-owned utilities, sets customer rates (e.g., a board of directors, commissioners or supervisors, or a city council).

EPA has determined that the legal authority of a municipal utility or other municipal corporation is specified in the State charter establishing the municipal corporation. Therefore, a dedicated fund for UST corrective actions may be established through a local government order by any municipal utility, or other municipal corporation, or special district whose State charter grants the authority to issue an order establishing such a fund. In their State charters, municipal corporations are granted express powers to conduct their primary business, and implied powers to carry out those actions that are incidental and essential to the conduct of their business. Because meeting statutory and regulatory requirements are both incidental and essential to the operation of a municipal utility, it would appear that, in general, the board of directors of a municipal utility would have the legal authority to establish an UST trust fund. Because the specific authorities of municipal utilities may vary from State to State, however, and because within a State, each charter may be unique, it may be appropriate for the board of a municipal utility to obtain the advice of legal counsel before voting to establish a dedicated fund for UST corrective actions. The Agency notes, for example, that public comments on the proposed rule claim that a New York statute expressly prohibits the creation of emergency response funds by school districts and Boards of Cooperative Education.

One commenter requested clarification of the requirement that dedicated funds cannot be commingled or otherwise used in normal operations, presumably because the term "normal operations" is not defined or described in the rule or preamble. The commenter also points out that, while the commingling requirement appeared in the preamble, it was not written into the rule itself.

EPA's intent in allowing local governments to establish a dedicated fund for UST corrective actions was to reduce the burden on and cost to local governments by not requiring a third-party trust fund. Whereas a third-party trust fund was authorized for use by non-governmental owners and operators, a third-party requirement for local governments was not considered necessary because of the experience of local governments in establishing and administering such funds. Nevertheless, EPA is concerned that funds reserved for meeting the costs of corrective actions and third-party liabilities associated with UST releases be easily identifiable and readily available.

On the issue of commingling funds, EPA has found that the investment options typically available to local governments offer minimally higher returns with larger deposits. Moreover, deposits larger than $100,000 would not be insured by the Federal Deposit Insurance Corporation, exposing a commingled fund to the risk of bank failure unless alternative insurance were obtained (e.g., from an agency or State government). Commingling funds may not be practical for local governments that seek to obtain higher returns on deposits by having a bank's trust department actively manage their assets, because the timing of cash needs from an operational fund and from a trust fund are so widely divergent that a prudent manager would select a different mix of investment instruments for the two funds, and consequently would establish separate funds.

Because of the minimum income gains potentially available through commingling funds, as well as insurance and asset management considerations, EPA has concluded that the potential gains from commingling accounts do not outweigh the associated costs. EPA has modified the language in the rule to reflect this concern: money held for the purposes of demonstrating financial responsibility must be held in a separate account dedicated either to UST responses in particular or to emergency and catastrophic events in general.

5. Combinations of Mechanisms

The mechanisms being provided today may be used by themselves or in combination with other mechanisms. Local governments qualifying for use of the bond rating or worksheet test mechanisms are not required to obtain additional evidence of financial responsibility, but may do so if they so choose. A guarantee or dedicated fund balance may be used to demonstrate financial responsibility for amounts not assured by other mechanisms.

B. Reporting by Owner or Operator

Each government demonstrating financial assurance using the mechanisms promulgated today must notify the implementing agency at the times specified in § 280.110.

C. Recordkeeping

Owners and operators are required to maintain evidence of all financial assurance mechanisms used to demonstrate financial responsibility under this subpart until the tank has been properly closed or, if corrective action is required, until corrective action has been completed and the tank has been properly closed as required by 40 CFR Part 280, Subpart G. In general, the recordkeeping requirements for the mechanisms being promulgated today are equivalent to those required for the mechanisms promulgated in the October 1988 rule. Because local governments are not uniformly required to submit data to third-party agencies, however, local governments using the worksheet test must maintain a copy of the underlying financial statements or other data used to support the use of the worksheet test. Also, local government owners and operators must maintain evidence of the authority that is used to establish dedicated funds for use in responding to UST releases. An owner or operator using the mechanisms promulgated today is required to maintain at his UST site or his office the following types of evidence for mechanisms used to demonstrate financial responsibility:

Bond Rating Test. Each local government using the bond rating test must maintain

(1) A letter signed by the chief financial officer (e.g., comptroller, controller, or treasurer) certifying the eligibility to use the bond rating test, and

(2) Originally signed and dated transmission from Moody's or Standard & Poor's, showing the amount, the type of bond and the bond rating assigned.

Such evidence must be on file on site or at the place of business no later than 120 days after the close of each fiscal year.

Worksheet Test. Each local government using the worksheet test must maintain

(1) A letter signed by the chief financial officer (e.g., comptroller, controller, or treasurer) certifying the accuracy of the calculations and the underlying data,

(2) A copy of the completed worksheet, and

(3) A copy of the underlying financial data (e.g., year-end financial statements) used to compute the worksheet.

Such evidence must be on file on site or at the place of business no later than 120 days after the close of each fiscal year.

Guarantee. Each local government using the guarantee must maintain

(1) A letter signed by the chief financial officer (e.g., comptroller, controller, or treasurer) certifying the use of the guarantee,

(2) An originally signed and dated guarantee contract, showing the addresses of all tanks for which financial assurance is guaranteed, the nature of the guarantee (third-party liability, corrective action, or both), and the limits of the guarantee,

(3) A letter signed by the chief financial officer (e.g., comptroller, controller, or treasurer) of the guarantor certifying (1) the eligibility to use the bond rating test (unless the guarantor is a State), (2) the eligibility to use the worksheet test (unless the guarantor is a State), or (3) the existence of a dedicated UST or emergency response trust fund meeting the requirements of § 280.107,

(4) For guarantors other than States, a copy of the documentation supporting

the bond rating or worksheet test, including (a) a copy of the originally signed and dated transmission from Moody's or Standard & Poor's to the guarantor, showing the issue size, the type of bond and the bond rating assigned, or (b) a copy of the completed worksheet and underlying financial data, and

(5) Originally signed duplicates of the standby trust funds worded as specified in this rule for guarantees, surety bonds, or letters of credit (as necessary).

Such evidence must be on file on site or at the place of business no later than 120 days after the close of each fiscal year.

Fund Balance. Each local government using the fund balance mechanism must maintain

(1) A letter signed by the chief financial officer (e.g., comptroller, controller, or treasurer) certifying the use of the fund balance mechanism,

(2) Originally-signed letter certification from the comptroller or treasurer worded as specified in the rule and letters or certificates from municipalities regarding coverage by municipal funds or other municipal assurances,

(3) A copy of the authorizing statute or resolution that clearly restricts use of the funds to the designated purposes,

(4) A financial statement showing the balance of cash or liquid investments in the fund, and

(5) A copy of either (a) the authorized bond resolution in an amount equalling or exceeding the unfunded portion of the fund or (b) State Attorney General's opinion showing that such authorization is unnecessary.

Such evidence must be on file on site or at the place of business no later than 120 days after the close of each fiscal year.

One commenter asserted that the proposed recordkeeping provisions were generally reasonable and did not represent an undue hardship to local government entities. Another commenter stated that the recordkeeping requirements of the proposed rule would be burdensome. The commenter indicated that requiring local government entities to be able to present evidence of financial capability upon request would be a suitable substitute for the proposed recordkeeping and reporting requirements.

The Agency emphasizes that there is no routine reporting requirement, but that the need to determine compliance with the requirements on an annual basis is considered to be a fundamental part of the rule. Records are to be retained by local governments and must be provided only when (1) a release occurs, (2) the local government becomes ineligible for a financial responsibility mechanism that it is using, (3) the local government installs a new tank, or (4) records are requested by the implementing agency.

D. Bankruptcy or Other Incapacity of the Owner or Operator

Any owner or operator named as a debtor in voluntary or involuntary bankruptcy proceedings (under Title 9 of the U.S. Code) is required to notify the Regional Administrator or the implementing agency within 10 days after commencement of such proceeding. In addition, any guarantor or indemnitor is required to notify the owner or operator by certified mail within 10 days after commencement of a voluntary or involuntary proceeding under Title 9 (Bankruptcy) of the U.S. Code that names such guarantor or indemnitor as debtor. Any owner who demonstrates financial responsibility using a third-party mechanism will be deemed to be without the required financial assurance in the event of a bankruptcy or incapacity of its provider of financial assurance, or a suspension or revocation of the authority of a provider to issue the mechanism relied upon (e.g., guarantee, indemnity contract, surety bond, insurance policy, risk retention group coverage policy, letter of credit, or State-required mechanism). Finally, municipalities are required to notify the Director of the implementing agency within 30 days of being notified that a provider of financial assurance (e.g., a guarantor) has declared bankruptcy or is otherwise incapable to cover assured costs, unless they are able to obtain alternative coverage.

V. Economic Impact Analysis

In conjunction with this rule, the Agency has performed three impact analyses: an Economic Impact Analysis, a Federalism Assessment, and a Paperwork Reduction Act estimate. Summaries of these analyses are presented below:

A. Economic Impact Analysis

This section describes the methodology and results of an Economic Impact Analysis of the rule. EPA estimates that about 2,300 local governmental entities will be able to demonstrate financial responsibility using the mechanisms promulgated today that would not be able to demonstrate financial responsibility using only the mechanisms allowed by the October 1988 rule. The Agency estimates that the use of these mechanisms will result in approximately 5,700 fewer USTs being closed because of a lack of financial assurance. The Agency estimates the total annualized cost savings to be about $4.5 million, with a present value of about $32 million over ten years, in constant 1987 dollars.

1. Compliance With Executive Order 12291

Executive Order 12291 (46 FR 13193, February 19, 1981) requires that a regulatory agency examine the potential impact of regulations. The regulatory agency must determine whether a new regulation will be "major." If it is, the regulatory agency must conduct a regulatory impact analysis (RIA). A major rule is defined as one that is likely to result in (1) an annual effect on the economy of $100 million or more; (2) a major increase in the costs or prices for consumers, individual industries, Federal, State, or local government agencies, or geographic regions; or (3) significant adverse effects on competition, employment, investment, productivity, innovation, or on the ability of U.S.-based enterprises to compete with foreign-based enterprises in domestic or export markets.

EPA has analyzed the local government financial responsibility rule. Based on this analysis, the Agency has concluded that this regulation will have an annual effect of less than $100 million. The rule is expected to reduce costs to the regulated community; these reductions are estimated to be less than $100 million on an annual basis. Accordingly, the regulation being promulgated today is not a major rule, as defined by Executive Order 12291. Nonetheless, the Agency is interested in the potential economic effects of the regulation and has developed an Economic Impact Analysis (EIA) to examine them. The following four sections summarize the results of the EIA: Section 2 describes the regulated community affected by this regulation; Section 3 presents some of the methods and assumptions used to produce the EIA; Section 4 discusses the regulation's cost impacts; and Section 5 describes its environmental impacts.

2. The Affected Community

EPA estimates that approximately 25,000 local government entities own more than 62,000 petroleum-containing USTs. For the purpose of this analysis, the regulated community is divided into eight categories: very large, large, medium, and small general purpose governments (i.e., cities, counties, and townships); and very large, large, medium, and small special purpose districts (including, for example, independent school districts and water districts). General purpose governments were grouped according to population: very large governments serve more than 50,000 persons; large governments serve more than 10,000 and fewer than 50,000 persons; medium governments serve more than 2,500 and fewer than 10,000 persons; and small governments serve fewer than 2,500 persons. These population categories are the same categories used in other EPA economic impact analyses, and have been used here at the suggestion of one commenter.

For this analysis, school districts were categorized by estimated population, assuming that the enrollment of a school district is one-sixth of the total population. Other special districts, which do not always have well-defined populations, were grouped according to annual revenues: very large districts have annual revenues of more than $100 million; large districts have annual revenues of more than $5 million and less than $100 million; medium districts have revenues of more than $200,000 and less than $5 million; and small districts have revenues of less than $200,000.

Table 1 shows the estimated total number of governments in each category, the number of UST-owning governments in each category, and the number of USTs owned. (A summary of the method used to develop these estimates is provided below.)

TABLE 1.—PROFILE OF LOCAL GOVERNMENTS

Category	Number of entities	Number of entities owning USTs	Total USTs owned
Very large governments	1,360	1,360	13,813
Large governments	4,351	3,870	9,580
Medium governments	7,054	3,643	5,126
Small governments	26,189	2,764	2,921
Very large special districts	862	840	3,148
Large special districts	4,930	4,785	17,359
Medium special districts	12,558	6,617	9,399
Small special districts	26,480	646	654
Total	83,784	24,525	62,000

Source: EPA Analysis.

EPA estimates that 1,360 very large general purpose local governments own approximately 13,800 USTs, 4,350 large general purpose governments own approximately 9,580 USTs, approximately 7,000 medium general purpose governments own approximately 5,100 USTs, and more than 26,000 small governments collectively own fewer than 3,000 USTs. That is, most small governments are not estimated to own any USTs. EPA estimates that about 860 very large districts (including school districts) own about 3,100 USTs, approximately 4,900 large districts own about 17,000 USTs, about 12,500 medium districts own roughly 9,400 USTs, and more than 26,000 small districts own about 650 USTs. All very large local government entities are assumed to own at least one UST. The Agency used probability theory and an estimate of the total number of USTs owned by all UST-owning entities to calculate the percentage of large, medium, and small government entities that own USTs.

3. Assumptions and Methodology Used in the EIA

The analysis uses several key assumptions to estimate the costs and other impacts of this regulation:

• The baseline used to estimate incremental costs and economic impacts of the self-test rule is the cost of complying with the financial responsibility rule published on October 26, 1988. EPA assumes that local government entities will comply using the options available under the October 1988 rule in the absence of the alternative mechanisms. Local governments unable to use the financial mechanisms available under the October 1988 rule are assumed to close their USTs, in compliance with the regulations.

• Cost impacts were evaluated on an annualized basis. To develop annual costs for insurance and UST closure, EPA calculated the equivalent annual payment having the same "present value" as the cost estimates developed for the UST technical standards regulations, using a ten-year period and a real discount rate of 7 percent.

• The estimated number of USTs owned by local governments is based on a derived relationship between the annual revenues of local governments and the number of USTs owned.

—The 1985 "Summary of State Reports on Releases From Underground Storage Tanks" provides data on the percentage of releases occurring in places of different populations.

—EPA assumes that release incidents are not biased towards places of different size and that the distribution of release incidents is the same as the distribution of USTs among local government entities.

—The analysis assumes that there are about 62,000 local government USTs, as estimated in the financial responsibility rule published in October 1988.

• The analysis uses budget data obtained from the 1982 Census of Governments to develop a relationship between budget and population and then between budget and number of USTs. (EPA used population statistics and budgets from 1982 to develop relationships between UST ownership, population, and budget, because these data provided information consistent with the UST ownership data used in this report. Estimated ownership patterns were not updated to reflect 1987 data. First, the relationship between population and UST ownership was assumed to remain stable from 1982 to 1987. Second, the relationship between constant-dollar budgets and UST-ownership was also assumed to remain stable. Third, the estimates of total UST ownership are based on 1987 data.)

• All local governments using insurance or mandatory State assurance funds to meet financial responsibility requirements for corrective action and compensation of third parties in the baseline are assumed to continue to use those mechanisms to comply with the financial responsibility requirements, rather than using the mechanisms promulgated today.

• All other local government entities that qualify for financial responsibility using the new mechanisms are assumed to incur costs ranging from $75 (for the bond rating test) to $253 (for the guarantee) per government per year to develop and maintain the required records and reports.

• Because local governments that do not qualify for financial responsibility under the worksheet test are assumed to be unable to obtain insurance or otherwise demonstrate financial responsibility, the analysis assumes that they close their UST systems and purchase fuel from retail petroleum dealers.

EPA estimated the fraction of local government entities that will be able to demonstrate financial assurance under the promulgated rules by assuming that governments not able to obtain insurance and not required to use State mechanisms will use the least onerous method for which they qualify:

• Local governments able to obtain insurance under the baseline are assumed to do so, rather than use the mechanisms being promulgated today, in order to minimize their exposure to potentially large UST-related costs.

- Local governments in States with mandatory assurance programs are assumed to use them rather than the mechanisms being promulgated today.
- Local governments with outstanding issues of investment grade bonds in amounts greater than $1 million (about 87 percent of all governments with investment grade ratings) are assumed to use the bond rating test. Analysis of data on Minnesota cities suggest that virtually all cities with populations of more than 10,000 have investment-rated general obligation bonds, and that virtually no cities with populations less than 2,000 have such bonds.
- Local governments not eligible to use the bond rating test are assumed to use the worksheet test, if they qualify. The Agency used the 1987 Census of Governments to estimate the fraction of governments with populations less than 200,000 or annual revenues less than $100 million (i.e., those that may not qualify to use the bond rating test) that qualify at the 10 percentile threshold.
- The Agency assumes that local governments with total fund balances greater than $4 million that do not qualify to use the worksheet test will establish a dedicated fund meeting the requirements. Data from the 1987 Census of Governments were used to estimate the percentage of governments having more than $4 million in funds that would not qualify under the worksheet test.
- The Agency assumes that 90 percent of school districts unable to demonstrate financial responsibility will obtain guarantees from surrounding general purpose governments. This assumption is based on the assumption that education will be deemed to be of sufficient importance that the general purpose governments served by school districts will act to insure that the USTs remain in operation.
- The Agency also assumes that half of all other special districts unable to otherwise demonstrate financial responsibility will be able to obtain guarantees from the general purpose governments served by the districts.
- All other general purpose governments and districts not able to demonstrate financial responsibility are assumed to close their USTs and purchase fuel from retail petroleum stations.

Figure 1 shows the estimated fraction of local governments using each financial assurance mechanism under today's rule. It should be noted that the assumed availability of guarantees represents just one of many plausible outcomes. Other possible outcomes range from (1) all governments failing to obtain insurance or qualify using one of the self-test mechanisms will be able to continue to operate their USTs, either by obtaining guarantees or by transferring ownership of their USTs to the State or to local governments able to demonstrate financial responsibility, (2) some other fraction of all governments, without regard to purpose, will be able to obtain guarantees, or (3) no governments will be able to obtain guarantees. The EIA discusses the sensitivity of the results to alternative assumptions about the availability of guarantees.

BILLING CODE 6560-50-M

Figure 1

Estimated Use of Financial Assurance Mechanisms

(Counties, Municipalities, Townships, School Districts, and Special Districts)[a]

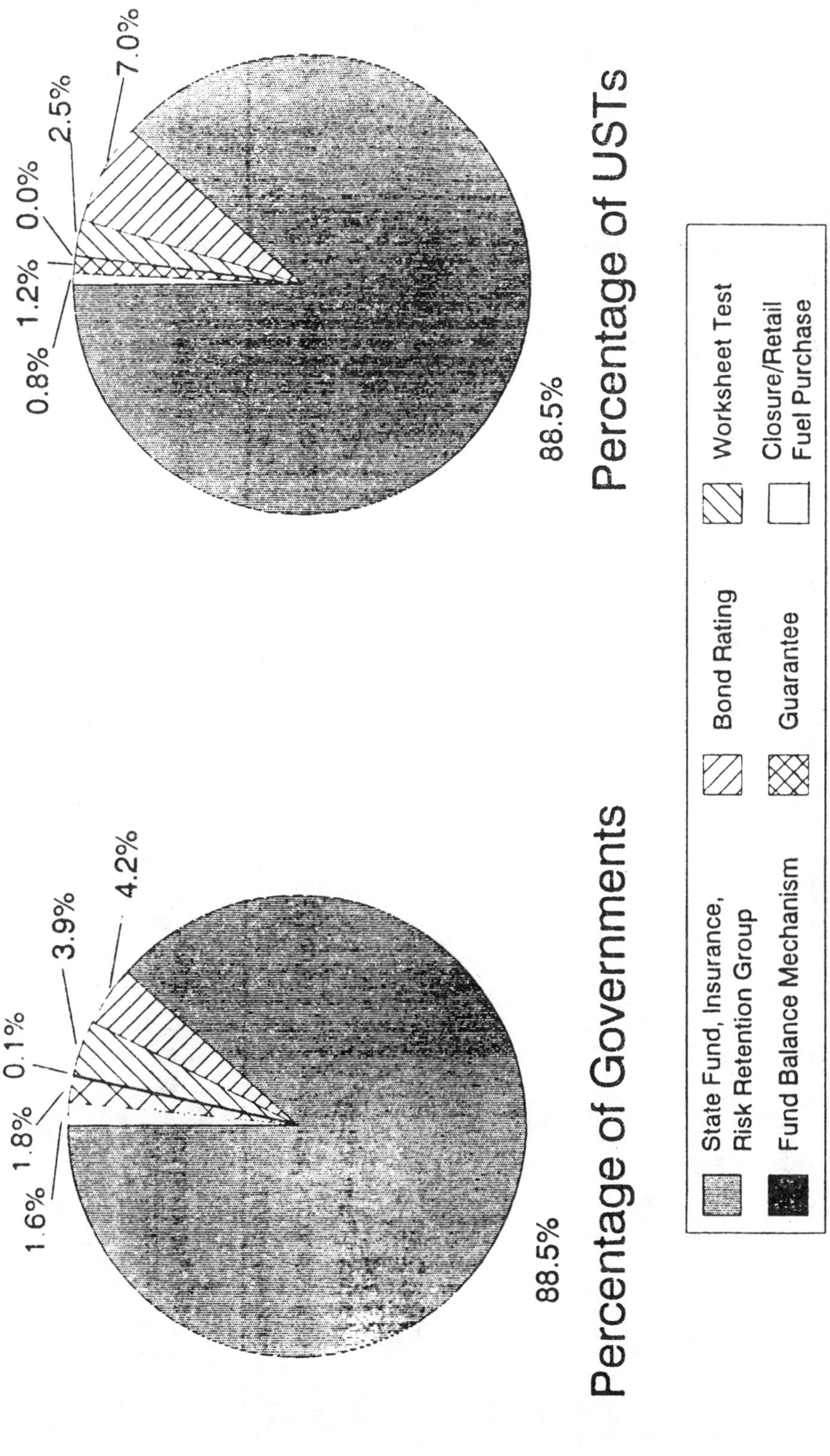

[a] Percentages may not add up to 100 percent due to rounding.

Source: EPA, "Economic Impact and Regulatory Flexibility Analysis of Additional Mechanisms for Local Government Entities to Demonstrate Financial Responsibility for Underground Storage Tanks." EPA Office of Underground Storage Tanks, 1992.

Table 2
Summary of Results of
Economic Impact Analysis

	General Purpose Governments				Special Districts			
	Very Large	Large	Medium	Small	Very Large	Large	Medium	Small
Number of UST-Owning Governments	1,360	3,870	3,643	2,764	840	4,785	6,617	646
Response Under October 1988 Rule								
Number of Governments Demonstrating Financial Responsibility	1,259	3,452	3,249	2,447	778	4,269	5,902	568
Number of USTs Covered By Financial Responsibility	12,791	8,546	4,573	2,585	2,915	15,485	8,384	575
Number of USTs Closed	1,022	1,034	553	336	233	1,874	1,015	80
Annual Cost ($ millions)	2.2	1.6	1.0	0.4	0.7	2.7	1.8	0.0
Response Under This Rule								
Additional Governments Demonstrating Financial Responsibility	101	396	365	283	62	509	501	47
Additional USTs Remaining in Operation	1,022	977	517	298	230	1,856	715	47
Annual Cost Savings ($ millions)	1.0	0.7	0.5	0.1	0.4	1.2	0.7	0.0

Source: EPA, Economic Impact Analysis of Additional Mechanisms for Local Government Entities to Demonstrate Financial Responsibility for Underground Storage Tanks (December, 1992).

BILLING CODE 6560-50-C

4. Cost Impacts

The cost impacts of the rule are measured as the difference between the costs of complying with the October 1988 financial responsibility rule and with this rule. Table 2 shows the estimated change from the baseline by type and size of local government.

The annualized total costs of complying with the October 1988 financial responsibility rule are estimated to be about $10 million, whereas the costs of complying with this rule are about $5.9 million. The rule, therefore, is estimated to result in an annual net savings of approximately $4.5 million.[17] Most of the savings result from fewer mandatory closures under the rule than in the baseline, with additional savings earned from the difference between wholesale and retail prices for fuel.[18] EPA estimates that the average annualized savings per local government entity (irrespective of UST ownership) will be about $54, with an average saving of about $185 per UST-owning government. Most of the total savings will be realized in the first year, because the revised compliance schedule requires that local governments without financial assurance close their USTs within one year of promulgation of this rule. The Agency estimates that very large general purpose governments will save about $1.0 million and very large special districts will save approximately $445,000 under the rule. Very large UST-owning governments are expected to realize savings of more than $700 each, or about 0.001 percent of their typical annual budgets. The Agency expects that all very large entities estimated to close their USTs in the absence of the rule will be able to demonstrate financial responsibility through the alternative mechanisms promulgated under today's rule.

Small general purpose governments save more relative to their budgets for two reasons: (1) The costs of closure are a larger percentage of their budgets; and (2) those entities with USTs spend proportionately more on fuel than medium and large entities and, therefore, save proportionately more by reduced fuel costs.[19] In the absence of the rule, EPA estimates that 85 percent of small governments and 90 percent of small districts would be unable to demonstrate financial responsibility and would be forced to close their USTs; under the rule, only an estimated one percent of small general purpose governments and five percent of small special districts are estimated to be forced to close their USTs.

5. Environmental Impacts

The rule has potential environmental impacts as well as economic impacts. As a result of the rule, more local governments may be able to demonstrate financial responsibility through the alternative mechanisms and more tanks may remain in operation. All local government USTs, however, are subject to the requirements established by the UST technical standards rule, which is designed to minimize the human health and environmental risks from underground storage tank operations. In addition, the rule requires local governments to demonstrate the ability to respond to UST releases to minimize potential environmental damages. EPA believes, therefore, that any negative human health and environmental effects resulting from the rule will be minimal.

B. Regulatory Flexibility Act

The Regulatory Flexibility Act generally requires all Federal agencies to review the impact of their regulations to determine whether they will have a significant economic impact on a substantial number of small entities. If so, the Agency must prepare a regulatory flexibility analysis. As discussed in the economic impact analysis, EPA has determined that this rule will provide a net benefit to small local governments by reducing their costs of compliance with the financial responsibility provisions of Subtitle I. For this reason, the Agency has concluded that a regulatory flexibility analysis is not necessary.

Nevertheless, EPA, as part of its on-going effort to be responsive to small entities affected by its regulations, has expanded its analysis of the economic effects associated with this rule. The additional findings support EPA's conclusions that local governments benefit from the mechanisms promulgated today. The analysis shows, in fact, that small local governments receive an even greater benefit as compared with their larger counterparts.

The Agency estimates that this rule will result in a net savings for local governments of $4.5 million per year. Most of the savings realized by local governments will be due to a significant decrease in the number of UST closures. In developing this rule, the Agency has sought to reduce the impacts on all governments regardless of size. The worksheet test, however, was specifically designed to recognize that even small local governments may be able to demonstrate the requisite financial strength needed to pay for an UST release. Based on an analysis presented in the EIA, approximately 89 percent of all small general purpose governments (including 98 percent of small counties, 91 percent of small municipalities, and 87 percent of small townships) will be able to demonstrate financial responsibility using the worksheet test.

The use of the governmental guarantee is also particularly suited to small local governments, such as school districts, which often have a substantial governmental relationship with a nearby city or county. EPA has estimated that approximately 59 percent of small school and special districts will be able to obtain guarantees.

Overall, the proportion of USTs owned by small local governments which will be unable to meet the financial responsibility requirement drops from about 12 percent to 2 percent as a result of the additional mechanisms promulgated today.

EPA is also considering adoption of another rule which would provide even more flexibility to small local governments. As described in the **Federal Register** on December 23, 1991 (56 FR 66369), an option considered but not yet proposed would extend the financial responsibility compliance date to 1998 for certain facilities that meet Federal economic criteria. This rule would be designed to benefit the small, possibly rural local governments which provide essential emergency or community services and which may be most in need of an additional extension. Despite these efforts, the Agency recognizes that some small local governments may still have difficulty complying with the financial responsibility and other requirements for underground storage tanks. Therefore, the Agency plans to conduct a study of the overall costs of these requirements and the potential impact of these costs on the ability of small

[17] The estimated costs for a municipal government using a state fund to demonstrate financial responsibility may not accurately reflect the true costs to society for providing state fund coverage. Most state funds impose a mandatory per-gallon fee on all petroleum products used in the state. Thus, all users of petroleum products in the state share in paying the costs associated with the state fund.

[18] The only entities that save under the rule are those that would have closed their USTs in the baseline and choose the self-test under the rule and those entities that use a third party mechanism in the baseline (other than insurance and state funds) and the less costly alternative mechanisms under the rule. Far more entities are in the former category than the latter.

[19] In general, small entities do not own USTs; the Agency estimates that fewer than 3,600 of the more than 50,000 small entities own USTs. EPA infers from this that those entities that do own USTs use them intensively, incurring fuel costs as high as seven percent of their annual budgets.

local governments to retain their USTs. The study would also look into the feasibility of alternative methods of providing fuel (for essential public services) without requiring the ownership of USTs by these governments. In addition, the study would assess the continued availability of State financial assistance and assurance funds and their effects on local governments that own USTs. EPA will use the findings from this study to help monitor the effects of the UST requirements on local governments and to help assess the need for any further Agency guidance or action.

It should be noted that UST closures do not necessarily result in a loss of availability of fuel. Local governments have several means of assuring a continuing fuel supply: Purchase of fuel from retail outlets, use of above-ground tanks (where fire codes permit), and "pooling" of fuel sources with neighboring governments (local or State) that are able to demonstrate financial responsibility. The concern expressed by local governments is the availability of fuel for emergency vehicles when retail stations may be closed. EPA understands that some local governments are participating in "card-lock" arrangements. Under a card-lock arrangement, participants are issued magnetic cards that can then unlock the fuel pump; fuel withdrawal is monitored and charged to the card owner. With this arrangement, no attendant is necessary, and there is 24-hour access to fuel. Some of these alternatives have the added benefit of removing the costs and liability of maintaining USTs from local governments.

One commenter contended that EPA's rationale for not developing a regulatory flexibility analysis for the proposed rule is flawed, arguing that because the proposed rule simply amends the financial responsibility requirement, which initially imposed a burden on the regulated community, the benefits of the proposed rule cannot be accurately assessed in isolation of these requirements. As discussed above, the Agency has used this rule to provide additional flexibility to local governments, and particularly small local governments.

EPA does not, however, consider a baseline of no financial responsibility requirement to be a reasonable baseline for analyzing this rule. After review with regard to the 1988 financial responsibility rule, the Agency determined that although most local governments have adequate stability and financial strength to respond to an UST release, not all local governments have the resources to meet their UST-related obligations. Given these concerns, EPA believes that exempting all local governments from the financial responsibility requirement would not be consistent with statutory intent.

C. Paperwork Reduction Act

The information collection requirements in this rule have been approved by the Office of Management and Budget (OMB) under the *Paperwork Reduction Act*, 44 U.S.C. 3501 *et seq.* and have been assigned control number 2050–0066. An Information Collection Request document has been prepared by EPA (ICR No. 1359.04) and a copy may be obtained from Sandy Farmer, Information Policy Branch, EPA, 401 M Street, SW. (PM–223Y), Washington, D.C. 20460 or by calling (202) 260–2740. The information collection requirements were approved by the Office of Management and Budget (OMB) under the Paperwork Reduction Act, 44 U.S.C. 3501 et seq. and assigned OMB control number 2050–0066. The Agency estimated the total annual burden on the regulated community of local governments to be 28,518 hours. The Agency estimated the burden on local government entities for reading the requirements to be 24,188 hours, the reporting or disclosure burden to be 218 hours, and the total recordkeeping burden to be 4,112 hours. The average burden for reading the requirements was estimated to be one hour. The average burden for reporting or disclosure was estimated to be two hours, while only 0.45 percent of all local government UST owners and operators were expected to be required to report each year. The average annual burden to maintain records of the alternative mechanisms was estimated to be 0.17 hours. These burden estimates included all aspects of the collection effort and included time for reviewing instructions, searching existing data sources, gathering and maintaining the data needed, and completing and reviewing the collection of information.

Send comments regarding the burden estimate or any other aspect of this collection of information, including suggestions for reducing this burden, to Chief, Information Policy Branch, PM–223Y, U.S. Environmental Protection Agency, 401 M Street, S.W., Washington, D.C. 20460; and to the Office of Information and Regulatory Affairs, Office of Management and Budget, Washington, D.C. 20503, marked "Attention: Jonathan Gledhill."

VI. Supporting Documents

In addition to supporting material found in the rulemaking docket, EPA has prepared the following supporting documents to support this rule:

EPA, "Background Document in Support of Financial Self-Test for Local Governments Subject to the Financial Responsibility Requirements of Subtitle I of the Resource Conservation and Recovery Act," U.S. Environmental Protection Agency, Office of Underground Storage Tanks (November, 1992).

EPA, "Economic Impact Analysis of Additional Mechanisms for Local Government Entities to Demonstrate Financial Responsibility for Underground Storage Tanks," EPA Office of Underground Storage Tanks (November, 1992).

EPA, "Response to Comments on the June 18, 1990 Proposed Rule to Provide Additional Mechanisms for Local Government Entities to Demonstrate Financial Responsibility for Underground Storage Tanks," EPA Office of Underground Storage Tanks (November, 1992).

List of Subjects in 40 CFR Part 280

Administrative practice and procedure, Environmental protection, Hazardous materials insurance, Hazardous substances, Insurance, Oil pollution, Penalties, Petroleum, Reporting and recordkeeping requirements, State program approval, Surety bonds, Underground storage tanks, Water pollution control, Water supply.

Dated: January 15, 1993.

William K. Reilly,

Administrator.

For the reasons set forth in the preamble, part 280 of title 40 of the Code of Federal Regulations is amended as follows:

PART 280—TECHNICAL STANDARDS AND CORRECTIVE ACTION REQUIREMENTS FOR OWNERS AND OPERATORS OF UNDERGROUND STORAGE TANKS

1. The authority citation for part 280 continues to read as follows:

Authority: 42 U.S.C. 6912, 6991, 6991a, 6991b, 6991c, 6991e, 6991f, and 6991h.

2. Section 280.92 is amended by removing the paragraph designations (a) through (o) and adding in alphabetical order three definitions reading as follows:

§ 280.92 Definition of terms.

* * * * *

Chief Financial Officer, in the case of local government owners and operators, means the individual with the overall authority and responsibility for the collection, disbursement, and use of funds by the local government.

* * * * *

Local government shall have the meaning given this term by applicable state law and includes Indian tribes. The term is generally intended to include: (1) Counties, municipalities, townships, separately chartered and operated special districts (including local government public transit systems and redevelopment authorities), and independent school districts authorized as governmental bodies by state charter or constitution; and (2) Special districts and independent school districts established by counties, municipalities, townships, and other general purpose governments to provide essential services.

* * * * *

Substantial governmental relationship means the extent of a governmental relationship necessary under applicable state law to make an added guarantee contract issued incident to that relationship valid and enforceable. A guarantee contract is issued "incident to that relationship" if it arises from a clear commonality of interest in the event of an UST release such as coterminous boundaries, overlapping constituencies, common ground-water aquifer, or other relationship other than monetary compensation that provides a motivation for the guarantor to provide a guarantee.

* * * * *

3. § 280.94 is amended by revising paragraphs (a) and (b) to read as follows:

§ 280.94 Allowable mechanisms and combinations of mechanisms.

(a) Subject to the limitations of paragraphs (b) and (c) of this section,

(1) An owner or operator, including a local government owner or operator, may use any one or combination of the mechanisms listed in §§ 280.95 through 280.103 to demonstrate financial responsibility under this subpart for one or more underground storage tanks, and

(2) A local government owner or operator may use any one or combination of the mechanisms listed in §§ 280.104 through 280.107 to demonstrate financial responsibility under this subpart for one or more underground storage tanks.

(b) An owner or operator may use a guarantee under § 280.96 or surety bond under § 280.98 to establish financial responsibility only if the Attorney(s) General of the state(s) in which the underground storage tanks are located has (have) submitted a written statement to the implementing agency that a guarantee or surety bond executed as described in this section is a legally valid and enforceable obligation in that state.

* * * * *

4. The following sections are redesignated according to the following table:

Old section no.	New section no.
§ 280.104	§ 280.108.
§ 280.105	§ 280.109.
§ 280.106	§ 280.110.
§ 280.107	§ 280.111.
§ 280.108	§ 280.112.
§ 280.109	§ 280.113.
§ 280.110	§ 280.114.
§ 280.111	§ 280.115.
§ 280.112	§ 280.116.

5. Newly designated §§ 280.109, 280.110, 280.111, 280.112, 280.114, and 280.115 are revised to read as follows:

§ 280.109 Cancellation or nonrenewal by a provider of financial assurance.

(a) Except as otherwise provided, a provider of financial assurance may cancel or fail to renew an assurance mechanism by sending a notice of termination by certified mail to the owner or operator.

(1) Termination of a local government guarantee, a guarantee, a surety bond, or a letter of credit may not occur until 120 days after the date on which the owner or operator receives the notice of termination, as evidenced by the return receipt.

(2) Termination of insurance or risk retention coverage, except for non-payment or misrepresentation by the insured, or state-funded assurance may not occur until 60 days after the date on which the owner or operator receives the notice of termination, as evidenced by the return receipt. Termination for non-payment of premium or misrepresentation by the insured may not occur until a minimum of 10 days after the date on which the owner or operator receives the notice of termination, as evidenced by the return receipt.

(b) If a provider of financial responsibility cancels or fails to renew for reasons other than incapacity of the provider as specified in § 280.114, the owner or operator must obtain alternate coverage as specified in this section within 60 days after receipt of the notice of termination. If the owner or operator fails to obtain alternate coverage within 60 days after receipt of the notice of termination, the owner or operator must notify the Director of the implementing agency of such failure and submit:

(1) The name and address of the provider of financial assurance;

(2) The effective date of termination; and

(3) The evidence of the financial assistance mechanism subject to the termination maintained in accordance with § 280.107(b).

§ 280.110 Reporting by owner or operator.

(a) An owner or operator must submit the appropriate forms listed in § 280.111(b) documenting current evidence of financial responsibility to the Director of the implementing agency:

(1) Within 30 days after the owner or operator identifies a release from an underground storage tank required to be reported under § 280.53 or § 280.61;

(2) If the owner or operator fails to obtain alternate coverage as required by this subpart, within 30 days after the owner or operator receives notice of:

(i) Commencement of a voluntary or involuntary proceeding under Title 11 (Bankruptcy), U.S. Code, naming a provider of financial assurance as a debtor,

(ii) Suspension or revocation of the authority of a provider of financial assurance to issue a financial assurance mechanism,

(iii) Failure of a guarantor to meet the requirements of the financial test,

(iv) Other incapacity of a provider of financial assurance; or

(3) As required by § 280.95(g) and § 280.109(b).

(b) An owner or operator must certify compliance with the financial responsibility requirements of this part as specified in the new tank notification form when notifying the appropriate state or local agency of the installation of a new underground storage tank under § 280.22.

(c) The Director of the Implementing Agency may require an owner or operator to submit evidence of financial assurance as described in § 280.111(b) or other information relevant to compliance with this subpart at any time.

(The information requirements in this section have been approved by the Office of Management and Budget and assigned OMB control number 2050–0066).

§ 280.111 Recordkeeping.

(a) Owners or operators must maintain evidence of all financial assurance mechanisms used to demonstrate financial responsibility under this subpart for an underground storage tank until released from the requirements of this subpart under § 208.113. An owner or operator must maintain such evidence at the underground storage tank site or the owner's or operator's place of work. Records maintained off-site must be made available upon request of the implementing agency.

(b) An owner or operator must maintain the following types of evidence of financial responsibility:

(1) An owner or operator using an assurance mechanism specified in §§ 280.95 through 280.100 or § 280.102 or §§ 280.104 through 280.107 must maintain a copy of the instrument worded as specified.

(2) An owner or operator using a financial test or guarantee, or a local government financial test or a local government guarantee supported by the local government financial test must maintain a copy of the chief financial officer's letter based on year-end financial statements for the most recent completed financial reporting year. Such evidence must be on file no later than 120 days after the close of the financial reporting year.

(3) An owner or operator using a guarantee, surety bond, or letter of credit must maintain a copy of the signed standby trust fund agreement and copies of any amendments to the agreement.

(4) A local government owner or operator using a local government guarantee under § 280.106(d) must maintain a copy of the signed standby trust fund agreement and copies of any amendments to the agreement.

(5) A local government owner or operator using the local government bond rating test under § 280.104 must maintain a copy of its bond rating published within the last twelve months by Moody's or Standard & Poor's.

(6) A local government owner or operator using the local government guarantee under § 280.106, where the guarantor's demonstration of financial responsibility relies on the bond rating test under § 280.104 must maintain a copy of the guarantor's bond rating published within the last twelve months by Moody's or Standard & Poor's.

(7) An owner or operator using an insurance policy or risk retention group coverage must maintain a copy of the signed insurance policy or risk retention group coverage policy, with the endorsement or certificate of insurance and any amendments to the agreements.

(8) An owner or operator covered by a state fund or other state assurance must maintain on file a copy of any evidence of coverage supplied by or required by the state under § 280.101(d).

(9) An owner or operator using a local government fund under § 280.107 must maintain the following documents:

(i) A copy of the state constitutional provision or local government statute, charter, ordinance, or order dedicating the fund, and

(ii) Year-end financial statements for the most recent completed financial reporting year showing the amount in the fund. If the fund is established under § 280.107(a)(3) using incremental funding backed by bonding authority, the financial statements must show the previous year's balance, the amount of funding during the year, and the closing balance in the fund.

(iii) If the fund is established under § 280.107(a)(3) using incremental funding backed by bonding authority, the owner or operator must also maintain documentation of the required bonding authority, including either the results of a voter referendum (under § 280.107(a)(3)(i)), or attestation by the State Attorney General as specified under § 280.107(a)(3)(ii).

(10) A local government owner or operator using the local government guarantee supported by the local government fund must maintain a copy of the guarantor's year-end financial statements for the most recent completed financial reporting year showing the amount of the fund.

(11)(i) An owner or operator using an assurance mechanism specified in §§ 280.95 through 280.107 must maintain an updated copy of a certification of financial responsibility worded as follows, except that instructions in brackets are to be replaced with the relevant information and the brackets deleted:

Certification of Financial Responsibility

[Owner or operator] hereby certifies that it is in compliance with the requirements of subpart H of 40 CFR part 280.

The financial assurance mechanism(s) used to demonstrate financial responsibility under subpart H of 40 CFR part 280 is (are) as follows:

[For each mechanism, list the type of mechanism, name of issuer, mechanism number (if applicable), amount of coverage, effective period of coverage and whether the mechanism covers "taking corrective action" and/or "compensating third parties for bodily injury and property damage caused by" either "sudden accidental releases" or "nonsudden accidental releases" or "accidental releases."]

[Signature of owner or operator]
[Name of owner or operator]
[Title]
[Date]
[Signature of witness or notary]
[Name of witness or notary]
[Date]

(ii) The owner or operator must update this certification whenever the financial assurance mechanism(s) used to demonstrate financial responsibility change(s).

(The information requirements in this section have been approved by the Office of Management and Budget and assigned OMB control number 2050-0066.)

§ 280.112 Drawing on financial assurance mechanisms.

(a) Except as specified in paragraph (d) of this section, the Director of the implementing agency shall require the guarantor, surety, or institution issuing a letter of credit to place the amount of funds stipulated by the Director, up to the limit of funds provided by the financial assurance mechanism, into the standby trust if:

(1)(i) The owner or operator fails to establish alternate financial assurance within 60 days after receiving notice of cancellation of the guarantee, surety bond, letter of credit, or, as applicable, other financial assurance mechanism; and

(ii) The Director determines or suspects that a release from an underground storage tank covered by the mechanism has occurred and so notifies the owner or operator or the owner or operator has notified the Director pursuant to subparts E or F of a release from an underground storage tank covered by the mechanism; or

(2) The conditions of paragraph (b)(1) or (b)(2) (i) or (ii) of this section are satisfied.

(b) The Director of the implementing agency may draw on a standby trust fund when:

(1) The Director makes a final determination that a release has occurred and immediate or long-term corrective action for the release is needed, and the owner or operator, after appropriate notice and opportunity to comply, has not conducted corrective action as required under 40 CFR part 280, subpart F; or

(2) The Director has received either:

(i) Certification from the owner or operator and the third-party liability claimant(s) and from attorneys representing the owner or operator and the third-party liability claimant(s) that a third-party liability claim should be paid. The certification must be worded as follows, except that instructions in brackets are to be replaced with the relevant information and the brackets deleted:

Certification of Valid Claim

The undersigned, as principals and as legal representatives of [insert: owner or operator] and [insert: name and address of third-party claimant], hereby certify that the claim of bodily injury [and/or] property damage caused by an accidental release arising from operating [owner's or operator's] underground storage tank should be paid in the amount of $[____________].

[Signatures]
Owner or Operator
Attorney for Owner or Operator
(Notary)
Date

[Signatures]
Claimant(s)
Attorney(s) for Claimant(s)
(Notary)
Date

or (ii) A valid final court order establishing a judgment against the owner or operator for bodily injury or property damage caused by an accidental release from an underground storage tank covered by financial assurance under this subpart and the Director determines that the owner or operator has not satisfied the judgment.

(c) If the Director of the implementing agency determines that the amount of corrective action costs and third-party liability claims eligible for payment under paragraph (b) of this section may exceed the balance of the standby trust fund and the obligation of the provider of financial assurance, the first priority for payment shall be corrective action costs necessary to protect human health and the environment. The Director shall pay third-party liability claims in the order in which the Director receives certifications under paragraph (b)(2)(i) of this section, and valid court orders under paragraph (b)(2)(ii) of this section.

(d) A governmental entity acting as guarantor under § 280.106(e), the local government guarantee without standby trust, shall make payments as directed by the Director under the circumstances described in § 280.112 (a), (b), and (c).

* * * * *

§ 280.114 Bankruptcy or other incapacity of owner or operator or provider of financial assurance.

(a) Within 10 days after commencement of a voluntary or involuntary proceeding under Title 11 (Bankruptcy), U.S. Code, naming an owner or operator as debtor, the owner or operator must notify the Director of the implementing agency by certified mail of such commencement and submit the appropriate forms listed in § 280.111(b) documenting current financial responsibility.

(b) Within 10 days after commencement of a voluntary or involuntary proceeding under Title 11 (Bankruptcy), U.S. Code, naming a guarantor providing financial assurance as debtor, such guarantor must notify the owner or operator by certified mail of such commencement as required under the terms of the guarantee specified in § 280.96.

(c) Within 10 days after commencement of a voluntary or involuntary proceeding under Title 11 (Bankruptcy), U.S. Code, naming a local government owner or operator as debtor, the local government owner or operator must notify the Director of the implementing agency by certified mail of such commencement and submit the appropriate forms listed in § 280.111(b) documenting current financial responsibility.

(d) Within 10 days after commencement of a voluntary or involuntary proceeding under Title 11 (Bankruptcy), U.S. Code, naming a guarantor providing a local government financial assurance as debtor, such guarantor must notify the local government owner or operator by certified mail of such commencement as required under the terms of the guarantee specified in § 280.106.

(e) An owner or operator who obtains financial assurance by a mechanism other than the financial test of self-insurance will be deemed to be without the required financial assurance in the event of a bankruptcy or incapacity of its provider of financial assurance, or a suspension or revocation of the authority of the provider of financial assurance to issue a guarantee, insurance policy, risk retention group coverage policy, surety bond, letter of credit, or state-required mechanism. The owner or operator must obtain alternate financial assurance as specified in this subpart within 30 days after receiving notice of such an event. If the owner or operator does not obtain alternate coverage within 30 days after such notification, he must notify the Director of the implementing agency.

(f) Within 30 days after receipt of notification that a state fund or other state assurance has become incapable of paying for assured corrective action or third-party compensation costs, the owner or operator must obtain alternate financial assurance.

§ 280.115 Replenishment of guarantees, letters of credit, or surety bonds.

(a) If at any time after a standby trust is funded upon the instruction of the Director of the implementing agency with funds drawn from a guarantee, local government guarantee with standby trust, letter of credit, or surety bond, and the amount in the standby trust is reduced below the full amount of coverage required, the owner or operator shall by the anniversary date of the financial mechanism from which the funds were drawn:

(1) Replenish the value of financial assurance to equal the full amount of coverage required, or

(2) Acquire another financial assurance mechanism for the amount by which funds in the standby trust have been reduced.

(b) For purposes of this section, the full amount of coverage required is the amount of coverage to be provided by § 280.93 of this subpart. If a combination of mechanisms was used to provide the assurance funds which were drawn upon, replenishment shall occur by the earliest anniversary date among the mechanisms.

6. New § 280.104 is added to read as follows:

§ 280.104 Local government bond rating test.

(a) A general purpose local government owner or operator and/or local government serving as a guarantor may satisfy the requirements of § 280.93 by having a currently outstanding issue or issues of general obligation bonds of $1 million or more, excluding refunded obligations, with a Moody's rating of Aaa, Aa, A, or Baa, or a Standard & Poor's rating of AAA, AA, A, or BBB. Where a local government has multiple outstanding issues, or where a local government's bonds are rated by both Moody's and Standard and Poor's, the lowest rating must be used to determine eligibility. Bonds that are backed by credit enhancement other than municipal bond insurance may not be considered in determining the amount of applicable bonds outstanding.

(b) A local government owner or operator or local government serving as a guarantor that is not a general-purpose local government and does not have the legal authority to issue general obligation bonds may satisfy the requirements of § 280.93 by having a currently outstanding issue or issues of revenue bonds of $1 million or more, excluding refunded issues and by also having a Moody's rating of Aaa, A, A, or Baa, or a Standard & Poor's rating of AAA, AA, A, or BBB as the lowest rating for any rated revenue bond issued by the local government. Where bonds are rated by both Moody's and Standard & Poor's, the lower rating for each bond must be used to determine eligibility. Bonds that are backed by credit enhancement may not be considered in determining the amount of applicable bonds outstanding.

(c) The local government owner or operator and/or guarantor must maintain a copy of its bond rating published within the last 12 months by Moody's or Standard & Poor's.

(d) To demonstrate that it meets the local government bond rating test, the chief financial officer of a general purpose local government owner or operator and/or guarantor must sign a letter worded exactly as follows, except that the instructions in brackets are to be replaced by the relevant information and the brackets deleted:

Letter from Chief Financial Officer

I am the chief financial officer of [insert: name and address of local government owner or operator, or guarantor]. This letter is in support of the use of the bond rating test to demonstrate financial responsibility for [insert: "taking corrective action" and/or "compensating third parties for bodily injury and property damage"] caused by [insert: "sudden accidental releases" and/or "nonsudden accidental releases"] in the amount of at least [insert: dollar amount] per occurrence and [insert: dollar amount] annual aggregate arising from operating (an) underground storage tank(s).

Underground storage tanks at the following facilities are assured by this bond rating test: [List for each facility: the name and address of the facility where tanks are assured by the bond rating test].

The details of the issue date, maturity, outstanding amount, bond rating, and bond rating agency of all outstanding bond issues that are being used by [name of local government owner or operator, or guarantor] to demonstrate financial responsibility are as follows: [complete table]

Issue date	Maturity date	Outstanding amount	Bond rating	Rating agency
				[Moody's or Standard & Poor's]

The total outstanding obligation of [insert amount], excluding refunded bond issues, exceeds the minimum amount of $1 million. All outstanding general obligation bonds issued by this government that have been rated by Moody's or Standard & Poor's are rated as at least investment grade (Moody's Baa or Standard & Poor's BBB) based on the most recent ratings published within the last 12 months. Neither rating service has provided notification within the last 12 months of downgrading of bond ratings below investment grade or of withdrawal of bond rating other than for repayment of outstanding bond issues.

I hereby certify that the wording of this letter is identical to the wording specified in 40 CFR Part 280.104(d) as such regulations were constituted on the date shown immediately below.

[Date] ______

[Signature] ______

[Name] ______

[Title] ______

(e) To demonstrate that it meets the local government bond rating test, the chief financial officer of local government owner or operator and/or guarantor other than a general purpose government must sign a letter worded exactly as follows, except that the instructions in brackets are to be replaced by the relevant information and the brackets deleted:

Letter from Chief Financial Officer

I am the chief financial officer of [insert: name and address of local government owner or operator, or guarantor]. This letter is in support of the use of the bond rating test to demonstrate financial responsibility for [insert: "taking corrective action" and/or "compensating third parties for bodily injury and property damage"] caused by [insert : "sudden accidental releases" and/or "nonsudden accidental releases"] in the amount of at least [insert: dollar amount] per occurrence and [insert: dollar amount] annual aggregate arising from operating (an) underground storage tank(s). This local government is not organized to provide general governmental services and does not have the legal authority under state law or constitutional provisions to issue general obligation debt.

Underground storage tanks at the following facilities are assured by this bond rating test: [List for each facility: the name and address of the facility where tanks are assured by the bond rating test].

The details of the issue date, maturity, outstanding amount, bond rating, and bond rating agency of all outstanding revenue bond issues that are being used by [name of local government owner or operator, or guarantor] to demonstrate financial responsibility are as follows: [complete table]

Issue date	Maturity date	Outstanding amount	Bond rating	Rating agency
				[Moody's or Standard & Poor's]

The total outstanding obligation of [insert amount], excluding refunded bond issues, exceeds the minimum amount of $1 million. All outstanding revenue bonds issued by this government that have been rated by Moody's or Standard & Poor's are rated as at least investment grade (Moody's Baa or Standard & Poor's BBB) based on the most recent ratings published within the last 12 months. The revenue bonds listed are not backed by third-party credit enhancement or are insured by a municipal bond insurance company. Neither rating service has provided notification within the last 12 months of downgrading of bond ratings below investment grade or of withdrawal of bond rating other than for repayment of outstanding bond issues.

I hereby certify that the wording of this letter is identical to the wording specified in 40 CFR Part 280.104(e) as such regulations were constituted on the date shown immediately below.

[Date] ______

[Signature] ______

[Name] ______

[Title] ______

(f) The Director of the implementing agency may require reports of financial condition at any time from the local government owner or operator, and/or local government guarantor. If the Director finds, on the basis of such reports or other information, that the local government owner or operator, and/or guarantor, no longer meets the local government bond rating test requirements of § 280.104, the local government owner or operator must obtain alternative coverage within 30 days after notification of such a finding.

(g) If a local government owner or operator using the bond rating test to provide financial assurance finds that it no longer meets the bond rating test requirements, the local government owner or operator must obtain alternative coverage within 150 days of the change in status.

7. New § 280.105 is added to read as follows:

§ 280.105 Local government financial test.

(a) A local government owner or operator may satisfy the requirements of § 280.93 by passing the financial test specified in this section. To be eligible to use the financial test, the local government owner or operator must have the ability and authority to assess and levy taxes or to freely establish fees and charges. To pass the local government financial test, the owner or operator must meet the criteria of paragraphs (b)(2) and (b)(3) of this section based on year-end financial statements for the latest completed fiscal year.

(b)(1) The local government owner or operator must have the following information available, as shown in the year-end financial statements for the latest completed fiscal year:

(i) *Total revenues:* Consists of the sum of general fund operating and non-operating revenues including net local taxes, licenses and permits, fines and forfeitures, revenues from use of money and property, charges for services, investment earnings, sales (property, publications, etc.), intergovernmental revenues (restricted and unrestricted), and total revenues from all other governmental funds including enterprise, debt service, capital projects, and special revenues, but excluding revenues to funds held in a trust or agency capacity. For purposes of this test, the calculation of total revenues shall exclude all transfers between funds under the direct control of the local government using the financial test (interfund transfers), liquidation of investments, and issuance of debt.

(ii) *Total expenditures:* Consists of the sum of general fund operating and non-operating expenditures including public safety, public utilities, transportation, public works, environmental protection, cultural and recreational, community development, revenue sharing, employee benefits and compensation, office management, planning and zoning, capital projects, interest payments on debt, payments for retirement of debt principal, and total expenditures from all other governmental funds including enterprise, debt service, capital projects, and special revenues. For purposes of this test, the calculation of total expenditures shall exclude all transfers between funds under the direct control of the local government using the financial test (interfund transfers).

(iii) *Local revenues:* Consists of total revenues (as defined in paragraph (b)(1)(i) of this section) minus the sum of all transfers from other governmental entities, including all monies received from Federal, state, or local government sources.

(iv) *Debt service:* Consists of the sum of all interest and principal payments on all long-term credit obligations and all interest-bearing short-term credit obligations. Includes interest and principal payments on general obligation bonds, revenue bonds, notes, mortgages, judgments, and interest bearing warrants. Excludes payments on non-interest-bearing short-term obligations, interfund obligations, amounts owed in a trust or agency capacity, and advances and contingent loans from other governments.

(v) *Total funds:* Consists of the sum of cash and investment securities from all funds, including general, enterprise, debt service, capital projects, and special revenue funds, but excluding employee retirement funds, at the end of the local government's financial reporting year. Includes Federal securities, Federal agency securities, state and local government securities, and other securities such as bonds, notes and mortgages. For purposes of this test, the calculation of total funds shall exclude agency funds, private trust funds, accounts receivable, value of real property, and other non-security assets.

(vi) *Population* consists of the number of people in the area served by the local government.

(2) The local government's year-end financial statements, if independently audited, cannot include an adverse auditor's opinion or a disclaimer of opinion. The local government cannot have outstanding issues of general obligation or revenue bonds that are rated as less than investment grade.

(3) The local government owner or operator must have a letter signed by the chief financial officer worded as specified in paragraph (c) of this section.

(c) To demonstrate that it meets the financial test under paragraph (b) of this section, the chief financial officer of the local government owner or operator, must sign, within 120 days of the close of each financial reporting year, as defined by the twelve-month period for which financial statements used to support the financial test are prepared, a letter worded exactly as follows, except that the instructions in brackets are to be replaced by the relevant information and the brackets deleted:

Letter From Chief Financial Officer

I am the chief financial officer of [insert: name and address of the owner or operator]. This letter is in support of the use of the local government financial test to demonstrate financial responsibility for [insert: "taking corrective action" and/or "compensating third parties for bodily injury and property damage"] caused by [insert: "sudden accidental releases" and/or "nonsudden accidental releases"] in the amount of at least [insert: dollar amount] per occurrence and [insert: dollar amount] annual aggregate arising from operating [an] underground storage tank[s].

Underground storage tanks at the following facilities are assured by this financial test [List for each facility: the name and address of the facility where tanks assured by this financial test are located. If separate mechanisms or combinations of mechanisms are being used to assure any of the tanks at this facility, list each tank assured by this financial test by the tank identification number provided in the notification submitted pursuant to 40 CFR Part 280.22 or the corresponding state requirements.]

This owner or operator has not received an adverse opinion, or a disclaimer of opinion from an independent auditor on its financial statements for the latest completed fiscal year. Any outstanding issues of general obligation or revenue bonds, if rated, have a Moody's rating of Aaa, Aa, A, or Baa or a Standard and Poor's rating of AAA, AA, A, or BBB; if rated by both firms, the bonds have a Moody's rating of Aaa, Aa, A, or Baa and a Standard and Poor's rating of AAA, AA, A, or BBB.

Worksheet for Municipal Financial Test

Part I: Basic Information

1. Total Revenues

a. Revenues (dollars) __________

Value of revenues excludes liquidation of investments and issuance of debt. Value includes all general fund operating and non-operating revenues, as well as all revenues from all other governmental funds including enterprise, debt service, capital projects, and special revenues, but excluding revenues to funds held in a trust or agency capacity.

b. Subtract interfund transfers (dollars)__________

c. Total Revenues (dollars)__________

2. Total Expenditures

a. Expenditures (dollars) __________

Value consists of the sum of general fund operating and non-operating expenditures including interest payments on debt, payments for retirement of debt principal, and total expenditures from all other governmental funds including enterprise, debt service, capital projects, and special revenues.

b. Subtract interfund transfers (dollars)__________

c. Total Expenditures (dollars)__________

3. Local Revenues

a. Total Revenues (from 1c) (dollars) __________

b. Subtract total intergovernmental transfers (dollars)__________

c. Local Revenues (dollars)__________

4. Debt Service

a. Interest and fiscal charges (dollars)__________

b. Add debt retirement (dollars)__________

c. Total Debt Service (dollars)__________

5. Total Funds (Dollars)__________

(Sum of amounts held as cash and investment securities from all funds, excluding amounts held for employee retirement funds, agency funds, and trust funds)

6. Population (Persons)__________

Part II: Application of Test

7. Total Revenues to Population

a. Total Revenues (from 1c)__________

b. Population (from 6)__________

c. Divide 7a by 7b __________

d. Subtract 417__________

e. Divide by 5,212__________
f. Multiply by 4.095__________

8. Total Expenses to Population

a. Total Expenses (from 2c)__________
b. Population (from 6)__________
c. Divide 8a by 8b __________
d. Subtract 524 __________
e. Divide by 5,401__________
f. Multiply by 4.095__________

9. Local Revenues to Total Revenues

a. Local Revenues (from 3c)__________
b. Total Revenues (from 1c)__________
c. Divide 9a by 9b __________
d. Subtract .695__________
e. Divide by .205__________
f. Multiply by 2.840 __________

10. Debt Service to Population

a. Debt Service (from 4d) __________
b. Population (from 6)__________
c. Divide 10a by 10b __________
d. Subtract 51 __________
e. Divide by 1,038__________
f. Multiply by −1.866__________

11. Debt Service to Total Revenues

a. Debt Service (from 4d)__________
b. Total Revenues (from 1c)__________
c. Divide 11a by 11b __________
d. Subtract .068 __________
e. Divide by .259 __________
f. Multiply by −3.533 __________

12. Total Revenues to Total Expenses

a. Total Revenues (from 1c)__________
b. Total Expenses (from 2c)__________
c. Divide 12a by 12b__________
d. Subtract .910 __________
e. Divide by .899 __________
f. Multiply by 3.458 __________

13. Funds Balance to Total Revenues

a. Total Funds (from 5) __________
b. Total Revenues (from 1c)__________
c. Divide 13a by 13b __________
d. Subtract .891 __________
e. Divide by 9.156__________
f. Multiply by 3.270 __________

14. Funds Balance to Total Expenses

a. Total Funds (from 5)__________
b. Total Expenses (from 2c)__________
c. Divide 14a by 14b__________
d. Subtract .866 __________
e. Divide by 6.409 __________
f. Multiply by 3.270 __________

15. Total Funds to Population __________

a. Total Funds (from 5) __________
b. Population (from 6)__________
c. Divide 15a by 15b __________
d. Subtract 270 __________
e. Divide by 4,548 __________
f. Multiply by 1.866 __________

16. Add 7f + 8f + 9f + 10f + 11f + 12f + 13f + 14f + 15f + 4.937__________

I hereby certify that the financial index shown on line 16 of the worksheet is greater than zero and that the wording of this letter is identical to the wording specified in 40 CFR Part 280.105(c) as such regulations were constituted on the date shown immediately below.

[Date]
[Signature]
[Name]
[Title]

(d) If a local government owner or operator using the test to provide financial assurance finds that it no longer meets the requirements of the financial test based on the year-end financial statements, the owner or operator must obtain alternative coverage within 150 days of the end of the year for which financial statements have been prepared.

(e) The Director of the implementing agency may require reports of financial condition at any time from the local government owner or operator. If the Director finds, on the basis of such reports or other information, that the local government owner or operator no longer meets the financial test requirements of § 280.105 (b) and (c), the owner or operator must obtain alternate coverage within 30 days after notification of such a finding.

(f) If the local government owner or operator fails to obtain alternate assurance within 150 days of finding that it no longer meets the requirements of the financial test based on the year-end financial statements or within 30 days of notification by the Director of the implementing agency that it no longer meets the requirements of the financial test, the owner or operator must notify the Director of such failure within 10 days.

8. New § 280.106 is added to read as follows:

§ 280.106 Local government guarantee.

(a) A local government owner or operator may satisfy the requirements of § 280.93 by obtaining a guarantee that conforms to the requirements of this section. The guarantor must be either the state in which the local government owner or operator is located or a local government having a "substantial governmental relationship" with the owner and operator and issuing the guarantee as an act incident to that relationship. A local government acting as the guarantor must:

(1) demonstrate that it meets the bond rating test requirement of § 280.104 and deliver a copy of the chief financial officer's letter as contained in § 280.104(c) to the local government owner or operator; or

(2) demonstrate that it meets the worksheet test requirements of § 280.105 and deliver a copy of the chief financial officer's letter as contained in § 280.105(c) to the local government owner or operator; or

(3) demonstrate that it meets the local government fund requirements of § 280.107(a), § 280.107(b), or § 280.107(c) and deliver a copy of the chief financial officer's letter as contained in § 280.107 to the local government owner or operator.

(b) If the local government guarantor is unable to demonstrate financial assurance under any of §§ 280.104, 280.105, 280.107(a), 280.107(b), or 280.107(c), at the end of the financial reporting year, the guarantor shall send by certified mail, before cancellation or non-renewal of the guarantee, notice to the owner or operator. The guarantee will terminate no less than 120 days after the date the owner or operator receives the notification, as evidenced by the return receipt. The owner or operator must obtain alternative coverage as specified in § 280.114(c).

(c) The guarantee agreement must be worded as specified in paragraph (d) or (e) of this section, depending on which of the following alternative guarantee arrangements is selected:

(1) If, in the default or incapacity of the owner or operator, the guarantor guarantees to fund a standby trust as directed by the Director of the implementing agency, the guarantee shall be worded as specified in paragraph (d) of this section.

(2) If, in the default or incapacity of the owner or operator, the guarantor guarantees to make payments as directed by the Director of the implementing agency for taking corrective action or compensating third parties for bodily injury and property damage, the guarantee shall be worded as specified in paragraph (e) of this section.

(d) If the guarantor is a state, the local government guarantee with standby trust must be worded exactly as follows, except that instructions in brackets are to be replaced with relevant information and the brackets deleted:

Local Government Guarantee With Standby Trust Made by a State

Guarantee made this [date] by [name of state], herein referred to as guarantor, to [the state implementing agency] and to any and all third parties, and obliges, on behalf of [local government owner or operator].

Recitals

(1) Guarantor is a state.

(2) [Local government owner or operator] owns or operates the following underground storage tank(s) covered by this guarantee: [List the number of tanks at each facility and the name(s) and address(es) of the facility(ies) where the tanks are located. If more than one instrument is used to assure different tanks at any one facility, for each tank covered by this instrument, list the tank identification number provided in the notification submitted pursuant to 40 CFR Part 280 or the corresponding state requirement, and the name and address of the facility.] This guarantee satisfies 40 CFR

Part 280, Subpart H requirements for assuring funding for [insert: "taking corrective action" and/or "compensating third parties for bodily injury and property damage caused by" either "sudden accidental releases" or "nonsudden accidental releases" or "accidental releases"; if coverage is different for different tanks or locations, indicate the type of coverage applicable to each tank or location] arising from operating the above-identified underground storage tank(s) in the amount of [insert dollar amount] per occurrence and [insert dollar amount] annual aggregate.

(3) Guarantor guarantees to [implementing agency] and to any and all third parties that:

In the event that [local government owner or operator] fails to provide alternative coverage within 60 days after receipt of a notice of cancellation of this guarantee and the [Director of the implementing agency] has determined or suspects that a release has occurred at an underground storage tank covered by this guarantee, the guarantor, upon instructions from the [Director] shall fund a standby trust fund in accordance with the provisions of 40 CFR part 280.112, in an amount not to exceed the coverage limits specified above.

In the event that the [Director] determines that [local government owner or operator] has failed to perform corrective action for releases arising out of the operation of the above-identified tank(s) in accordance with 40 CFR part 280, subpart F, the guarantor upon written instructions from the [Director] shall fund a standby trust fund in accordance with the provisions of 40 CFR part 280.112, in an amount not to exceed the coverage limits specified above.

If [owner or operator] fails to satisfy a judgment or award based on a determination of liability for bodily injury or property damage to third parties caused by ["sudden" and/or "nonsudden"] accidental releases arising from the operation of the above-identified tank(s), or fails to pay an amount agreed to in settlement of a claim arising from or alleged to arise from such injury or damage, the guarantor, upon written instructions from the [Director], shall fund a standby trust in accordance with the provisions of 40 CFR part 280.112 to satisfy such judgment(s), award(s), or settlement agreement(s) up to the limits of coverage specified above.

(4) Guarantor agrees to notify [owner or operator] by certified mail of a voluntary or involuntary proceeding under Title 11 (Bankruptcy), U.S. Code naming guarantor as debtor, within 10 days after commencement of the proceeding.

(5) Guarantor agrees to remain bound under this guarantee notwithstanding any modification or alteration of any obligation of [owner or operator] pursuant to 40 CFR part 280.

(6) Guarantor agrees to remain bound under this guarantee for so long as [local government owner or operator] must comply with the applicable financial responsibility requirements of 40 CFR part 280, Subpart H for the above identified tank(s), except that guarantor may cancel this guarantee by sending notice by certified mail to [owner or operator], such cancellation to become effective no earlier than 120 days after receipt of such notice by [owner or operator], as evidenced by the return receipt.

(7) The guarantor's obligation does not apply to any of the following:

(a) Any obligation of [local government owner or operator] under a workers' compensation, disability benefits, or unemployment compensation law or other similar law;

(b) Bodily injury to an employee of [insert: local government owner or operator] arising from, and in the course of, employment by [insert: local government owner or operator];

(c) Bodily injury or property damage arising from the ownership, maintenance, use, or entrustment to others of any aircraft, motor vehicle, or watercraft;

(d) Property damage to any property owned, rented, loaded to, in the care, custody, or control of, or occupied by [insert: local government owner or operator] that is not the direct result of a release from a petroleum underground storage tank;

(e) Bodily damage or property damage for which [insert owner or operator] is obligated to pay damages by reason of the assumption of liability in a contract or agreement other than a contract or agreement entered into to meet the requirements of 40 CFR part 280.93.

(8) Guarantor expressly waives notice of acceptance of this guarantee by [the implementing agency], by any or all third parties, or by [local government owner or operator].

I hereby certify that the wording of this guarantee is identical to the wording specified in 40 CFR part 280.106(d) as such regulations were constituted on the effective date shown immediately below.

Effective date: ——————

[Name of guarantor]

[Authorized signature for guarantor]

[Name of person signing]

[Title of person signing]

Signature of witness or notary:

If the guarantor is a local government, the local government guarantee with standby trust must be worded exactly as follows, except that instructions in brackets are to be replaced with relevant information and the brackets deleted:

Local Government Guarantee With Standby Trust Made by a Local Government

Guarantee made this [date] by [name of guaranteeing entity], a local government organized under the laws of [name of state], herein referred to as guarantor, to [the state implementing agency] and to any and all third parties, and obliges, on behalf of [local government owner or operator].

Recitals

(1) Guarantor meets or exceeds [select one: the local government bond rating test requirements of 40 CFR Part 280.104, the local government financial test requirements of 40 CFR Part 280.105, or the local government fund under 40 CFR Part 280.107(a), 280.107(b), or 280.107(c)].

(2) [Local government owner or operator] owns or operates the following underground storage tank(s) covered by this guarantee: [List the number of tanks at each facility and the name(s) and address(es) of the facility(ies) where the tanks are located. If more than one instrument is used to assure different tanks at any one facility, for each tank covered by this instrument, list the tank identification number provided in the notification submitted pursuant to 40 CFR Part 280 or the corresponding state requirement, and the name and address of the facility.] This guarantee satisfies 40 CFR Part 280, Subpart H requirements for assuring funding for [insert: "taking corrective action" and/or "compensating third parties for bodily injury and property damage caused by" either "sudden accidental releases" or "nonsudden accidental releases" or "accidental releases"; if coverage is different for different tanks or locations, indicate the type of coverage applicable to each tank or location] arising from operating the above-identified underground storage tank(s) in the amount of [insert dollar amount] per occurrence and [insert: dollar amount] annual aggregate.

(3) Incident to our substantial governmental relationship with [local government owner or operator], guarantor guarantees to [implementing agency] and to any and all third parties that:

In the event that [local government owner or operator] fails to provide alternative coverage within 60 days after receipt of a notice of cancellation of this guarantee and the [Director of the implementing agency] has determined or suspects that a release has occurred at an underground storage tank covered by this guarantee, the guarantor, upon instructions from the [Director] shall fund a standby trust fund in accordance with the provisions of 40 CFR Part 280.112, in an amount not to exceed the coverage limits specified above.

In the event that the [Director] determines that [local government owner or operator] has failed to perform corrective action for releases arising out of the operation of the above-identified tank(s) in accordance with 40 CFR Part 280, Subpart F, the guarantor upon written instructions from the [Director] shall fund a standby trust fund in accordance with the provisions of 40 CFR Part 280.112, in an amount not to exceed the coverage limits specified above.

If [owner or operator] fails to satisfy a judgment or award based on a determination of liability for bodily injury or property damage to third parties caused by ["sudden" and/or "nonsudden"] accidental releases arising from the operation of the above-identified tank(s), or fails to pay an amount agreed to in settlement of a claim arising from or alleged to arise from such injury or damage, the guarantor, upon written instructions from the [Director], shall fund a standby trust in accordance with the provisions of 40 CFR Part 280.112 to satisfy such judgment(s), award(s), or settlement agreement(s) up to the limits of coverage specified above.

(4) Guarantor agrees that, if at the end of any fiscal year before cancellation of this guarantee, the guarantor fails to meet or exceed the requirements of the financial responsibility mechanism specified in paragraph (1), guarantor shall send within 120 days of such failure, by certified mail,

notice to [local government owner or operator], as evidenced by the return receipt.

(5) Guarantor agrees to notify [owner or operator] by certified mail of a voluntary or involuntary proceeding under Title 11 (Bankruptcy), U.S. Code naming guarantor as debtor, within 10 days after commencement of the proceeding.

(6) Guarantor agrees to remain bound under this guarantee notwithstanding any modification or alteration of any obligation of [owner or operator] pursuant to 40 CFR Part 280.

(7) Guarantor agrees to remain bound under this guarantee for so long as [local government owner or operator] must comply with the applicable financial responsibility requirements of 40 CFR Part 280, Subpart H for the above identified tank(s), except that guarantor may cancel this guarantee by sending notice by certified mail to [owner or operator], such cancellation to become effective no earlier than 120 days after receipt of such notice by [owner or operator], as evidenced by the return receipt.

(8) The guarantor's obligation does not apply to any of the following:

(a) Any obligation of [local government owner or operator] under a workers' compensation, disability benefits, or unemployment compensation law or other similar law;

(b) Bodily injury to an employee of [insert: local government owner or operator] arising from, and in the course of, employment by [insert: local government owner or operator];

(c) Bodily injury or property damage arising from the ownership, maintenance, use, or entrustment to others of any aircraft, motor vehicle, or watercraft;

(d) Property damage to any property owned, rented, loaned to, in the care, custody, or control of, or occupied by [insert: local government owner or operator] that is not the direct result of a release from a petroleum underground storage tank;

(e) Bodily damage or property damage for which [insert: owner or operator] is obligated to pay damages by reason of the assumption of liability in a contract or agreement other than a contract or agreement entered into to meet the requirements of 40 CFR Part 280.93.

(9) Guarantor expressly waives notice of acceptance of this guarantee by [the implementing agency], by any or all third parties, or by [local government owner or operator].

I hereby certify that the wording of this guarantee is identical to the wording specified in 40 CFR Part 280.106(d) as such regulations were constituted on the effective date shown immediately below.

Effective date: ______

[Name of guarantor]
[Authorized signature for guarantor]
[Name of person signing]
[Title of person signing]
Signature of witness or notary:

______

(e) If the guarantor is a state, the local government guarantee without standby trust must be worded exactly as follows, except that instructions in brackets are to be replaced with relevant information and the brackets deleted:

Local Government Guarantee Without Standby Trust Made by a State

Guarantee made this [date] by [name of state], herein referred to as guarantor, to [the state implementing agency] and to any and all third parties, and obliges, on behalf of [local government owner or operator].

Recitals

(1) Guarantor is a state.

(2) [Local government owner or operator] owns or operates the following underground storage tank(s) covered by this guarantee: [List the number of tanks at each facility and the name(s) and address(es) of the facility(ies) where the tanks are located. If more than one instrument is used to assure different tanks at any one facility, for each tank covered by this instrument, list the tank identification number provided in the notification submitted pursuant to 40 CFR Part 280 or the corresponding state requirement, and the name and address of the facility.] This guarantee satisfies 40 CFR Part 280, Subpart H requirements for assuring funding for [insert: "taking corrective action" and/or "compensating third parties for bodily injury and property damage caused by" either "sudden accidental releases" or "nonsudden accidental releases" or "accidental releases"; if coverage is different for different tanks or locations, indicate the type of coverage applicable to each tank or location] arising from operating the above-identified underground storage tank(s) in the amount of [insert: dollar amount] per occurrence and [insert: dollar amount] annual aggregate.

(3) Guarantor guarantees to [implementing agency] and to any and all third parties and obliges that:

In the event that [local government owner or operator] fails to provide alternative coverage within 60 days after receipt of a notice of cancellation of this guarantee and the [Director of the implementing agency] has determined or suspects that a release has occurred at an underground storage tank covered by this guarantee, the guarantor, upon written instructions from the [Director] shall make funds available to pay for corrective actions and compensate third parties for bodily injury and property damage in an amount not to exceed the coverage limits specified above.

In the event that the [Director] determines that [local government owner or operator] has failed to perform corrective action for releases arising out of the operation of the above-identified tank(s) in accordance with 40 CFR Part 280, Subpart F, the guarantor upon written instructions from the [Director] shall make funds available to pay for corrective actions in an amount not to exceed the coverage limits specified above.

If [owner or operator] fails to satisfy a judgment or award based on a determination of liability for bodily injury or property damage to third parties caused by ["sudden" and/or "nonsudden"] accidental releases arising from the operation of the above-identified tank(s), or fails to pay an amount agreed to in settlement of a claim arising from or alleged to arise from such injury or damage, the guarantor, upon written instructions from the [Director], shall make funds available to compensate third parties for bodily injury and property damage in an amount not to exceed the coverage limits specified above.

(4) Guarantor agrees to notify [owner or operator] by certified mail of a voluntary or involuntary proceeding under Title 11 (Bankruptcy), U.S. Code naming guarantor as debtor, within 10 days after commencement of the proceeding.

(5) Guarantor agrees to remain bound under this guarantee notwithstanding any modification or alteration of any obligation of [owner or operator] pursuant to 40 CFR Part 280.

(6) Guarantor agrees to remain bound under this guarantee for so long as [local government owner or operator] must comply with the applicable financial responsibility requirements of 40 CFR Part 280, Subpart H for the above identified tank(s), except that guarantor may cancel this guarantee by sending notice by certified mail to [owner or operator], such cancellation to become effective no earlier than 120 days after receipt of such notice by [owner or operator], as evidenced by the return receipt. If notified of a probable release, the guarantor agrees to remain bound to the terms of this guarantee for all charges arising from the release, up to the coverage limits specified above, notwithstanding the cancellation of the guarantee with respect to future releases.

(7) The guarantor's obligation does not apply to any of the following:

(a) Any obligation of [local government owner or operator] under a workers' compensation disability benefits, or unemployment compensation law or other similar law;

(b) Bodily injury to an employee of [insert local government owner or operator] arising from, and in the course of, employment by [insert: local government owner or operator];

(c) Bodily injury or property damage arising from the ownership, maintenance, use, or entrustment to others of any aircraft, motor vehicle, or watercraft;

(d) Property damage to any property owned, rented, loaded to, in the care, custody, or control of, or occupied by [insert: local government owner or operator] that is not the direct result of a release from a petroleum underground storage tank;

(e) Bodily damage or property damage for which [insert: owner or operator] is obligated to pay damages by reason of the assumption of liability in a contract or agreement other than a contract or agreement entered into to meet the requirements of 40 CFR Part 280.93.

(8) Guarantor expressly waives notice of acceptance of this guarantee by [the implementing agency], by any or all third parties, or by [local government owner or operator].

I hereby certify that the wording of this guarantee is identical to the wording specified in 40 CFR Part 280.106(e) as such regulations were constituted on the effective date shown immediately below.

Effective date: ______

[Name of guarantor]
[Authorized signature for guarantor]
[Name of person signing]
[Title of person signing]
Signature of witness or notary:

______

If the guarantor is a local government, the local government guarantee without standby trust must be worded exactly as follows, except that instructions in brackets are to be replaced with relevant information and the brackets deleted:

Local Government Guarantee Without Standby Trust Made by a Local Government

Guarantee made this [date] by [name of guaranteeing entity], a local government organized under the laws of [name of state], herein referred to as guarantor, to [the state implementing agency] and to any and all third parties, and obliges, on behalf of [local government owner or operator].

Recitals

(1) Guarantor meets or exceeds [select one: the local government bond rating test requirements of 40 CFR part 280.104, the local government financial test requirements of 40 part CFR 280.105, the local government fund under 40 CFR part 280.107(a), 280.107(b), or 280.107(c).

(2) [Local government owner or operator] owns or operates the following underground storage tank(s) covered by this guarantee: [List the number of tanks at each facility and the name(s) and address(es) of the facility(ies) where the tanks are located. If more than one instrument is used to assure different tanks at any one facility, for each tank covered by this instrument, list the tank identification number provided in the notification submitted pursuant to 40 CFR Part 280 or the corresponding state requirement, and the name and address of the facility.] This guarantee satisfies 40 CFR part 280, subpart H requirements for assuring funding for [insert: "taking corrective action" and/or "compensating third parties for bodily injury and property damage caused by" either "sudden accidental releases" or "nonsudden accidental releases" or "accidental releases"; if coverage is different for different tanks or locations, indicate the type of coverage applicable to each tank or location] arising from operating the above-identified underground storage tank(s) in the amount of [insert: dollar amount] per occurrence and [insert: dollar amount] annual aggregate.

(3) Incident to our substantial governmental relationship with [local government owner or operator], guarantor guarantees to [implementing agency] and to any and all third parties and obliges that:

In the event that [local government owner or operator] fails to provide alternative coverage within 60 days after receipt of a notice of cancellation of this guarantee and the [Director of the implementing agency] has determined or suspects that a release has occurred at an underground storage tank covered by this guarantee, the guarantor, upon written instructions from the [Director] shall make funds available to pay for corrective actions and compensate third parties for bodily injury and property damage in an amount not to exceed the coverage limits specified above.

In the event that the [Director] determines that [local government owner or operator] has failed to perform corrective action for releases arising out of the operation of the above-identified tank(s) in accordance with 40 CFR part 280, Subpart F, the guarantor upon written instructions from the [Director] shall make funds available to pay for corrective actions in an amount not to exceed the coverage limits specified above.

If [owner or operator] fails to satisfy a judgment or award based on a determination of liability for bodily injury or property damage to third parties caused by ["sudden" and/or "nonsudden"] accidental releases arising from the operation of the above-identified tank(s), or fails to pay an amount agreed to in settlement of a claim arising from or alleged to arise from such injury or damage, the guarantor, upon written instructions from the [Director], shall make funds available to compensate third parties for bodily injury and property damage in an amount not to exceed the coverage limits specified above.

(4) Guarantor agrees that if at the end of any fiscal year before cancellation of this guarantee, the guarantor fails to meet or exceed the requirements of the financial responsibility mechanism specified in paragraph (1), guarantor shall send within 120 days of such failure, by certified mail, notice to [local government owner or operator], as evidenced by the return receipt.

(5) Guarantor agrees to notify [owner or operator] by certified mail of a voluntary or involuntary proceeding under Title 11 (Bankruptcy), U.S. Code naming guarantor as debtor, within 10 days after commencement of the proceeding.

(6) Guarantor agrees to remain bound under this guarantee notwithstanding any modification or alteration of any obligation of [owner or operator] pursuant to 40 CFR part 280.

(7) Guarantor agrees to remain bound under this guarantee for so long as [local government owner or operator] must comply with the applicable financial responsibility requirements of 40 CFR Part 280, Subpart H for the above identified tank(s), except that guarantor may cancel this guarantee by sending notice by certified mail to [owner or operator], such cancellation to become effective no earlier than 120 days after receipt of such notice by [owner or operator], as evidenced by the return receipt. If notified of a probable release, the guarantor agrees to remain bound to the terms of this guarantee for all charges arising from the release, up to the coverage limits specified above, notwithstanding the cancellation of the guarantee with respect to future releases.

(8) The guarantor's obligation does not apply to any of the following:

(a) Any obligation of [local government owner or operator] under a workers' compensation disability benefits, or unemployment compensation law or other similar law;

(b) Bodily injury to an employee of [insert: local government owner or operator] arising from, and in the course of, employment by [insert: local government owner or operator];

(c) Bodily injury or property damage arising from the ownership, maintenance, use, or entrustment to others of any aircraft, motor vehicle, or watercraft;

(d) Property damage to any property owned, rented, loaded to, in the care, custody, or control of, or occupied by [insert: local government owner or operator] that is not the direct result of a release from a petroleum underground storage tank;

(e) Bodily damage or property damage for which [insert: owner or operator] is obligated to pay damages by reason of the assumption of liability in a contract or agreement other than a contract or agreement entered into to meet the requirements of 40 CFR Part 280.93.

(9) Guarantor expressly waives notice of acceptance of this guarantee by [the implementing agency], by any or all third parties, or by [local government owner or operator],

I hereby certify that the wording of this guarantee is identical to the wording specified in 40 CFR Part 280.106(e) as such regulations were constituted on the effective date shown immediately below.

Effective date: ______

[Name of guarantor]

[Authorized signature for guarantor]

[Name of person signing]

[Title of person signing]

Signature of witness or notary:

---

9. New § 280.107 is added to read as follows:

§ 280.107 Local government fund.

A local government owner or operator may satisfy the requirements of § 280.93 by establishing a dedicated fund account that conforms to the requirements of this section. Except as specified in paragraph (b), a dedicated fund may not be commingled with other funds or otherwise used in normal operations. A dedicated fund will be considered eligible if it meets one of the following requirements:

(a) The fund is dedicated by state constitutional provision, or local government statute, charter, ordinance, or order to pay for taking corrective action and for compensating third parties for bodily injury and property damage caused by accidental releases arising from the operation of petroleum underground storage tanks and is funded for the full amount of coverage required under § 280.93, or funded for part of the required amount of coverage and used in combination with other mechanism(s) that provide the remaining coverage; or

(b) The fund is dedicated by state constitutional provision, or local government statute, charter, ordinance, or order as a contingency fund for general emergencies, including taking corrective action and compensating third parties for bodily injury and property damage caused by accidental releases arising from the operation of petroleum underground storage tanks, and is funded for five times the full amount of coverage required under § 280.93, or funded for part of the required amount of coverage and used in combination with other mechanism(s) that provide the remaining coverage. If the fund is funded for less than five times the amount of coverage required under § 280.93, the amount of financial responsibility demonstrated by the fund may not exceed one-fifth the amount in the fund; or

(c) The fund is dedicated by state constitutional provision, or local government statute, charter, ordinance or order to pay for taking corrective action and for compensating third parties for bodily injury and property damage caused by accidental releases arising from the operation of petroleum underground storage tanks. A payment is made to the fund once every year for seven years until the fund is fully-funded. This seven year period is hereafter referred to as the "pay-in-period." The amount of each payment must be determined by this formula:

$$\frac{TF - CF}{Y}$$

Where TF is the total required financial assurance for the owner or operator, CF is the current amount in the fund, and Y is the number of years remaining in the pay-in-period, and;

(1) The local government owner or operator has available bonding authority, approved through voter referendum (if such approval is necessary prior to the issuance of bonds), for an amount equal to the difference between the required amount of coverage and the amount held in the dedicated fund. This bonding authority shall be available for taking corrective action and for compensating third parties for bodily injury and property damage caused by accidental releases arising from the operation of petroleum underground storage tanks, or

(2) The local government owner or operator has a letter signed by the appropriate state attorney general stating that the use of the bonding authority will not increase the local government's debt beyond the legal debt ceilings established by the relevant state laws. The letter must also state that prior voter approval is not necessary before use of the bonding authority.

(d) To demonstrate that it meets the requirements of the local government fund, the chief financial officer of the local government owner or operator and/or guarantor must sign a letter worded exactly as follows, except that the instructions in brackets are to be replaced by the relevant information and the brackets deleted:

Letter from Chief Financial Officer

I am the chief financial officer of [insert: name and address of local government owner or operator, or guarantor]. This letter is in support of the use of the local government fund mechanism to demonstrate financial responsibility for [insert: "taking corrective action" and/or "compensating third parties for bodily injury and property damage"] caused by [insert: "sudden accidental releases" and/or "nonsudden accidental releases"] in the amount of at least [insert: dollar amount] per occurrence and [insert: dollar amount] annual aggregate arising from operating (an) underground storage tank(s).

Underground storage tanks at the following facilities are assured by this local government fund mechanism: [List for each facility: the name and address of the facility where tanks are assured by the local government fund].

[Insert: "The local government fund is funded for the full amount of coverage required under § 280.93, or funded for part of the required amount of coverage and used in combination with other mechanism(s) that provide the remaining coverage." or "The local government fund is funded for ten times the full amount of coverage required under § 280.93, or funded for part of the required amount of coverage and used in combination with other mechanisms(s) that provide the remaining coverage," or "A payment is made to the fund once every year for seven years until the fund is fully-funded and [name of local government owner or operator] has available bonding authority, approved through voter referendum, of an amount equal to the difference between the required amount of coverage and the amount held in the dedicated fund" or "A payment is made to the fund once every year for seven years until the fund is fully-funded and I have attached a letter signed by the State Attorney General stating that (1) the use of the bonding authority will not increase the local government's debt beyond the legal debt ceilings established by the relevant state laws and (2) that prior voter approval is not necessary before use of the bonding authority"].

The details of the local government fund are as follows:

Amount in Fund (market value of fund at close of last fiscal year): ______

[If fund balance is incrementally funded as specified in § 280.107(c), insert:

Amount added to fund in the most recently completed fiscal year: ______

Number of years remaining in the pay-in period: ______]

A copy of the state constitutional provision, or local government statute, charter, ordinance or order dedicating the fund is attached.

I hereby certify that the wording of this letter is identical to the wording specified in 40 CFR 280.107(d) as such regulations were constituted on the date shown immediately below.

[Date]

[Signature]

[Name]

[Title]

[FR Doc. 93-2824 Filed 2-17-93; 8:45 am]

BILLING CODE 6560-50-P

APPENDIX J

Minimum Required Amounts of Financial Responsibility for Petroleum Underground Storage Tanks

(40 C.F.R. § 280.93)

MINIMUM AMOUNT	CLASS OF OWNER OR OPERATOR
a) Per Occurrence	
$1 million	Tanks located at petroleum marketing facilities, or which handle an average of more than 10,000 gallons of product per month
$500,000.00	All other petroleum USTs
b) Annual Aggregate Amounts	
$1 million	1 to 100 petroleum USTs
$2 million	101 or more petroleum USTs

APPENDIX K

State Financial Assurance Funds

State	Agency with Primary Authority
Alaska	Department of Environmental Conservation
Alabama	Department of Environmental Management
Arizona	Department of Environmental Quality
California	Waste Resources Control Board
Colorado	Department of Health
Connecticut	Department of Environmental Protection
Delaware	Department of Natural Resources and Environmental Control
Florida	Department of Environmental Regulation
Georgia	Department of Natural Resources
Iowa	Underground Storage Tank Board
Idaho	State Insurance Fund
Illinois	State Comptroller
Indiana	Department of Environmental Management
Kansas	Department of Health and Environment
Kentucky	Petroleum Storage Tank Environment Assurance Fund Commission
Louisiana	Department of Environmental Quality
Massachusetts	Department of Public Safety
Maine	Department of Environmental Protection
Michigan	Department of Management and Budget
Minnesota	Department of Commerce
Missouri	Department of Natural Resources

State	Agency with Primary Authority
Mississippi	Department of Environmental Quality
Montana	Petroleum Tank Release Compensation Board
North Carolina	Department of Environment, Health and Natural Resources
North Dakota	Commissioner of Insurance
Nebraska	Department of Environmental Control
New Hampshire	Department of Environmental Services
New Mexico	Environment Department
Nevada	Division of Environmental Protection
New York	Office of the State Comptroller
Ohio	Petroleum Underground Storage Tank Release Compensation Board
Oklahoma	Oklahoma Corporation Commission
Pennsylvania	Insurance Department
South Carolina	Department of Health and Environmental Control
South Dakota	Department of Commerce and Regulation
Tennessee	Department of Environment and Conservation
Texas	Texas Water Commission
Utah	Department of Environmental Quality
Virginia	State Water Control Board
Vermont	Department of Environmental Conservation
Washington	Pollution Liability Insurance Agency
Wisconsin	Department of Industry, Labor and Human Relations
Wyoming	Department of Environmental Quality

APPENDIX L

U.S. EPA Penalty Guidance for Violations of UST Regulations (November 1990) and Guidance for Federal Field Citation Enforcement (October 6, 1993)

U.S. EPA PENALTY GUIDANCE FOR VIOLATIONS OF UST REGULATIONS

November 1990

Office of Underground Storage Tanks
U.S. Environmental Protection Agency

NOTICE

The procedures set forth in this document are intended solely for the guidance of the U.S. EPA. They are not intended, and cannot be relied on, to create rights, substantive or procedural, enforceable by any party in litigation with the United States government. The U.S. EPA reserves its right to act at variance with this guidance and to change it at any time without public notice.

TABLE OF CONTENTS

TABLE OF CONTENTS
(Continued)

APPENDICES

LIST OF EXHIBITS

CHAPTER 1. INTRODUCTION TO UST PENALTY GUIDANCE

This document provides guidance to U.S. Environmental Protection Agency (EPA) Regional Offices on calculating civil penalties against owner/operators of underground storage tanks (USTs) who are in violation of the UST technical standards and financial responsibility regulations. The methodology described in this guidance seeks to ensure that UST civil penalties, which can be as high as $10,000 for each tank for each day of violation, are assessed in a fair and consistent manner, and that such penalties serve to deter potential violators and assist in achieving compliance.

This penalty document is part of a series of enforcement documents which includes: (1) the Agency's UST/LUST Enforcement Procedures Guidance Manual (OSWER Directive 9610.11, July 1990), which provides guidance to U.S. EPA Regional personnel on taking enforcement actions against violations of the UST technical requirements; and (2) the draft "Interim Enforcement Response Strategy for Violations of UST Financial Responsibility Requirements," which provides guidance on taking enforcement actions against violations of the financial responsibility requirements. Although these enforcement documents are intended primarily for U.S. EPA Regional enforcement staff, State and local UST implementing agencies may find it useful to adapt some of the concepts and methodologies for their own UST enforcement programs.

This chapter briefly describes the U.S. EPA's authorities for taking enforcement action and assessing civil penalties. It also provides an overview of the enforcement actions that may be taken in response to UST violations, and indicates how the assessment of penalties fits into the enforcement framework.

1.1 U.S. EPA PENALTY AUTHORITY

The U.S. EPA's authority for assessing civil penalties for violations of UST requirements is provided by Subtitle I of the Resource Conservation and Recovery Act (RCRA). Under the Hazardous and Solid Waste Amendments of 1984, Congress added Subtitle I to RCRA in response to the growing environmental and health problems created by releases from USTs. The statutory framework for the national UST program is set forth in Sections 9002 through 9004 of Subtitle I.

Under Section 9006 of Subtitle I, EPA is authorized to take enforcement actions and assess penalties against violators of requirements promulgated under Subtitle I, including technical standards and financial responsibility requirements.[1] In particular, Section 9006(a) provides the authority to issue administrative orders requiring compliance within a reasonable specified time period. All such orders will be processed within the Agency according to the Consolidated Rules of Practice (CROP).[2] Pursuant to Section 9006(d), a Section 9006 compliance order may assess a civil penalty, provided that the penalty does not exceed $10,000 for each tank for each day of violation of the technical standards

---

[1] These are contained in two separate rules: the UST Technical Standards Rule, 40 CFR Part 280, Subparts A through G (promulgated September 23, 1988) and the UST Financial Responsibility Rule, 40 CFR Part 280, Subpart H (promulgated October 26, 1988).

[2] 40 CFR Part 22, "The Consolidated Rules of Practice Governing the Administrative Assessment of Civil Penalties and the Revocation or Suspension of Permits." The CROP was extended to cover administrative enforcement actions under Section 9006 (see 53 FR 5373, February 24, 1988).

and financial responsibility rules.[3] This document presents guidance for determining the appropriate civil penalty amount for an administrative complaint and order, and discusses use of penalties in field citations.

In addition to administrative enforcement actions, EPA may initiate judicial enforcement actions under Section 9006 to compel compliance with Subtitle I's statutory and regulatory requirements. EPA's judicial enforcement actions are processed through Federal courts and are reserved for violations of administrative orders. Under such actions, EPA is authorized to seek judicial penalties of up to $25,000 for each day of continued noncompliance with an administrative order issued under Section 9006 or a corrective action order issued under Section 9003. In these cases, Agency personnel should seek the maximum penalty.[4]

1.2 OVERVIEW OF THE UST ENFORCEMENT PROCESS

The UST/LUST Enforcement Procedures Guidance Manual (OSWER Directive 9610.11, July 1990) describes the range of enforcement actions that may be taken in response to an UST violation. These enforcement options vary from initial responses, such as warning letters or notices of violation (NOVs), which encourage compliance, to more stringent actions, such as administrative orders and judicial injunctions, which compel compliance and, if appropriate, penalize violators. Exhibit 1 presents the various enforcement actions that may be taken once a violation of an UST requirement is identified. In general, enforcement personnel will take the least costly enforcement action that appears necessary to achieve compliance and create a strong deterrent, and will escalate the severity of the enforcement response if the initial action fails.

As shown in Exhibit 1, there are two approaches to taking enforcement actions. Under the "traditional" approach, enforcement personnel may initially respond to a discovered violation by issuing a warning letter or NOV to inform the owner/operator of the violation, explain what actions need to be taken, and indicate possible consequences if the owner/operator fails to achieve compliance. If necessary, enforcement personnel may then meet with the owner/operator to negotiate an agreed-upon course of action for the owner/operator to follow to achieve compliance. However, for recalcitrant violators, or where violations pose a threat to human health and the environment, enforcement personnel will typically issue administrative complaints or take judicial action. To provide a deterrent effect, an administrative complaint may include an initial penalty target figure. Upon receipt of the complaint, a violator may pay the penalty specified, request an informal settlement conference, and/or request an administrative hearing. Regardless of the violator's response, the outcome generally will be a final penalty that the violator must pay or else face judicial prosecution. Exhibit 1 shows where the target and final penalties appear in the enforcement process.

As an alternative to the traditional approach, enforcement personnel may initiate an enforcement response using field citations (see Chapter 5). Field citations, similar to traffic tickets, are modified compliance orders issued by inspectors on-site at a facility when violations are discovered. However, the use of field citations is generally limited to first-time violators when compliance is expected and when the violation does not pose an immediate threat to human health and the environment. A typical

[3] This $10,000 limit also applies to violations of the Interim Prohibition provisions and any requirement of an approved State program. For violations of the May 1985 (statutory) notification requirements, the penalty may not exceed $10,000 for each tank.

[4] This guidance is in no way intended to limit the penalty amounts sought in civil judicial actions. In settling judicial cases, however, the Agency may use the narrative penalty assessment criteria set forth in this guidance to determine or justify the penalty amount that the Agency agrees to accept in settlement.

Exhibit 1

Overview of Enforcement Response Options

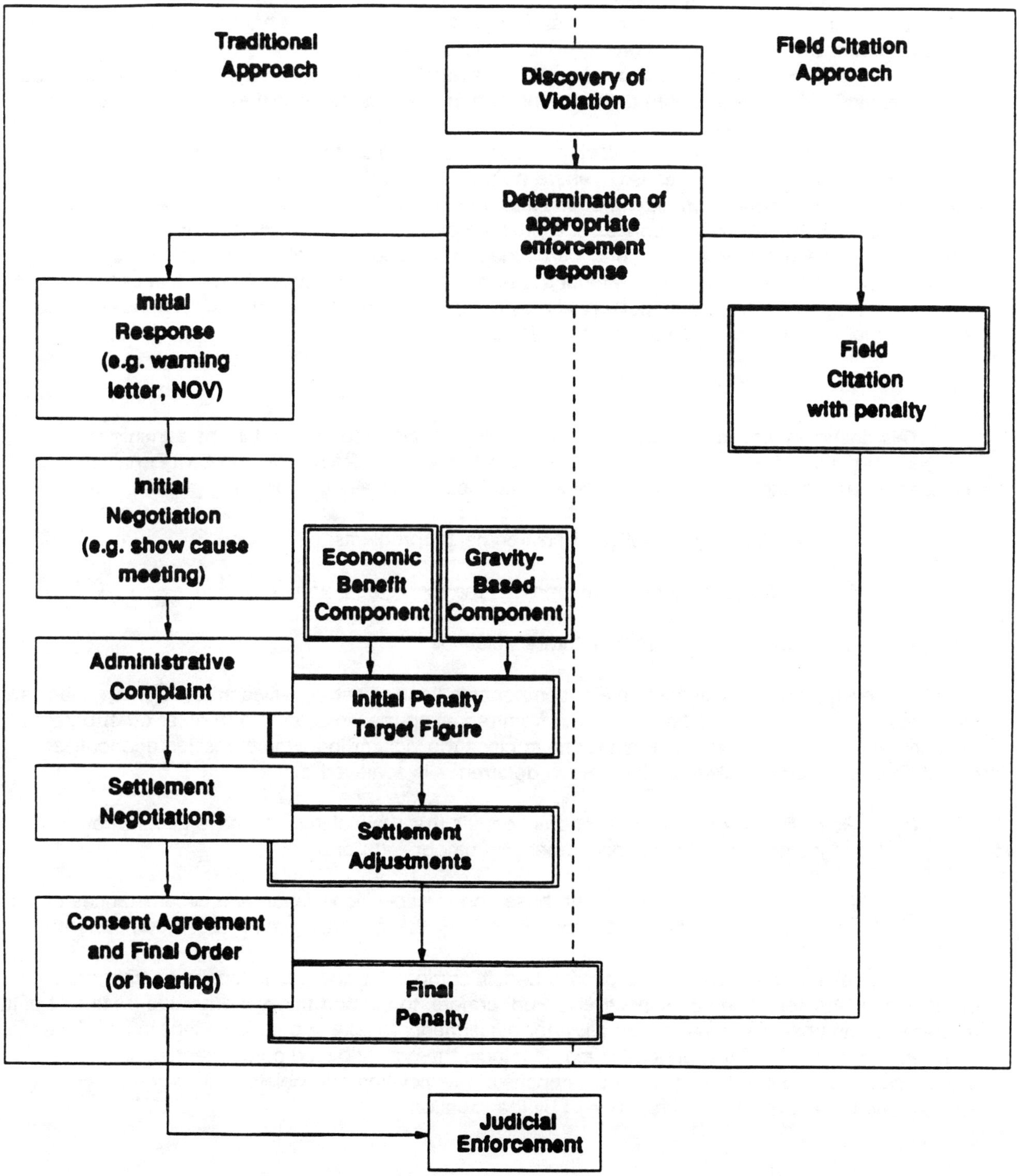

NOTE: This exhibit presents an overview of enforcement options only, and does not mandate a certain order of action. Actual enforcement actions may begin at any point in the process.

f0n003-3

field citation will not only require that the violator take actions to achieve compliance, but will also assess a pre-established, non-negotiable penalty. This penalty is usually fairly low (e.g., $100) to encourage prompt payment and response. In paying the citation penalty, the violator gives up the right to appeal and consents to the requirements specified; thus, the citation is analogous to the final penalty that results from settlement negotiations. This alternative path to arriving at a penalty is also shown in Exhibit 1. If the owner/operator fails to respond to the field citation, enforcement personnel may resort to enforcement actions under the traditional approach or may initiate judicial actions.

Under the UST program's franchise approach, States will undertake most of the enforcement actions. However, in certain cases (e.g., where an owner/operator is particularly recalcitrant or the State lacks sufficient enforcement authority), Federal assistance may be needed. In such cases, the Regional office may omit initial, informal responses and proceed directly with administrative or judicial actions. However, U.S. EPA enforcement also may be needed at the beginning of an enforcement case in certain circumstances (e.g., in States without active enforcement programs or on Indian Lands). In such cases, Regional enforcement personnel may begin with either the traditional responses or may determine that it is appropriate to use field citations.

1.3 UST PENALTY ASSESSMENT FRAMEWORK

This document provides guidance on calculating penalties to be used in the administrative enforcement actions described above. Consistent with the U.S. EPA's Policy on Civil Penalties, penalties assessed under this methodology are intended to achieve the following goals:[5]

- Encourage timely resolution of environmental problems;
- Support fair and equitable treatment of the regulated community; and
- Deter potential violators from future violations.

Exhibit 2 provides an overview of the major components used to set penalties at levels that will achieve these goals. Specifically, to deter the violator from repeating the violation and to deter other potential violators from failing to comply, the penalty must place the violator in a worse position economically than if he or she had complied on time. Such deterrence is achieved by:

(1) Removing any significant economic benefit that the violator may have gained from noncompliance (the "economic benefit component"); and

(2) Charging an additional amount, based on the specific violation and circumstances of the case, to penalize the violator for not obeying the law (the "gravity-based component").

The procedures for determining the economic benefit component and gravity-based component are discussed in Chapters 2 and 3, respectively. Furthermore, to support fair and equitable treatment of the regulated community, the penalty must allow for adjustments to take into account legitimate differences between similar cases. Thus, under this methodology, the gravity-based component incorporates adjustments that reflect the specific circumstances of the violation, the violator's background and actions, and the environmental threat posed by the situation.

---

[5] The "EPA Policy on Civil Penalties" (EPA General Enforcement Policy #GM-21, February 1984) and the "Framework for Statute-Specific Approaches to Penalty Assessment" (EPA General Enforcement Policy #GM-22, February 1984) establish a consistent Agency-wide approach to the assessment of civil penalties.

Exhibit 2

Process for Assessing UST Civil Penalties

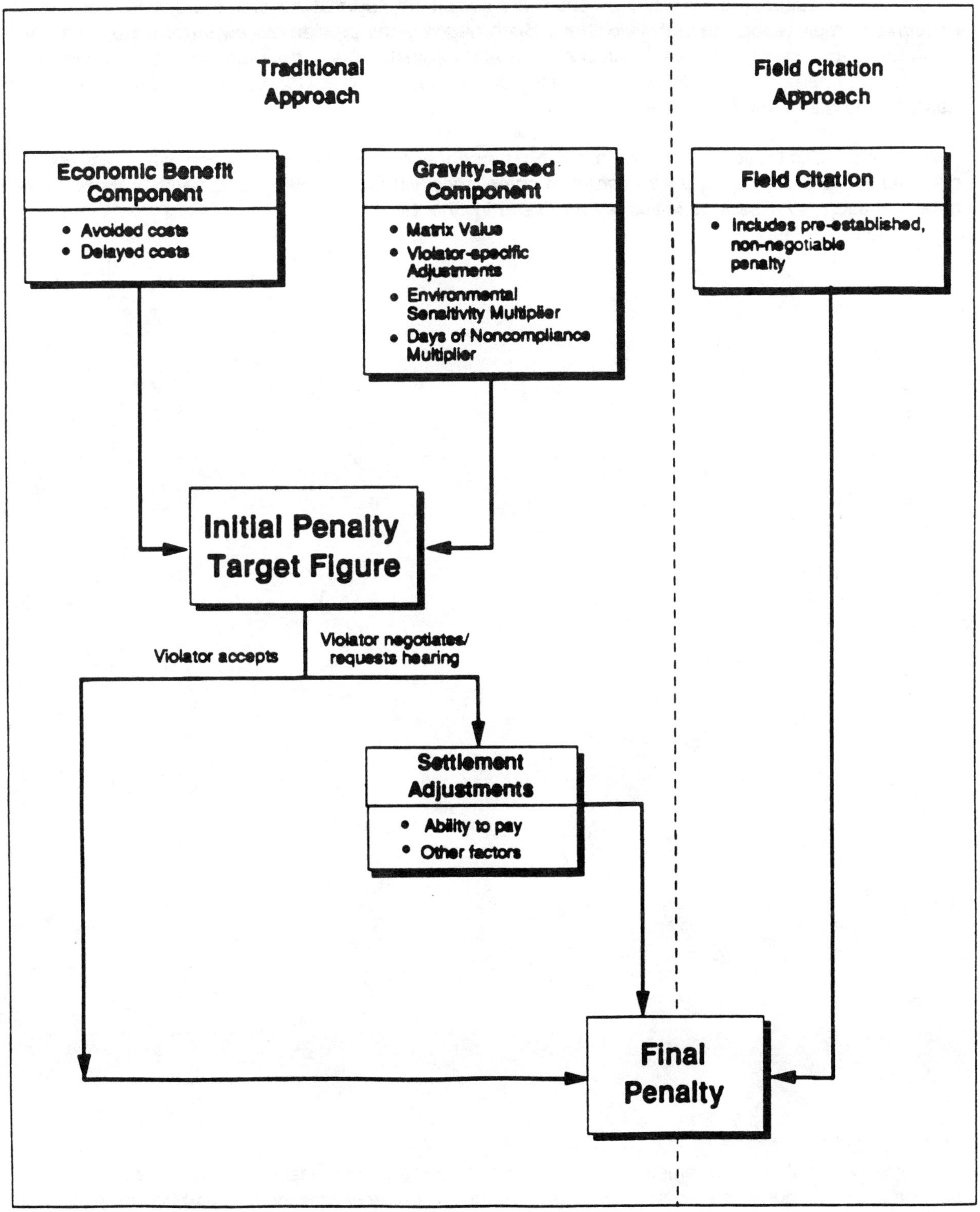

10n003-2

The sum of the economic benefit component and the gravity-based component yields the initial penalty target figure that is assessed in the administrative complaint.[6] For each case that involves more than one violation, the Regional case team will need to decide on the number of counts addressed in the complaint. Each count should be accompanied by an appropriate penalty calculation, and the sum of these penalties will be the initial penalty target figure assessed in the complaint. Once a complaint is issued, the Agency may enter into settlement negotiations with the owner/operator to encourage timely resolution of the violation. Such negotiations provide the owner/operator with the opportunity to present evidence to support downward adjustments in the penalty. The process of adjusting the penalty during settlement negotiations is addressed in Chapter 4. The outcome of such negotiations will be the final penalty.

For specific types of cases, enforcement personnel may issue field citations, which assess penalties while encouraging a swift return to compliance without a drawn-out appeals process. The use of field citations to assess penalties is addressed in Chapter 5.

[6] However, it should be remembered that the sum of the gravity-based component plus the economic benefit component cannot be greater than the statutory maximum of $10,000 for each tank for each day of violation of the technical standards and financial responsibility regulations.

Chapter 2

Determining the Economic Benefit Component

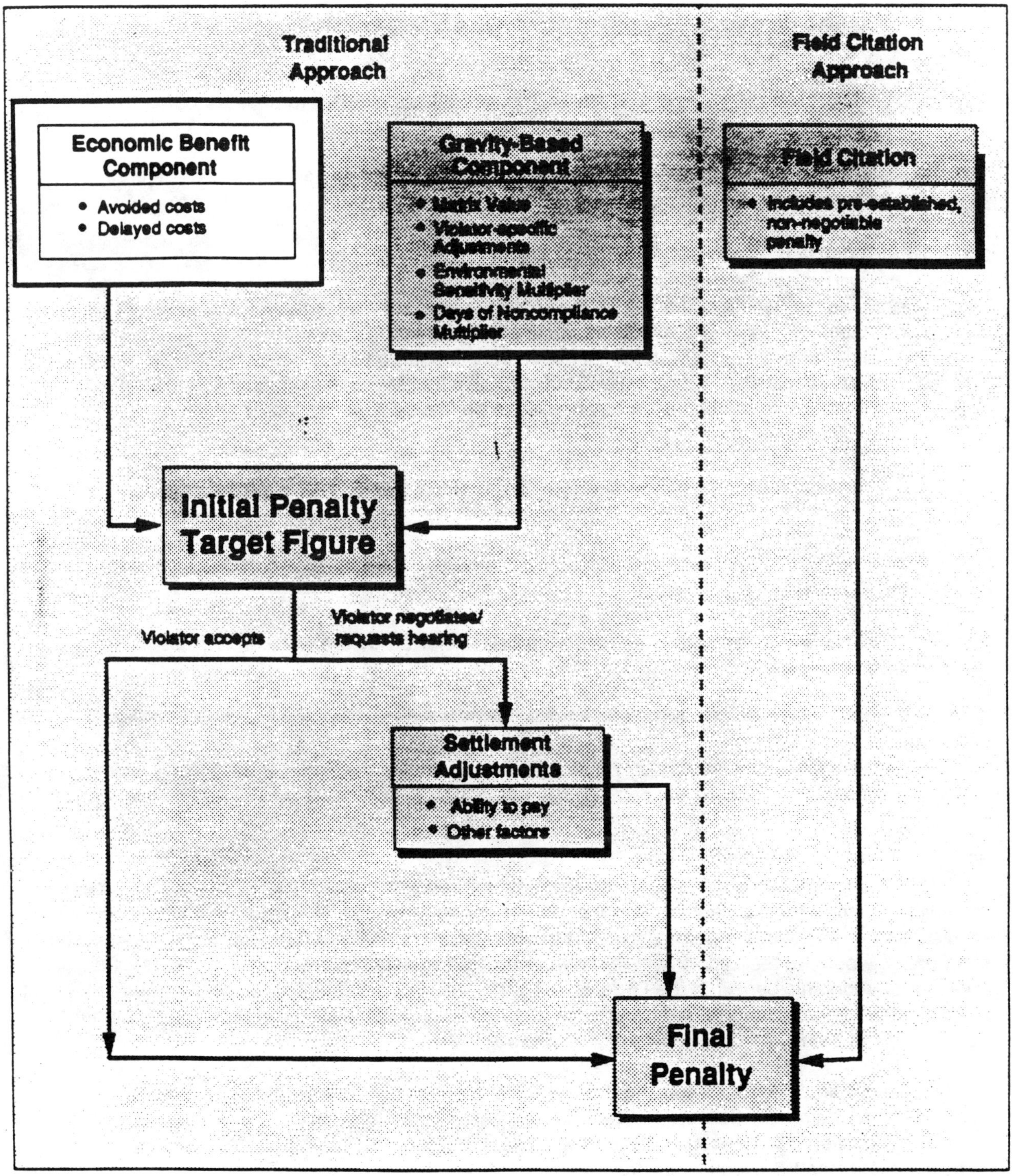

f0n003-4

CHAPTER 2. DETERMINING THE ECONOMIC BENEFIT COMPONENT

As explained in the preceding chapter, to ensure that the penalty deters potential violators, the initial penalty target figure assessed in the complaint must include two fundamental components:

- Economic Benefit Component, which removes any significant profit from noncompliance; and
- Gravity-Based Component, which imposes an assessment to penalize current and/or past noncompliance.

This chapter discusses the process for determining the economic benefit component. The gravity-based component is discussed in Chapter 3.

2.1 DEFINITION OF ECONOMIC BENEFIT COMPONENT

The economic benefit component represents the economic advantage that a violator has gained by delaying capital and/or non-depreciable costs and by avoiding operational and maintenance costs associated with compliance.[7] The total economic benefit component is based on the benefit from two sources: (1) avoided costs; and (2) delayed costs. All penalties assessed must include the full economic benefit unless the benefit is determined to be "incidental," i.e., less than $100.

Economic Benefit Component = Avoided Costs + Delayed Costs

Avoided costs are the periodic, operation and maintenance expenditures that should have been incurred, but were not.

Delayed costs are the expenditures that have been deferred by the violation, but will be incurred to achieve compliance.

The Agency-wide penalty policy prescribes the use of two methods for calculating a violator's economic benefit from noncompliance:[8] (1) the rule-of-thumb approach; and (2) the software program

[7] This policy does not outline a methodology for the recovery, as a measure of economic benefit, of profits proximately attributable to illegal or non-compliant activities. Because the Federal UST regulations do not include a permitting process, the Agency is not presently aware of situations where such profits would be realized, or where we would expect to seek recovery of such profits as a measure of economic benefit in the Federal UST program. Should EPA determine that the recovery of such profits is appropriate in a particular case, the Agency will calculate such profits in a manner consistent with the RCRA Civil Penalty Policy (October 1990).

[8] Revised guidelines for calculating the economic benefit from noncompliance are incorporated into a memorandum from Courtney Price (Assistant Administrator for Enforcement and Compliance Monitoring) entitled, "Guidance for Calculating the Economic Benefit of Noncompliance for a Civil Penalty Assessment" (November 5, 1984).

called BEN.[9] The rule-of-thumb approach (described in the sections that follow) should be used for making an initial estimate of the economic benefit of noncompliance. If the initial estimate is less than $10,000, the rule-of-thumb calculation may be used as a basis for the economic benefit assessed in the penalty. If, however, the estimate indicates that the economic benefit is greater than $10,000, the BEN model should be used. The BEN model should also be used if the violator rejects the rule-of-thumb calculation.

The BEN model, which is accessible by computer from anywhere in the country, uses a financial analysis technique known as "discounting" to determine the net present value of economic gains from noncompliance. BEN determines the economic benefit for an individual violator based on 12 specific factors, or inputs, including the violator's initial capital investment, nondepreciable expenditures, and operation and maintenance costs. For some inputs, such as income tax rate, annual inflation rate, and discount rate, BEN will provide standard values if the user does not have actual figures. This use of standard values allows for national consistency in determining economic benefit. Because the majority of UST violations will be associated with an economic benefit of less than $10,000, the rule-of-thumb approach will be used in most cases.

The procedures for calculating the economic benefit of noncompliance using the rule-of-thumb approach are described below. Because of the fundamental differences between avoided and delayed costs, the process for determining the economic benefit component will depend on the type of cost involved. The sections that follow describe methods for calculating each type of cost.

2.2 AVOIDED COSTS

Avoided costs are the operation and maintenance expenditures that are averted by the violator's failure to comply. These are considered to be avoided because they will never be incurred even if the violator comes into compliance. For example, a violator who has failed to maintain product inventory records in the past never will have to make up for the costs saved, even if he is directed to start maintaining inventory records now. Other examples of avoided costs include: (1) failure to conduct a required periodic test; (2) failure to obtain financial assurance by the phase-in date; and (3) failure to conduct periodic maintenance of equipment. The violator's benefit from avoided costs is generally expressed as the avoided expenditures plus the interest potentially earned on the money not spent.

DETERMINING AVOIDED COSTS

$$\text{Avoided Costs} = \text{Avoided Expenditures} + \frac{\text{Avoided Expenditures} \times \text{Interest} \times \text{Number of Days}}{365 \text{ Days}} \times (1 - \text{Marginal Tax Rate})$$

Avoided Expenditures are estimated using local, comparable costs.

Interest is the equity discount rate provided in the BEN model (currently 18.1 percent).

Number of Days is from the date of noncompliance to the date of compliance.

365 Days is the number of days in a year.

Marginal Tax Rate is based on corporate tax rates or financial responsibility compliance class.

[9] For information, contact the BEN/ABEL Coordinator in the Office of Enforcement at the U.S. EPA Headquarters by phoning (202) 475-6777 or FTS 475-6777.

To determine the value of the interest, compounded annually, the equity discount rate should be used. This represents the risk-free rate (T-bill) plus the cost of financing for pollution control equipment. This rate can be obtained by calling the EPA Office of Enforcement or by accessing the BEN computer model.[10] As of the beginning of FY91, the equity discount rate was 18.1 percent. When used in the formula, this number should be expressed as a decimal and not a percentage (e.g., 0.181, instead of 18.1%).

The marginal tax rate (MTR) used in calculating the avoided costs will vary depending on the size of the business. Exhibit 3 provides a list of appropriate tax rates based on the facility or company's taxable income. As with the interest rate, this number should be expressed as a decimal, not a percentage (e.g., 0.15 instead of 15%). To determine the taxable income, enforcement staff should contact EPA's National Enforcement Investigations Center (NEIC) to determine whether the business in violation is listed in the Dun and Bradstreet Business Information Report data base.[11] The data base provides information on the annual incomes of a large number of companies across the country, including the smaller, "Mom and Pop" businesses. Although most of the incomes listed in the data base are those reported to Dun and Bradstreet, the data base also includes some estimated incomes for companies that have not reported.

If information on annual income cannot be obtained from NEIC, enforcement staff may use the company's financial responsibility compliance class as a basis for determining the appropriate marginal tax rate, as follows:

MARGINAL TAX RATES BASED ON FINANCIAL RESPONSIBILITY COMPLIANCE CLASS

Compliance Class [a]	Tax Rate
FR Classes 1 & 2	0.34 (34%)
FR Class 3	0.25 (25%)
FR Class 4	0.15 (15%)

[a] Compliance class is determined as follows: Class 1 - large petroleum marketing firms with 1,000 or more USTs or any firm with net worth over $20 million; Class 2 - large and medium-sized petroleum marketing firms with 100 to 999 USTs; Class 3 - smaller petroleum marketing firms with 13 to 99 USTs; and Class 4 - very small marketing firms with 1 to 12 USTs or less than 100 USTs at one site, all other firms with net worth of less than $20 million, and municipalities.

In the absence of specific information on the violator's FR compliance class, enforcement staff should assume that the violator is in FR Class 4 (which will result in the highest penalty).

---

[10] To obtain the equity discount rate from the Office of Enforcement, or to access BEN, call the BEN/ABEL coordinator at (202) 475-6777 or FTS 475-6777.

[11] For information from the Dun and Bradstreet data base call NEIC at (303) 236-3219 or FTS 8-776-3219. Using information on the violator's name and location (city and State), NEIC staff can search the data base for information on the company's annual income.

Exhibit 3

Applicable Tax Rates for Determining Avoided Costs

MARGINAL TAX RATE BASED ON FEDERAL CORPORATE TAX RATES
(from 1989 U.S. Master Tax Guide):

Taxable income over	Not over	Tax rate
$0	$50,000	15%
$50,000	$75,000	25%
$75,000	$100,000	34%
$100,000	$335,000	39%*
$335,000	....	34%

*An additional 5% tax is applied to income between $100,000 and $335,000 to phase out the benefits of the graduated rates in that income range.

The marginal tax rate is applied to each increment of income specified above (e.g., for an income of $75,000, 15% is applied to the first $50,000 and 25% to the next $25,000). The weighted average tax rates below have been calculated for each $10,000 increment in income to reflect the actual tax burden at each income level. These values will facilitate the determination of penalty amounts by eliminating the need to calculate the tax burden on each increment of marginal taxable income. To find the weighted tax rate, round the estimated taxable income to the nearest $10,000 and use the tax rate indicated in the table.

WEIGHTED AVERAGE TAX RATES BY INCOME LEVEL**

Taxable Income not greater than	Tax Rate	Taxable Income not greater than	Tax Rate
$50,000	0.15	$200,000	0.31
$60,000	0.17	$210,000	0.31
$70,000	0.18	$220,000	0.31
$80,000	0.19	$230,000	0.32
$90,000	0.21	$240,000	0.32
$100,000	0.22	$250,000	0.32
$110,000	0.24	$260,000	0.33
$120,000	0.25	$270,000	0.33
$130,000	0.26	$280,000	0.33
$140,000	0.27	$290,000	0.33
$150,000	0.28	$300,000	0.33
$160,000	0.29	$310,000	0.34
$170,000	0.29	$320,000	0.34
$180,000	0.30	$330,000	0.34
$190,000	0.30	≥ $340,000	0.34

**This table includes the additional 5% tax applied to incomes between $100,000 and $335,000.

2.3 DELAYED COSTS

Delayed costs are the capital expenditures and one-time non-depreciable costs that have been deferred because the violator failed to comply with the requirements. Examples of delayed costs include: (1) failure to install required equipment, such as cathodic protection; and (2) failure to clean up a spill. These expenditures are considered only to be delayed, and not avoided altogether, because the violator will eventually have to incur these costs to come into compliance. The benefit from delayed costs is generally expressed as only the return on investment that could have been earned on the money not spent.

DETERMINING DELAYED COSTS

$$\text{Delayed Costs} = \frac{\text{Delayed Expenditures} \times \text{Interest} \times \text{Number of Days}}{\text{365 Days}}$$

Delayed Expenditures are estimated using local, comparable costs.

Interest is the equity discount rate used in the BEN model (currently 18.1 percent).

Number of Days is from the date of noncompliance to the date of compliance.

365 Days is the number of days in a year.

For delayed costs there is no computation of the tax rate. Although there may be a modest tax consequence for the violator because of delayed costs, this effect was deemed to be insignificant. Furthermore, such a tax consequence only would be incurred if the violation were to span more than one of the violator's tax years.

Chapter 3

Determining the Gravity-Based Component

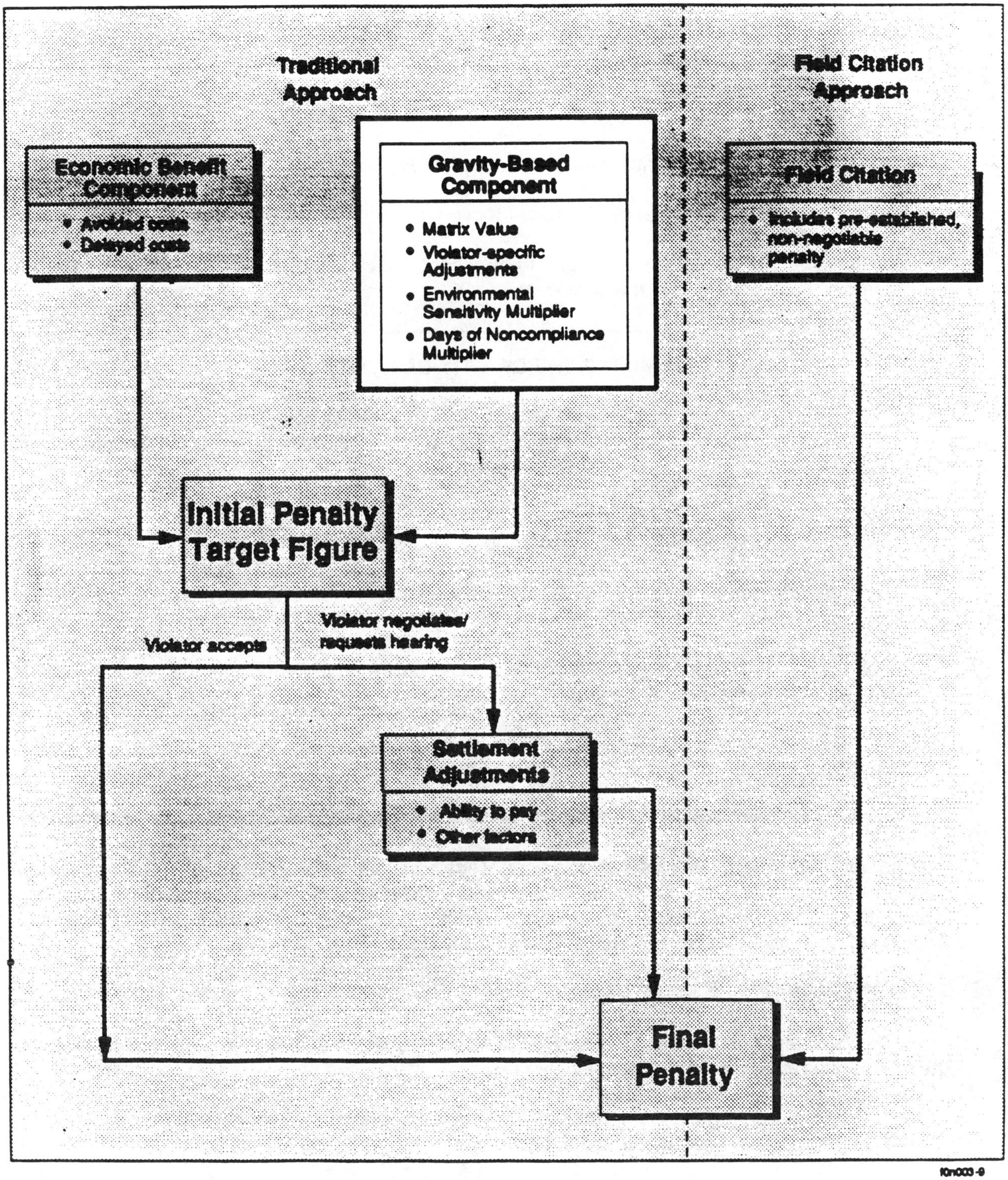

CHAPTER 3. DETERMINING THE GRAVITY-BASED COMPONENT

The second component of a penalty, and the one that serves to deter potential violators, is the gravity-based component. The purpose of the gravity-based component is to ensure that violators are economically disadvantaged relative to owner/operators of those facilities in compliance, and to penalize current and/or past noncompliance. The gravity-based component consists of four elements:

- Matrix Value (Section 3.1);
- Violator-Specific Adjustments to the Matrix Value (Section 3.2);
- Environmental Sensitivity Multiplier (Section 3.3); and
- Days of Noncompliance Multiplier (Section 3.4).

The gravity-based component is then added to the economic benefit component to arrive at the initial penalty target figure assessed in the complaint.

DETERMINING THE GRAVITY-BASED COMPONENT

Gravity-Based Component	=	Matrix Value	x	Violator-Specific Adjustments	x	Environmental Sensitivity Multiplier	x	Days of Noncompliance Multiplier

Matrix Value is based on potential for harm and deviation from the requirement.

Violator-Specific Adjustments to the matrix value are based on violator's cooperation, willfulness, history of noncompliance, and other factors.

Environmental Sensitivity Multiplier (ESM) is a value based on the environmental sensitivity associated with the location of the facility.

Days of Noncompliance Multiplier (DNM) is a value based on the number of days of noncompliance.

If the complaint results in settlement negotiations, certain factors used to adjust the matrix value may be re-assessed during negotiations to determine whether a downward adjustment in the gravity-based component is appropriate. In general, it is the violator's responsibility to provide evidence in support of reducing the penalty assessment during the settlement stage (see Chapter 4).

3.1 DETERMINING THE MATRIX VALUE

The first step in determining the gravity-based component is determining the initial matrix value. The matrix value is based on the following two criteria:

- Extent of deviation from requirement - An assessment of the extent to which the violation deviates from the UST statutory or regulatory requirements.

- Actual or potential harm - An assessment of the likelihood that the violation could (or did) result in harm to human health or the environment and/or has (or had) an adverse effect on the regulatory program.

A matrix has been developed in which these two criteria form the axes (Exhibit 4). Three gravity levels apply to each of these criteria -- major, moderate, and minor -- and form the grid of the matrix. Thus, the matrix has nine cells, each of which contains a penalty amount. The specific cell to be used in determining the matrix value is identified by selecting a gravity level for both factors. As a guide to determining the appropriate gravity level, Appendix A provides a list of selected violations of the Federal UST requirements and the associated deviation from the requirements and potential for harm.

Based on the type of violation (see Appendix A), penalties will be assessed on a per-tank basis if the specific requirement or violation is clearly associated with one tank (e.g., tank upgrading). If the requirement addresses the entire facility (e.g., recordkeeping practices), the penalty will be assessed on a per-facility basis. For requirements that address piping, the unit of assessment will depend on whether the piping is associated with one tank or with more than one tank. Appendix A indicates the suggested unit of assessment for specific violations.

3.1.1 Extent of Deviation from Requirements

The first factor in determining the matrix value is the extent of deviation from the requirements. The categories for extent of deviation from the requirements are the following:

- Major - The violator deviates from the requirements of the regulation or statute to such an extent that there is substantial noncompliance. An example is installing a bare steel tank without cathodic protection.

- Moderate - The violator significantly deviates from the requirement of the regulation or statute, but to some extent has implemented the requirement as intended. An example is installing improperly constructed cathodic protection.

- Minor - The violator deviates slightly from the regulatory or statutory requirements, but most of the requirements are met. An example is failing to keep every maintenance record on properly constructed cathodic protection.

3.1.2 Potential for Harm

The second criterion for determining the matrix value of a violation is the extent to which the owner/operator's actions resulted in, or were likely to result in, a situation that could cause harm to human health or the environment. When determining this factor, it is the potential in each situation that is important, not solely whether the harm has actually occurred. Violators should not be rewarded with lower penalties simply because no harm has occurred. The potential extent of this harm, if it were to occur, is addressed by the environmental sensitivity multiplier, discussed in Section 3.3 of this chapter.

The potential-for-harm factor will also be applied to violations of administrative requirements (e.g., recordkeeping and notification requirements) that are integral to the regulatory program. For violations of these requirements, enforcement personnel should consider the "importance" of the requirement violated. For example, failure to submit tank notification data may be considered to have significant potential for harm because the Agency has few other sources of information on the location of USTs.

Exhibit 4

Matrix Values for Determining the Gravity-Based Component of a Penalty

Extent of Deviation from Requirement

Potential for Harm	Major	Moderate	Minor
Major	1,500	1,000	500
Moderate	750	500	250
Minor	200	100	50

NOTE: These amounts constitute the matrix value only. They are <u>not</u> the initial penalty target figure. The initial penalty target figure is calculated as follows:

$$\text{Initial Penalty Target Figure} = \text{Economic Benefit Component} + \left(\textit{MATRIX VALUE} \times \text{Violator-Specific Adjustments} \times \text{Environmental Sensitivity Multiplier} \times \text{Days of Noncompliance Multiplier} \right)$$

f0n003-1

For purpose of this guidance, the categories for potential for harm are the following:

- Major - The violation causes or may cause a situation resulting in a substantial or continuing risk to human health and the environment and/or may have a substantial adverse effect on the regulatory program. Examples are: (1) improperly installing a fiberglass reinforced plastic tank (because a catastrophic release may result); or (2) failing to provide adequate release detection by the specified phase-in date (because without release detection a release may go unnoticed for a lengthy period of time with detrimental consequences).

- Moderate - The violation causes or may cause a situation resulting in a significant risk to human health and the environment and/or may have a significant adverse effect on the regulatory program. An example would be installing a tank that fails to meet tank corrosion protection standards (because it could result in a release, although the use of release detection is expected to minimize the potential for continuing harm from the release).

- Minor - The violation causes or may cause a situation resulting in a relatively low risk to human health and the environment and/or may have a minor adverse effect on the regulatory program. An example would be failing to provide certification of UST installation (assuming that the installation was done correctly).

3.2 VIOLATOR-SPECIFIC ADJUSTMENTS

In general, adjustments to the matrix value may be made at both the pre-negotiation and settlement stages of penalty assessment to address the unique facts of each case and to resolve the case quickly. Prior to settlement negotiations, enforcement personnel have the discretion to use any relevant information to adjust the matrix value upwards or downwards. These adjustments are solely at the discretion of EPA enforcement personnel.

Specifically, to ensure that penalties are assessed in a fair and consistent manner, and take into account case-specific differences, enforcement personnel have the option of adjusting the matrix value based on any information known about the violator's: (1) degree of cooperation or noncooperation; (2) degree of willfulness or negligence; (3) history of noncompliance; and (4) other unique factors.

VIOLATOR-SPECIFIC ADJUSTMENTS TO THE MATRIX VALUE

Adjustment Factor	Range of Percentage Adjustment
✓ Degree of Cooperation/Noncooperation	Between 50% increase and 25% decrease
Degree of Willfulness or Negligence	Between 50% increase and 25% decrease
History of Noncompliance	Up to 50% increase only
Other Unique Factors	Between 50% increase and 25% decrease

The sections that follow discuss these four adjustment factors. In addition, the matrix value should be adjusted to reflect the environmental sensitivity and the days of noncompliance, which are discussed in Sections 3.3 and 3.4, respectively. Subsequent adjustments made during the settlement stage, including adjustments for inability to pay, are discussed in Chapter 4.

To ensure that the penalty maintains a deterrent effect, enforcement staff should consider adjustments toward increased penalties in all cases (i.e., make upwards adjustments to the matrix value). It is up to the violator to present information during settlement that mitigates use of such upward adjustments. However, to ensure that penalties are calculated fairly and consistently, any upwards adjustment may be made only if the circumstances of the case warrant such adjustments. Furthermore, for any adjustments made to the matrix value, justification must be provided on the penalty assessment worksheet (see Appendix B).

3.2.1 Degree of Cooperation/Noncooperation

The first factor that may be considered in adjusting the matrix value is the violator's cooperation or good faith efforts in response to enforcement actions. In adjusting for the violator's degree of cooperation or noncooperation, enforcement staff may consider making upward adjustments by as much as 50 percent and downward adjustments by as much as 25 percent of the matrix value.

In order to have the matrix value reduced, the owner/operator must demonstrate cooperative behavior by going beyond what is minimally required to comply with requirements that are closely related to the initial harm addressed. For example, an owner/operator may indicate a willingness to establish an environmental auditing program to check compliance at other UST facilities, if appropriate, or may demonstrate efforts to accelerate compliance with other UST regulations for which the phase-in deadline has not yet passed.[12] Because compliance with the regulation is expected from the regulated community, no downward adjustment may be made if the good faith efforts to comply primarily consist of coming into compliance. That is, there should be no "reward" for doing now what should have been done in the first place. On the other hand, lack of cooperation with enforcement officials can result in an increase of up to 50 percent of the matrix value.

3.2.2 Degree of Willfulness or Negligence

The second adjustment that may be made to the matrix value is for willfulness or negligence, which takes into account the owner/operator's culpability and intentions in committing the violation. In assessing the degree of willfulness or negligence, the following factors may be considered:

- How much control the violator had over events constituting the violation (e.g., whether the violation could have been prevented or was beyond the owner/operator's control, as in the case of a natural disaster);

- The foreseeability of the events constituting the violation;

- Whether the violator made any good faith efforts to comply and/or took reasonable precautions against the events constituting the violation; and

- Whether the violator knew or should have known of the hazards associated with the conduct; and

- Whether the violator knew of the legal requirement that was violated (resulting in an upward adjustment only).[13]

---

[12] For information on establishing environmental auditing programs, see "EPA Policy on the Inclusion of Environmental Auditing Provisions in Enforcement Settlements," U.S. EPA, Office of Enforcement and Compliance Monitoring, November 1986.

[13] Lack of knowledge of the legal requirements may not be used as a basis to reduce the matrix value. Rather, informed violation of the law should serve to increase the matrix value.

In certain circumstances, the amount of control that the violator has over how quickly the violation is remedied also can be relevant. Specifically, if correction of a violation is delayed by factors that the violator clearly can show were not reasonably foreseeable and out of his or her control, the penalty assigned for the duration of noncompliance may be reduced (see Section 3.4), although the original penalty for noncompliance should not be. In assessing the degree of willfulness, enforcement staff may consider making upward adjustments by as much as 50 percent and downward adjustments by as much as 25 percent of the matrix value.

3.2.3 History of Noncompliance

The third factor to be considered in adjusting the matrix value is the violator's history of noncompliance. Previous violations of any environmental regulation are usually considered clear evidence that the violator was not deterred by previous interaction with enforcement staff and enforcement actions. Unless the current violation was caused by factors entirely out of the control of the violator, prior violations should be taken as an indication that the matrix value should be adjusted upwards. When assessing the history of noncompliance, some of the factors that may be considered are:

- Number of previous violations;
- Seriousness of the previous violations;
- Time period over which previous violations occurred;
- Similarity of the previous violations;
- Enforcement tools utilized (e.g., whether the owner/operator's previous behavior required use of more stringent enforcement actions); and
- Violator's response to the previous violation(s) with respect to correction of the problem.

For purposes of this document, a "prior violation" includes any act or omission for which an accountable enforcement action has occurred (e.g., an inspection that found a violation, a notice of violation, an administrative or judicial complaint, or a consent order). A prior violation of the same or a related requirement would constitute a similar violation.

In cases of large corporations that have many divisions and/or subsidiaries, if the same corporation is involved in the current violation the adjustments for history of noncompliance will apply. In addition, enforcement staff should be wary of a company that changes operators or shifts responsibility for compliance to different persons or organizational units as a way of avoiding increased penalties. A consistent pattern of noncompliance by several divisions or subsidiaries of a corporation may be found, even though the facilities are at different locations. Again, in these situations, enforcement staff may make only upward adjustments to the matrix value by as much as 50 percent.

3.2.4 Other Unique Factors

This guidance allows an adjustment for unanticipated factors that may arise on a case-by-case basis. As with the previous factors, enforcement staff may want to make upward adjustments to the matrix value by as much as 50 percent and downward adjustments by as much as 25 percent for such reasons.

3.3 ENVIRONMENTAL SENSITIVITY MULTIPLIER (ESM)

In addition to the violator-specific adjustments discussed above, enforcement personnel may make a further adjustment to the matrix value based on potential site-specific impacts that could be caused by the violation. The environmental sensitivity multiplier takes into account the adverse environmental effects that the violation may have had, given the sensitivity of the local area to damage posed by a potential or actual release. This factor differs from the potential-for-harm factor (discussed in Section 3.1.2) which takes into account the probability that a release or other harmful action would occur because of the violation. The environmental sensitivity multiplier addressed here looks at the actual or potential impact that such a release, once it did occur, would have on the local environment and public health.

To calculate the environmental sensitivity multiplier, enforcement personnel must first determine the sensitivity of the environment. For purposes of this document, the environmental sensitivity will be either low, moderate, or high. Factors to consider in determining the appropriate sensitivity level include:

- Amount of petroleum or hazardous substance potentially or actually released (e.g., size of the tanks and number of tanks at the facility that were involved in the violation, as they relate to the potential volume of materials released);
- Toxicity of petroleum or hazardous substance released;
- Potential hazards presented by the release or potential release, such as explosions or other human health hazards;
- Geologic features of the site that may affect the extent of the release and may make remediation difficult;
- Actual or potential human or environmental receptors, including:
    - Likelihood that release may contaminate a nearby river or stream;
    - Number of drinking water wells potentially affected;
    - Proximity to environmentally sensitive areas, such as wetlands; and
    - Proximity to sensitive populations, such as children (e.g., in schools).
- Ecological or aesthetic value to environmentally sensitive areas.

Thus, a "low" sensitivity value may be given in a case where one tank containing petroleum is located in clay soil in a semi-residential area where all drinking water is supplied by municipal systems, and where little wildlife is expected to be affected. A moderate sensitivity value may be given if: several tanks were in violation; the geology of the site would allow for some movement of a plume of released substance; and several drinking water wells could have been affected. A high sensitivity value may be given if: a number of tanks (or very large tanks) were involved; there were several potential receptors of the released substance through drinking water wells or contact with contaminated surface water; and the contamination would be difficult to remediate. Each level of sensitivity is given a corresponding multiplier value, as provided below.

DETERMINING THE ENVIRONMENTAL SENSITIVITY MULTIPLIER

Environmental Sensitivity Multiplier (ESM) is based on the potential or actual environmental impact at the site, and is given a corresponding value as follows:

Environmental Sensitivity	ESM
Low	1.0
Moderate	1.5
High	2.0

3.4 DAYS OF NONCOMPLIANCE MULTIPLIER

The final adjustment that may be made to the matrix value takes into account the number of days of noncompliance. To determine the amount of the adjustment, locate the days of noncompliance multiplier (or DNM) in the table below that corresponds to the duration of the violation:

DETERMINING THE DAYS OF NONCOMPLIANCE MULTIPLIER

Days of Noncompliance Multiplier (DNM) is based on the number of days of noncompliance:

Days of Noncompliance	DNM
0 - 90	1.0
91 - 180	1.5
181 - 270	2.0
271 - 365	2.5
Each additional 6 months or fraction thereof	add 0.5

The DNM is then multiplied by the adjusted matrix value and environmental sensitivity multiplier to obtain the gravity-based component of the penalty, as follows:

DETERMINING THE GRAVITY-BASED COMPONENT

Gravity-Based Component = Matrix Value x Violator-Specific Adjustments x Environmental Sensitivity Multiplier x Days of Noncompliance Multiplier

The economic benefit component is added to the gravity-based component to form the initial penalty target figure to be assessed in the complaint. As discussed previously, this figure cannot exceed $10,000 for each tank for each day of violation.

Chapter 4

Settlement Adjustments

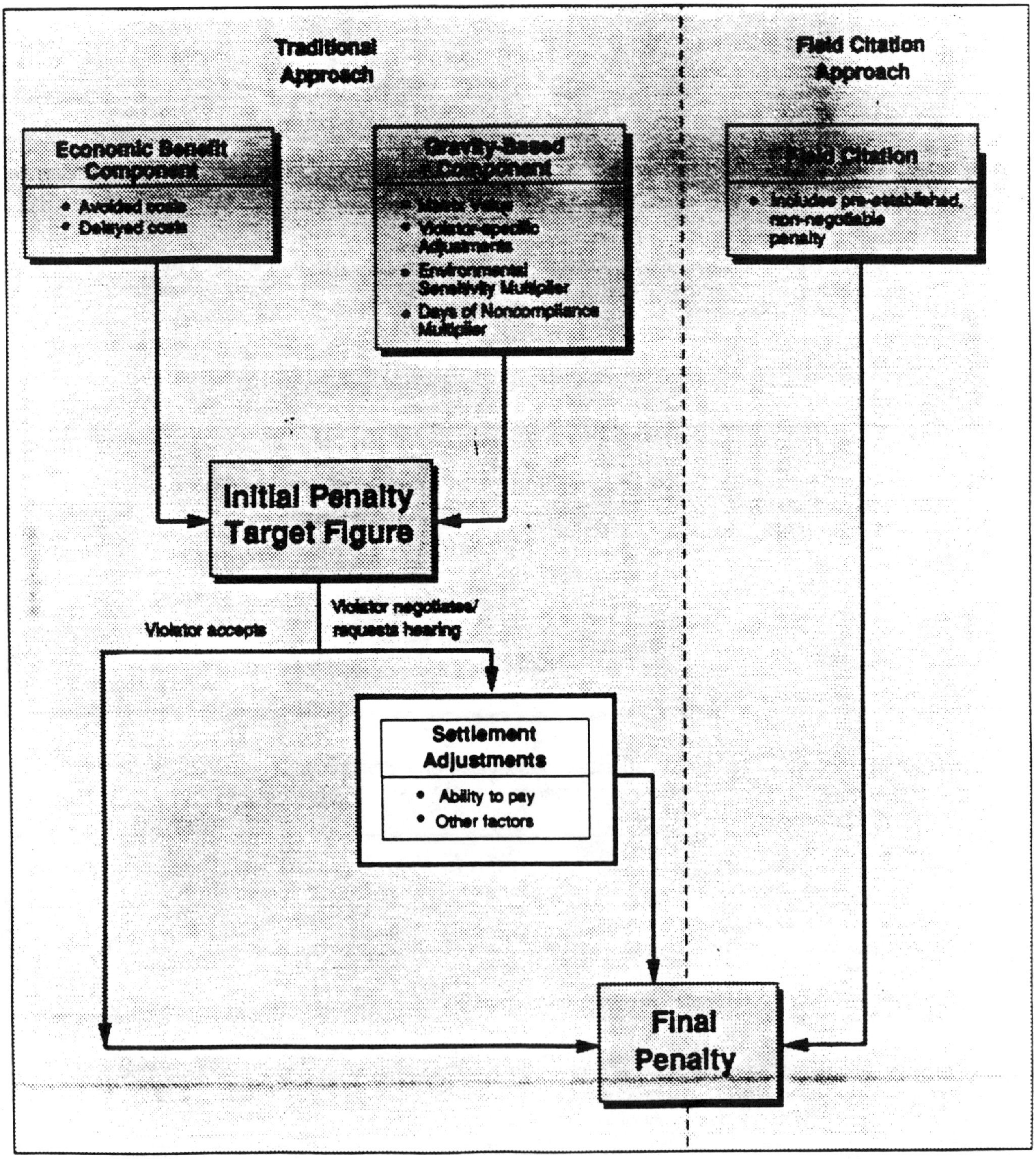

f0n003-7

CHAPTER 4. SETTLEMENT ADJUSTMENTS

After the initial penalty target figure has been presented to the potential violator in a complaint, additional adjustments may be made as part of a settlement compromise. All such adjustments are entirely within the discretion of Agency personnel. The burden is always on the owner/operator to provide evidence supporting any reduction of the penalty.

In response to a complaint, the owner/operator may request an informal conference and/or a hearing to settle the penalty and violation. The Federal Consolidated Rules of Practice (CROP) procedures for administrative actions at 40 CFR Part 22 provide for a settlement conference and a right to a public hearing, giving the owner/operator the opportunity to present data to support a penalty adjustment. At a minimum, enforcement personnel may consider adjustments based on the four violator-specific adjustment factors discussed in Chapter 3, including:

- Degree of cooperation/noncooperation;
- Degree of willfulness or negligence;
- History of noncompliance; and
- Other unique factors.

The settlement adjustment is usually not made to the economic benefit component unless new and better information about the economic benefits is made available. The Agency should maintain a record that includes a statement of the reasons for adjusting the penalty.

In addition to the adjustment factors listed above, and because of the nature of the UST regulated community, one factor that commonly will be discussed during negotiations is the owner/operator's inability to pay. An adjustment may need to be made for inability to pay to ensure fair and equitable treatment of the regulated community. It is important, however, that this reduction not allow the regulated community to regard violations of environmental requirements as a way to save money. Furthermore, a penalty should not be reduced when a violator refuses to correct a violation, has a history of noncompliance, or in cases with egregious violations, e.g., failure to abate a release that is contaminating drinking-water supplies.

The Agency should assume that the owner/operator is able to pay unless the owner/operator demonstrates otherwise. The inability to pay adjustment should be based on the amount of the initial penalty target figure and the financial condition of the business, but it is the owner/operator's responsibility to provide evidence of inability to pay. The owner/operator may provide evidence, such as tax returns, to document his or her claims. In cases when the owner/operator fails to demonstrate inability to pay, the Agency should determine whether the owner/operator is unwilling to pay, in which case no adjustments to the initial penalty target figure should be made. In cases where the owner/operator can successfully demonstrate: (1) that the company is unable to pay; or (2) that payment of all or a portion of the penalty will preclude the violator from achieving compliance, the following options may be considered:

- An installment payment plan with interest;

- A delayed payment schedule with interest;

- An in-kind mitigation activity performed by the owner/operator;

- An environmental auditing program implemented by the owner/operator; or

- Reduction of up to 80 percent of the gravity-based component.

A reduction of the gravity-based component should be considered only after determining that the other four options are not feasible.[14]

In order to evaluate a violator's claim regarding inability to pay, two sources of information are available to determine the likelihood that a company can afford to pay a certain civil penalty:

National Enforcement Investigation Center (NEIC). The NEIC of EPA's Office of Enforcement has developed the Superfund Financial Assessment System that can determine a company's ability to pay. For publicly owned companies, specific financial data is available from NEIC. If investigating a private company, enforcement staff can report financial data to NEIC and it will be keyed into NEIC's computerized economic computer model for analysis.[15]

ABEL. EPA's Office of Enforcement developed the "ABEL" model as part of an ongoing effort to evaluate the financial health of firms involved in enforcement proceedings. The ABEL model has been used by EPA, Regions, and States to evaluate a firm's claim regarding inability to pay based on 21 inputs gathered from the company's Federal income tax returns from the previous 3 years. Enforcement staff may access ABEL by computer dial-up on a personal computer with a modem and an ABEL user ID number.[16] In addition, OUST has developed a PC-based model called ABELPRO which is a simplified version of ABEL that is run on a PC using a LOTUS spreadsheet or Macintosh Excel.[17]

---

[14] The Agency is currently developing cross-media guidance on environmental mitigation projects which, when final, will supersede the "Alternative Payments" section of the Agency's February 16, 1984 penalty policy (#GM-22). Until the revised Agency guidance is finalized, the Agency's 1984 penalty policy should be consulted for additional guidance.

[15] For further information, contact the NEIC at (303) 236-5100 or FTS 8-776-5100.

[16] To obtain the ABEL User's Manual and user ID numbers for computer hookup, contact the BEN/ABEL Coordinator at the U.S. EPA Headquarters, by phoning (202) 475-6777 or FTS 475-6777.

[17] For information, contact the appropriate Regional Desk Officer at U.S. EPA Headquarters' Office of Underground Storage Tanks.

Chapter 5

Use of Field Citations

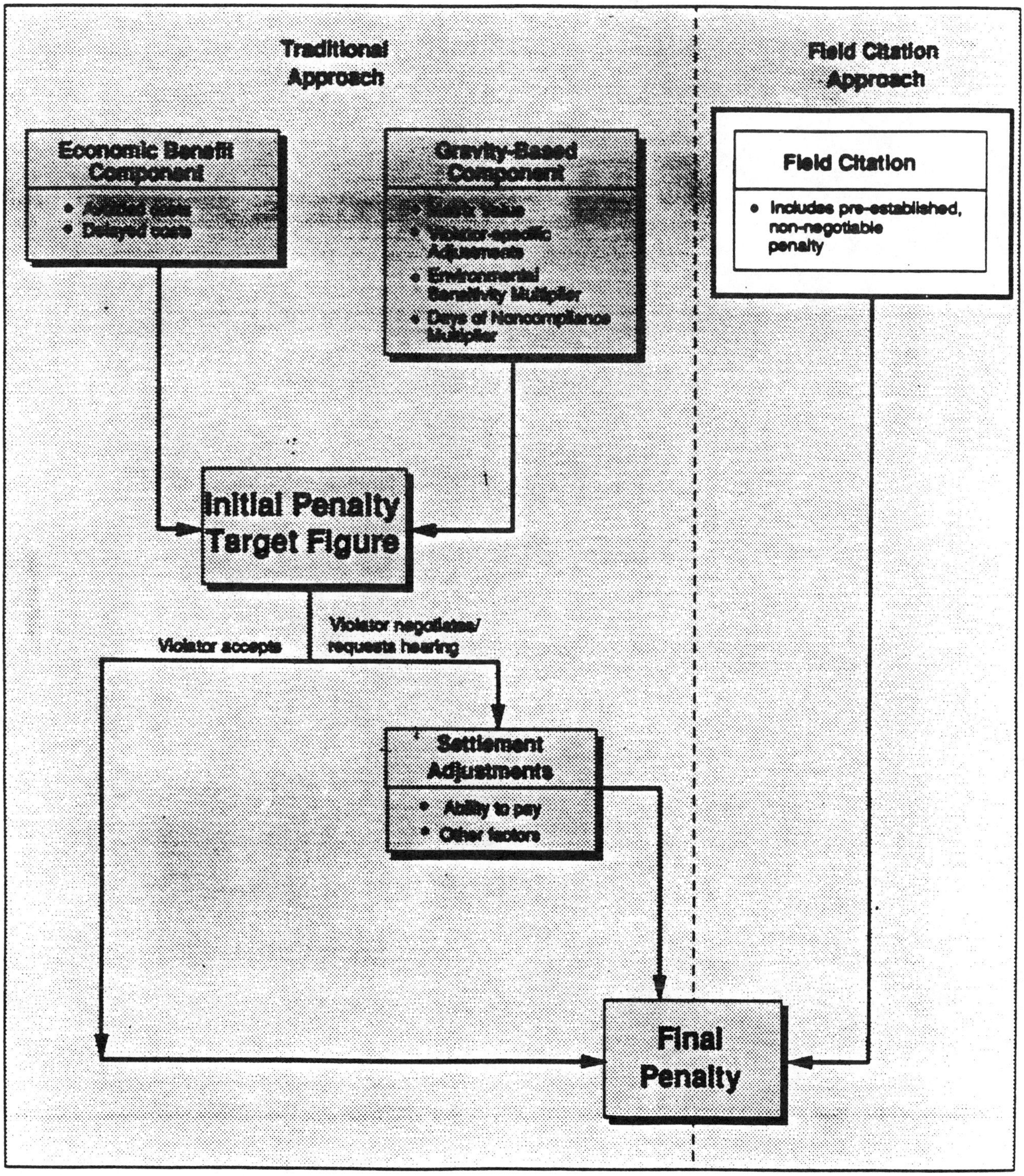

CHAPTER 5. USE OF FIELD CITATIONS

<Reserved>

The Office of Underground Storage Tanks (OUST) has been exploring the use of field citations as an alternative means of assessing civil penalties and obtaining compliance with UST requirements. Once the manner in which field citations will be used in the Federal UST program has been determined, this policy will be revised to reflect how field citations fit into the UST penalty policy.

APPENDIX A:

MATRIX VALUES FOR SELECTED VIOLATIONS OF FEDERAL UNDERGROUND STORAGE TANK REGULATIONS

APPENDIX A:
MATRIX VALUES FOR SELECTED VIOLATIONS OF FEDERAL UNDERGROUND STORAGE TANK REGULATIONS*

Regulatory Citation	Violation	Unit Assessment[1/]	Deviation from Requirement	Potential for Harm	Matrix Value
SUBPART B -- UST SYSTEMS: DESIGN, CONSTRUCTION, INSTALLATION, AND NOTIFICATION					
	§280.20 Performance standards for new UST systems				
§280.20(a)(1)	Installation of an improperly constructed fiberglass-reinforced plastic tank	(T)	Major	Major	$1500
§280.20(a)(2)	Installation of an improperly designed and constructed metal tank that fails to meet corrosion protection standards	(T)	Major	Moderate	$750
§280.20(a)(2)(i)	Installation of a metal tank with unsuitable dielectric coating	(T)	Major	Moderate	$750
§280.20(a)(2)(ii)	Installation of an improperly designed cathodic protection system for a metal tank	(T)	Moderate	Moderate	$500
§280.20(a)(2)(iii)	Improper installation of cathodic protection system for a metal tank	(T)	Moderate	Moderate	$500
§280.20(a)(2)(iv)	Improper operation and maintenance of tank cathodic protection system	(T)	Major	Moderate	$750
§280.20(a)(3)	Installation of an improperly constructed steel-fiberglass-reinforced-plastic tank	(T)	Major	Moderate	$750
§280.20(b)(1)	Installation of improperly constructed fiberglass-reinforced plastic piping	(P)	Major	Major	$1500
§280.20(b)(2)	Failure to provide any cathodic protection for metal piping	(P)	Major	Moderate	$750
§280.20(b)(2)(i)	Installation of piping with unsuitable dielectric coating	(T)	Major	Moderate	$750
§280.20(b)(2)(ii)	Installation of improperly designed cathodic protection for metal piping	(P)	Moderate	Moderate	$500
§280.20(b)(2)(iii)	Improper installation of cathodic protection system for piping	(P)	Moderate	Moderate	$500
§280.20(b)(2)(iv)	Improper operation and maintenance of cathodic protection system for metal piping	(P)	Major	Moderate	$750

[1/] Unit assessment refers to whether the penalty should be applied per tank (T) or per facility (F). Where the violation applies to piping (P), the assessment will depend on whether the piping is associated with one tank or more than one tank.

*** NOTE: THIS LIST OF SELECTED VIOLATIONS IS NOT INTENDED TO BE EXHAUSTIVE AND, THEREFORE, MAY NOT INCLUDE ALL POSSIBLE VIOLATIONS.**

Regulatory Citation	Violation *	Unit Assess-ment[1/]	Deviation from Requirement	Potential for Harm	Matrix Value
SUBPART B -- UST SYSTEMS: DESIGN, CONSTRUCTION, INSTALLATION, AND NOTIFICATION (Continued)					
§280.20(c)(1)	Failure to install any spill prevention system	(T)	Major	Major	$1500
§280.20(c)(1)(i)	Installation of inadequate spill prevention equipment in a new tank	(T)	Major	Major	$1500
§280.20(c)(1)	Failure to install any overfill prevention system	(T)	Major	Moderate	$750
§280.20(c)(1)(ii)	Installation of inadequate overfill prevention equipment in a new tank	(T)	Major	Moderate	$750
§280.20(d)	Failure to install tank in accordance with accepted codes and standards	(T)	Varies[2/]	Varies[2/]	see matrix
§280.20(d)	Failure to install piping in accordance with accepted codes and standards	(P)	Varies[2/]	Varies[2/]	see matrix
§280.20(e)	Failure to provide any certification of UST installation	(F)	Moderate	Minor	$100
§280.20(e)(1)-(6)	Failure to provide complete certification of UST installation	(F)	Minor	Minor	$50
	280.21 Upgrading of existing UST systems				
§280.21(b)	Failure to meet all tank upgrade standards	(T)	Major	Major	$1500
§280.21(b)(1)(i)	Improper installation of interior lining for tank upgrade requirements	(T)	Major	Major	$1500
§280.21(b)(1)(ii)	Failure to meet interior lining inspection requirements for tank upgrade	(T)	Major	Moderate	$750
§280.21(b)(2)(i)	Failure to ensure that tank is structurally sound before installing cathodic protection	(T)	Major	Moderate	$750
§280.21(b)(2)(ii)	Failure to provide any monthly monitoring of cathodic protection for tank upgrade requirement	(T/F)	Major	Major	$1500
§280.21(b)(2)(ii)	Failure to provide continuous monthly monitoring of cathodic protection for tank upgrade requirement	(T/F)	Moderate	Minor	$100

[2/] Deviation from requirement and potential for harm will vary depending upon specific code or standard violated.

* NOTE: THIS LIST OF SELECTED VIOLATIONS IS NOT INTENDED TO BE EXHAUSTIVE AND, THEREFORE, MAY NOT INCLUDE ALL POSSIBLE VIOLATIONS.

Regulatory Citation	Violation	Unit Assess-ment[1/]	Deviation from Requirement	Potential for Harm	Matrix Value
SUBPART B -- UST SYSTEMS: DESIGN, CONSTRUCTION, INSTALLATION, AND NOTIFICATION (Continued)					
§280.21(b)(2)(iii)	Failure to meet tightness test requirements for a tank upgraded with cathodic protection	(T/F)	Major	Moderate	$750
§280.21(b)(2)(iv)	Failure to meet requirements for testing for corrosion holes for a tank upgraded with cathodic protection	(T/F)	Major	Moderate	$750
§280.21(c)	Failure to install any cathodic protection for metal piping upgrade requirements	(P)	Major	Major	$1500
§280.21(c)	Failure to meet tightness test requirements for cathodically protected metal piping	(P)	Major	Moderate	$750
§280.21(d)	Failure to provide spill prevention system for an existing tank	(T)	Major	Major	$1500
§280.21(d)	Failure to provide overfill prevention system for an existing tank	(T)	Major	Moderate	$750
	280.22 Notification requirements				
§280.22(a)	Failure to notify state or local agency within 30 days of bringing an UST system into use	(T)	Major	Major	$1500
§280.22(a)	Failure to notify designated state or local agency of existing tank	(T)	Major	Major	$1500
§280.22(c)	Failure to identify on the submitted notification form all known tanks at that site	(F)	Major	Moderate	$750
§280.22(c)	Failure to submit a separate notification form for all notified tanks that are located at a separate place of operation	(F)	Major	Minor	$200
§280.22(e)-(f)	Failure to provide complete certification of all requirements on the notification form	(F)	Moderate	Minor	$100
§280.22(g)	Failure to inform tank purchaser of notification requirements	(T)	Major	Major	$1500

*** NOTE: THIS LIST OF SELECTED VIOLATIONS IS NOT INTENDED TO BE EXHAUSTIVE AND, THEREFORE, MAY NOT INCLUDE ALL POSSIBLE VIOLATIONS.**

Regulatory Citation	Violation	Unit Assess-ment[1/]	Deviation from Requirement	Potential for Harm	Matrix Value
SUBPART C -- GENERAL OPERATING REQUIREMENTS					
	280.30 Spill and overfill control				
§280.30(a)	Failure to take necessary precautions to prevent overfill/spillage during the transfer of product	(F)	Major	Major	$1500
§280.30(b)	Failure to report a spill/overfill	(F)	Major	Major	$1500
§280.30(b)	Failure to investigate and clean up a spill/overfill	(F)	Major	Major	$1500
	280.31 Operation and maintenance of corrosion protection				
§280.31(a)	Failure to operate and maintain corrosion protection system continuously	(F/T)	Major	Major	$1500
§280.31(b)(1)	Failure to ensure that cathodic protection system is tested within 6 months of installation	(F/T)	Major	Major	$1500
§280.31(b)(1)	Failure to ensure that cathodic protection system is tested every 3 years thereafter	(T/F)	Major	Moderate	$750
§280.31(b)(1)	Failure to meet one 3-year test for cathodic protection system	(T/F)	Moderate	Minor	$100
§280.31(b)(2)	Failure to inspect cathodic protection system in accordance with accepted codes	(T/F)	Major	Moderate	$750
§280.31(c)	Failure to inspect impressed current systems every 60 days	(T/F)	Major	Moderate	$750
§280.31(d)	Failure to maintain any records of cathodic protection inspections	(T/F)	Major	Moderate	$750
§280.31(d)	Failure to maintain every record of cathodic protection inspections	(T/F)	Moderate	Minor	$100
	280.32 Compatibility				
§280.32	Failure to ensure that UST system is made of or lined with materials compatible with substance stored	(T/P)	Major	Major	$1500

*** NOTE: THIS LIST OF SELECTED VIOLATIONS IS NOT INTENDED TO BE EXHAUSTIVE AND, THEREFORE, MAY NOT INCLUDE ALL POSSIBLE VIOLATIONS.**

Regulatory Citation	Violation	Unit Assess-ment[1/]	Deviation from Requirement	Potential for Harm	Matrix Value
SUBPART C -- GENERAL OPERATING REQUIREMENTS (Continued)					
	280.33 Repairs allowed				
§280.33(a)	Failure to repair UST system in accordance with accepted codes and standards	(T)	Varies[2/]	Varies[2/]	see matrix
§280.33(b)	Failure to repair fiberglass-reinforced UST in accordance with accepted codes and standards	(T)	Varies[2/]	Varies[2/]	see matrix
§280.33(c)	Failure to replace metal piping that has released product	(P)	Major	Major	$1500
§280.33(c)	Failure to repair fiberglass-reinforced piping in accordance with manufacturers specifications	(P)	Major	Major	$1500
§280.33(d)	Failure to ensure that repaired tank systems are tightness tested within 30 days of completion of repair	(T)	Major	Moderate	$750
§280.33(e)	Failure to test cathodic protection system within 6 months of repair of an UST system	(T)	Major	Moderate	$750
§280.33(f)	Failure to maintain records of each repair to an UST system	(T)	Major	Major	$1500
	280.34 Reporting and recordkeeping				
(For violations of reporting and recordkeeping, see appropriate regulatory section (e.g., reporting of releases will be under Subpart D).					
SUBPART D -- RELEASE DETECTION					
	280.40 General requirements for all UST systems				
§280.40(a)(1)	Failure to provide release detection method capable of detecting a release from tank or piping that routinely contains product	(T/F)	Major	Major	$1500
§280.40(a)(2)	Failure to install, calibrate, operate, or maintain release detection method in accordance with manufacturer's instructions	(T/F)	Major	Major	$1500

*** NOTE: THIS LIST OF SELECTED VIOLATIONS IS NOT INTENDED TO BE EXHAUSTIVE AND, THEREFORE, MAY NOT INCLUDE ALL POSSIBLE VIOLATIONS.**

Regulatory Citation	Violation	Unit Assess-ment[1/]	Deviation from Requirement	Potential for Harm	Matrix Value
	SUBPART D -- RELEASE DETECTION (Continued)				
§280.40(a)(3)	Failure to provide a release detection method that meets the performance requirements in §280.43 or §280.44	(F)	Major	Major	$1500
§280.40(b)	Failure to notify implementing agency when release detection indicates release	(F)	Major	Major	$1500
§280.40(c)	Failure to provide any release detection method by phase-in date	(F)	Major	Major	$1500
§280.40(d)	Failure to close any UST system that cannot meet release detection requirements.	(F)	Major	Major	$1500
	280.41 Requirements for petroleum UST systems				
§280.41(a)	Failure to monitor tanks at least every 30 days, if appropriate	(T)	Major	Major	$1500
§280.41(a)(1)	Failure to conduct tank tightness testing every 5 years, if appropriate	(T)	Major	Major	$1500
§280.41(a)(2)	Failure to conduct annual tank tightness testing, if appropriate	(T)	Major	Major	$1500
§280.41(b)	Failure to use any underground piping monitoring method	(P)	Major	Major	$1500
	280.42 Requirements for hazardous substance UST systems				
§280.42(a)	Failure to provide release detection for an existing hazardous substance tank system	(F)	Major	Major	$1500
§280.42(b)	Failure to provide adequate release detection for a new hazardous substance UST system	(F)	Major	Major	$1500
§280.42(b)(1)	Failure to provide adequate secondary containment of tank for a hazardous substance UST	(T)	Major	Major	$1500
§280.42(b)(2)	Failure to provide adequate double-walled tank/adequate lining for a hazardous substance UST	(T)	Major	Major	$1500

*** NOTE: THIS LIST OF SELECTED VIOLATIONS IS <u>NOT</u> INTENDED TO BE EXHAUSTIVE AND, THEREFORE, MAY NOT INCLUDE ALL POSSIBLE VIOLATIC**

Regulatory Citation	Violation	Unit Assess-ment[1/]	Deviation from Requirement	Potential for Harm	Matrix Value
SUBPART D -- RELEASE DETECTION (Continued)					
§280.42(b)(3)	Failure to provide adequate external liners for a hazardous substance UST	(T)	Major	Major	$1500
§280.42(b)(4)	Failure to provide adequate secondary containment of piping for a hazardous substance UST	(T)	Major	Major	$1500
	280.44 Methods of release detection for piping				
§280.44	Failure to provide any release detection for underground piping	(P)	Major	Major	$1500
§280.44(a)	Failure to provide adequate line leak detector system for underground piping	(P)	Major	Major	$1500
§280.44(b)	Failure to provide adequate line tightness testing system for underground piping system	(P)	Major	Major	$1500
§280.44(c)	Inadequate use of applicable tank release detection methods	(P)	Major	Major	$1500
	280.45 Release detection recordkeeping				
§280.45	Failure to maintain any records of release detection monitoring	(F)	Major	Major	$1500
§280.45	Failure to maintain every record of release detection monitoring	(F)	Moderate	Minor	$100
§280.45(a)	Failure to document all release detection performance claims for 5 years after installation	(F)	Moderate	Minor	$100
§280.45(b)	Failure to maintain any results of sampling, testing or monitoring for release detection for at least 1 year	(F)	Major	Major	$1500
§280.45(b)	Failure to maintain every result of sampling, testing or monitoring for release detection for at least 1 year	(F)	Moderate	Minor	$100
§280.45(b)	Failure to retain results of tightness testing until next test is conducted	(F)	Major	Major	$1500
§280.45(c)	Failure to document any calibration, maintenance, and repair of release detection	(F)	Major	Major	$1500

* NOTE: THIS LIST OF SELECTED VIOLATIONS IS NOT INTENDED TO BE EXHAUSTIVE AND, THEREFORE, MAY NOT INCLUDE ALL POSSIBLE VIOLATIONS.

Regulatory Citation	Violation	Unit Assess-ment[1/]	Deviation from Requirement	Potential for Harm	Matrix Value
SUBPART D -- RELEASE DETECTION (Continued)					
§280.45(c)	Failure to document every calibration, maintenance, and repair of release detection	(F)	Moderate	Moderate	$500
SUBPART E -- RELEASE REPORTING, INVESTIGATION, AND CONFIRMATION					
	280.50 Reporting of suspected release				
§280.50(a)-(c)	Failure to report a suspected release within 24 hours to the implementing agency	(F)	Major	Major	$1500
	280.52 Release investigation and confirmation steps				
§280.52(a)-(b)	Failure to investigate and confirm a release (if appropriate) using accepted procedures	(F)	Major	Major	$1500
	280.53 Reporting and cleanup of spills and overfills				
§280.53(a)	Failure to report a spill/overfill (if appropriate) to implementing agency within 24 hours (or other specified time period)	(F)	Major	Major	$1500
§280.53(b)	Failure to contain and immediately clean up a spill/overfill of less than 25 gallons	(F)	Major	Major	$1500
§280.53(b)	Failure to contain and immediately clean up a hazardous substance spill/overfill	(F)	Major	Major	$1500
SUBPART F -- RELEASE RESPONSE AND CORRECTIVE ACTION					
§280.61	Failure to take initial response actions within specified time period after a release is confirmed	(F)	Major	Major	$1500

* NOTE: THIS LIST OF SELECTED VIOLATIONS IS NOT INTENDED TO BE EXHAUSTIVE AND, THEREFORE, MAY NOT INCLUDE ALL POSSIBLE VIOLATIONS.

Regulatory Citation	Violation	Unit Assess-ment[1/]	Deviation from Requirement	Potential for Harm	Matrix Value
SUBPART F -- RELEASE RESPONSE AND CORRECTIVE ACTION (Continued)					
§280.62	Failure to submit report on initial abatement measures within 20 days (or other specified time) of release confirmation	(F)	Major	Major	$1500
§280.63	Failure to submit report on initial site characterization within 45 days (or other specified time) of release confirmation	(F)	Major	Major	$1500
§280.64	Failure to submit report on free report removal within 45 days (or other specified time) of release confirmation	(F)	Major	Major	$1500
SUBPART G -- OUT-OF-SERVICE UST SYSTEMS AND CLOSURE					
	280.70 Temporary closure				
§280.70(a)	Failure to continue operation and maintenance of cathodic protection system in a temporarily closed tank system	(F/T)	Major	Moderate	$750
§280.70(a)	Failure to continue operation and maintenance of release detection in a temporarily closed tank system	(F/T)	Major	Major	$1500
§280.70(b)	Failure to comply with temporary closure requirements for a tank system for 3 or more months	(F/T)	Major	Moderate	$750
§280.70(c)	Failure to permanently close or upgrade a temporarily closed tank system after 12 months	(F/T)	Major	Major	$1500
	280.71 Permanent closure and changes-in-service				
§280.71(a)	Failure to notify implementing agency of a closure or change-in-service	(F/T)	Major	Major	$1500
§280.71(b)	Failure to remove all liquids and sludges for tank closure	(F/T)	Major	Major	$1500
§280.71(b)	Failure to remove closed tank from the ground or fill tank with an inert solid for tank closure	(F/T)	Major	Moderate	$750
§280.71(c)	Failure to empty and clean tank system and conduct a site assessment prior to a change-in-service	(F/T)	Major	Major	$1500

*** NOTE: THIS LIST OF SELECTED VIOLATIONS IS <u>NOT</u> INTENDED TO BE EXHAUSTIVE AND, THEREFORE, MAY NOT INCLUDE ALL POSSIBLE VIOLATIONS.**

Regulatory Citation	Violation	Unit Assess-ment[1/]	Deviation from Requirement	Potential for Harm	Matrix Value
	SUBPART G -- OUT-OF-SERVICE UST SYSTEMS AND CLOSURE (Continued)				
	280.72 Assessing the site at closure or change-in-service				
§280.72(a)	Failure to measure (if required) for the presence of a release before a permanent closure	(T/F)	Major	Major	$1500
§280.72(b)	If contaminated soil, contaminated ground water, or free product is discovered, failure to begin corrective action	(T/F)	Major	Major	$1500
	280.74 Closure records				
§280.74	Failure to maintain closure records for at least 3 years	(F)	Major	Major	$1500
§280.74	Failure to maintain change-in-service records for at least 3 years	(F)	Major	Major	$1500
	SUBPART H -- FINANCIAL RESPONSIBILITY				
§280.93(a)	Failure to comply with financial responsibility requirements by the required phase-in time	(F)	Major	Moderate	$750
§280.93(a)(1)-(2)	Failure to meet the requirement for per-occurrence coverage of insurance.	(F)	Major	Moderate	$750
§280.93(b)(1)-(2)	Failure to meet the requirement for annual aggregate coverage of insurance.	(F)	Major	Moderate	$750
§280.93(f)	Failure to review and adjust financial assurance after acquiring new or additional USTs	(F)	Major	Moderate	$750
§280.94	Use of an unapproved mechanism or combination of mechanisms to demonstrate financial responsibility	(F)	Major	Moderate	$750
§280.95	Use of falsified financial documents to pass financial test of self-insurance	(F)	Major	Moderate	$750
§280.106(a)(1)	Failure to report evidence of financial responsibility to the implementing agency within 30 days of detecting a known or suspected release	(F)	Moderate	Minor	$100

*** NOTE: THIS LIST OF SELECTED VIOLATIONS IS <u>NOT</u> INTENDED TO BE EXHAUSTIVE AND, THEREFORE, MAY NOT INCLUDE ALL POSSIBLE VIOLATIONS.**

Regulatory Citation	Violation	Unit Assess-ment[1/]	Deviation from Requirement	Potential for Harm	Matrix Value
SUBPART H -- FINANCIAL RESPONSIBILITY (Continued)					
§280.106(a)(2)	Failure to report evidence of financial responsibility to the implementing agency when new tanks are installed	(F)	Moderate	Minor	$100
§280.106(b)	Failure to report evidence of financial responsibility to the implementing agency if the provider becomes incapable of providing financial assurance and the owner or operator is unable to obtain alternate coverage within 30 days.	(F)	Moderate	Minor	$100
§280.107	Failure to maintain copies of the financial assurance mechanism(s) used to comply with financial responsibility rule and certification that the mechanism is in compliance with the requirements of the rule at the UST site or place of business	(F)	Moderate	Minor	$100

* NOTE: THIS LIST OF SELECTED VIOLATIONS IS NOT INTENDED TO BE EXHAUSTIVE AND, THEREFORE, MAY NOT INCLUDE ALL POSSIBLE VIOLATIONS.

APPENDIX B:

UST PENALTY COMPUTATION WORKSHEET

(This page is blank)

UST PENALTY COMPUTATION WORKSHEET

Assessments for each violation should be determined on separate worksheets and totaled. (If more space is needed, attach separate sheet.)

PART 1 - BACKGROUND

Company name____________________

Regulation violated____________________

____

____

Previous violations____________________

____

Date of requirement____________________ Date of inspection____________________

Date of compliance____________________ Explanation (if appropriate):

1. Days of noncompliance____________________

2. Number of tanks____________________

PART 2 - ECONOMIC BENEFIT COMPONENT

Avoided Expenditures____________________ Basis:____________________

Delayed Expenditures____________________ Basis: ____________________

Weighted Tax Rate____________________ Source: ____________________

Interest Rate____________________ Source: ____________________

$$\text{AVOIDED COSTS} = \left[\text{Avoided Expenditures} + \frac{\text{Avoided Expenditures} \times \text{Interest} \times \text{Number of Days}}{\text{365 Days}}\right] \times (1 - \text{Weighted Tax Rate})$$

3. Calculated Avoided Cost:____________________

UST PENALTY COMPUTATION WORKSHEET

DELAYED COSTS = $\frac{\text{Delayed Expenditures} \times \text{Interest} \times \text{Number of Days}}{\text{365 Days}}$

4. Calculated Delayed Cost:____________________

5. Economic Benefit Component:____________________(carry figure to Line 16).
   (Line 3 + Line 4)

PART 3 - MATRIX VALUE FOR THE GRAVITY-BASED COMPONENT

Potential for Harm:____________________ Extent of Deviation____________________

6. Matrix Value (MV):____________________ (from document page 16 or Appendix A)

7. Per-tank MV:____________________ (if violation is per facility, the amount on Line 7 will be the same as the amount on Line 6)
   (Line 2 x Line 6)

PART 4 - VIOLATOR-SPECIFIC ADJUSTMENTS TO MATRIX VALUE

	Percentage Change (+ or -)	x Matrix Value	= Dollar Adjustment (+ or -)	Justification for Adjustment:
8. Degree of cooperation/ noncooperation	______	______	______	
9. Degree of willfulness or negligence:	______	______	______	
10. History of noncompliance:	______	______	______	
11. Unique factors:	______	______	______	
12. Adjusted Matrix Value (Line 7 + Lines 8-11)			______	

UST PENALTY COMPUTATION WORKSHEET

PART 5 - GRAVITY-BASED COMPONENT

Level of
Environmental Sensitivity__________________ Justification:

13. ESM (from document Page 21)__________

14. DNM (from document Page 21)__________

GRAVITY-BASED COMPONENT = Adjusted Matrix Value x Environmental Sensitivity Multiplier x Days of Noncompliance Multiplier

15. Gravity-Based Component: _______________
(Line 12 x Line 13 x Line 14)

PART 6 - INITIAL PENALTY TARGET FIGURE

16. Economic Benefit Component ______________
(from Line 5)

17. Gravity-Based Component_________________
(from Line 15)

18. Initial Penalty Target Figure_________________
(Line 16 + Line 17)

SIGNATURE_________________________________ DATE_______________

(This page is blank)

APPENDIX C:

UST PENALTY COMPUTATION EXAMPLES

EXAMPLE 1

BACKGROUND

Inspection Date: April 12, 1990

Facility Name and Description: Ed's Gas and Go is a small gas station in a semi-rural part of the county. The facility has 4 tanks, apparently installed prior to 1965. Judging from the condition of the facility and adjacent store, Ed's income appears to be less than $50,000 per year.

Violations: During the inspection, the inspector observed that Ed failed to provide a method of release detection by the December 22, 1989 deadline, in violation of 40 CFR section 280.40(c).

Owner/Operator Response: Ed claimed no knowledge of the requirements for release detection. After being informed of methods for meeting the requirement, he indicated that he would use annual tank tightness testing and monthly inventory control, in accordance with 40 CFR section 280.41(a)(2). Ed began to conduct adequate monthly inventory control and arranged to have his tanks tested within 10 days.

Previous Actions at Facility: Previously, Ed had been given a warning letter for failure to comply with the notification requirements, but had complied upon receipt of the letter. No other previous violations were identified.

Current Status at Site: The inspector observed that given the age of the tanks, and Ed's previous inability to detect any releases, there was a good chance for a release to occur and go unnoticed for a significant length of time. However, Ed's subsequent tightness tests indicated that the tanks were tight. The geology in the area is fractured shale. There are no drinking water wells or sensitive wildlife receptors within a 5-mile radius of the site.

PENALTY CALCULATION DATA

Violation: 40 CFR section 280.40(c)

Days of violation: 120 days from date of noncompliance (December 22, 1989) to date of compliance (April 22, 1990, which was 10 days after the inspection).

Avoided expenditures: $2.50 per day = $300 for 120 days (estimated cost for labor needed to conduct daily inventory control, based on 1/2 hour labor at $5.00 per hour)

Delayed expenditures: $520 x 4 tanks = $2,080, where the average cost for a tank tightness test is $520. This is considered a delayed expenditure because it was necessary to achieve compliance in this time frame.

Interest rate: 18.1% (the equity discount rate used in the BEN model for 1990).

Tax rate: 15% (the weighted average tax rate for a facility with less than $50,000 annual income).

[NOTE: The numbers used to determine avoided and delayed expenditures were chosen for convenience only. They do not necessarily represent true costs in any State or Region in the country.]

UST PENALTY COMPUTATION WORKSHEET

Assessments for each violation should be determined on separate worksheets and totaled. (If more space is needed, attach separate sheet.)

PART 1 - BACKGROUND

Company name: Ed's Gas and Go

Regulation violated: 40 CFR section 280.40(c) - Failure to provide release detection by December 22, 1989 phase-in date.

Previous violations: Notification violation (1986) - warning letter issued.

Date of requirement: 12/22/89

Date of inspection: 4/12/90

Date of compliance: 4/22/89

Explanation (if appropriate): date of compliance is 10 days after inspection.

1. Days of noncompliance: 120

2. Number of tanks: 4

PART 2 - ECONOMIC BENEFIT COMPONENT

Avoided Expenditures: $300

Basis: $2.50 per day for monitoring

Delayed Expenditures: $2080

Basis: $520 per tank for tightness test

Weighted Tax Rate: 0.15 (15%)

Source: MTR for income < $50,000/year

Interest Rate: 0.181 (18%)

Source: BEN model (equity discount rate)

$$\textbf{AVOIDED COSTS} = \left[\text{Avoided Expenditures} + \frac{\text{Avoided Expenditures} \times \text{Interest} \times \text{Number of Days}}{\text{365 Days}}\right] \times (1 - \text{Weighted Tax Rate})$$

$$AC = \left[\$300 + \frac{(\$300 \times .181 \times 120)}{365}\right] \times [1 - .15] = \$270$$

3. Calculated Avoided Cost: $270

UST PENALTY COMPUTATION WORKSHEET

DELAYED COSTS = Delayed Expenditures x Interest x Number of Days / 365 Days

$$DC = \frac{\$2080 \times .181 \times 120}{365} = \$124$$

4. Calculated Delayed Cost: $124

5. Economic Benefit Component: $394 (carry figure to Line 16).
(Line 3 + Line 4)

PART 3 - MATRIX VALUE FOR THE GRAVITY-BASED COMPONENT

Potential for Harm: Major

Extent of Deviation: Major

6. Matrix Value (MV): $1500 (from document page 16 or Appendix A)

7. Per-tank MV: $6000 (if violation is per facility, the amount on Line 7 will be the same as the amount on Line 6)
(Line 2 x Line 6)

PART 4 - VIOLATOR-SPECIFIC ADJUSTMENTS TO MATRIX VALUE

	Percentage Change (+ or -)	x	Matrix Value	=	Dollar Adjustment (+ or -)	Justification for Adjustment:
8. Degree of cooperation/ noncooperation	0		$6000		0	Complied as required following inspection.
9. Degree of willfulness or negligence:	0		$6000		0	Did not knowingly violate requirements.
10. History of noncompliance:	+5%		$6000		+$300	Warning letter issued for previous violation.
11. Unique factors:	0		$6000		0	
12. Adjusted Matrix Value (Line 7 + Lines 8-11)					$6300	

UST PENALTY COMPUTATION WORKSHEET

PART 5 - GRAVITY-BASED COMPONENT

Level of
Environmental Sensitivity Moderate

Justification: Any release is not likely to have impact on nearby drinking-water sources. Potential impact on the environment would be minimal, although fractured shale would complicate remediation.

13. ESM (from document Page 21) 1.5

14. DNM (from document Page 21) 1.5

GRAVITY-BASED COMPONENT = Adjusted Matrix Value x Environmental Sensitivity Multiplier x Days of Noncompliance Multiplier

GBC = $6300 x 1.5 x 1.5 = $14,175

15. Gravity-Based Component: $14,175
(Line 12 x Line 13 x Line 14)

PART 6 - INITIAL PENALTY TARGET FIGURE

16. Economic Benefit Component $394
(from Line 5)

17. Gravity-Based Component $14,175
(from Line 15)

18. Initial Penalty Target Figure $14,569
(Line 16 + Line 17)

SIGNATURE ______________________ DATE ______________

EXAMPLE 2

BACKGROUND

Inspection Date: March 20, 1992

Facility Name and Description: Johnson's Petromart, located at Prairie View Lane, is one of eight facilities in a convenience store chain that spans three counties. This facility has a total of 5 USTs, and there are a total of 34 USTs at the 8 facilities. Based on an examination of the parent company's tax returns, it was determined that the company's taxable income was $280,000.

Violations: During the inspection, the inspector observed that the facility had no records of financial assurance coverage as required by the April 26, 1991 deadline. Subsequently, the inspector requested records for each of the 8 Johnson facilities. Upon further investigation, the inspector determined that the owner of the chain, Jack Johnson, had acquired private insurance (the owner did not qualify to self-insure) for the other 7 facilities. At the remaining facility, however, neither the owner nor the operator had obtained the required coverage, thereby constituting a violation of 40 CFR section 280.93(a). This facility is among the oldest in the Johnson's chain and is operated with 4 bare steel UST systems and one cathodically protected UST system. The other 7 facilities were opened subsequent to the interim prohibition and installed USTs that meet the Federal design, construction, and installation requirements. Therefore, obtaining insurance for these USTs was easier than for the facility in violation. The insurance company had indicated that it would be willing to ensure the remaining facility provided that the tanks were retrofitted with spill/overfill protection and cathodic protection.

Owner/Operator Response: Jack Johnson argued that it was the responsibility of the operator to upgrade his USTs so as to make them insurable. The operator of the facility claimed that he lacked the resources to upgrade his USTs and believed that the responsibility for meeting the FR requirements was the owner's. The enforcement staff determined that the owner was aware of his responsibility to insure the USTs at all of his facilities and that only he had the means to do so. The Agency attempted to enter into compliance negotiations with Jack Johnson, but to no avail. The Agency planned to issue an administrative complaint on July 1, 1992.

Previous Actions at Facility: Previously, one of the Johnson's facilities had been issued a warning letter for failure to notify the Agency after bringing a new UST into operation. The owner had complied after receiving the letter. Three other facilities had been issued warning letters for failure to maintain all of the required monitoring records for release detection.

Current Status at Site: At the time of the most recent inspection, it was determined that the facility in violation of the FR requirements had an adequate method of release detection, and no releases were determined to have occurred. The geology in the area of the facility is clay. The facility is located in a semi-residential/commercial area; however, there are no drinking water wells or sensitive wildlife receptors within a 3-mile radius of the site.

PENALTY CALCULATION DATA

Violation: 40 CFR section 280.93(a)

Days of violation: 430 days from date of noncompliance (April 26, 1991) to date of compliance (which, for purposes of assessing the penalty, was determined to be July 1, 1992, to coincide with the date of the administrative complaint).

Avoided expenditures: $27.40 per day = $11,781 for 430 days (estimated insurance premium, based on an annual premium of $2,000 per UST for 5 USTs)

Delayed expenditures: $15,000 x 4 = $60,000 (where the average cost for system retrofit is $15,000). This is considered a delayed cost because retrofitting would enable Johnson's to achieve compliance with the financial responsibility requirement.

Interest rate: 18.1% (the equity discount rate used in the BEN model for 1990).

Tax rate: 33% (the weighted average rate for a facility with $280,000 in taxable income).

[NOTE: The numbers used to determine avoided and delayed expenditures were chosen for convenience only. They do not necessarily represent true costs in any State or Region in the country.]

UST PENALTY COMPUTATION WORKSHEET

Assessments for each violation should be determined on separate worksheets and totaled. (If more space is needed, attach separate sheet.)

PART 1 - BACKGROUND

Company name Johnson's Petromart

Regulation violated 40 CFR section 280.93(a) - Failure to provide full financial coverage by compliance deadline.

Previous violations Notification violation (1989) - warning letter issued; release detection violation (1991) - warning letter issued.

Date of requirement 4/26/91

Date of inspection 3/20/92

Date of compliance 7/1/92

Explanation (if appropriate): date of compliance is considered to be date complaint is issued.

1. Days of noncompliance 430

2. Number of tanks 5 (or 4)*

*(only 4 need to be retrofit)

PART 2 - ECONOMIC BENEFIT COMPONENT

Avoided Expenditures $11,781

Basis: $27.40 per day insurance (5 tanks)

Delayed Expenditures $60,000

Basis: $15,000 per UST retrofit (4 tanks)

Weighted Tax Rate 0.33 (33%)

Source: MTR for $280,000 income

Interest Rate 0.181 (18.1%)

Source: BEN model (equity discount rate)

$$\text{AVOIDED COSTS} = \left[\text{Avoided Expenditures} + \frac{\text{Avoided Expenditures} \times \text{Interest} \times \text{Number of Days}}{365 \text{ Days}}\right] \times (1 - \text{Weighted Tax Rate})$$

$$AC = \left[\$11{,}781 + \frac{\$11{,}481 \times .181 \times 430}{365}\right] \times (1 - .33) = \$9576$$

3. Calculated Avoided Cost: $9576

UST PENALTY COMPUTATION WORKSHEET

DELAYED COSTS = Delayed Expenditures x Interest x Number of Days / 365 Days

$$DC = \frac{\$60{,}000 \times .181 \times 430}{365} = \$12{,}794$$

4. Calculated Delayed Cost: $12,794

5. Economic Benefit Component: $22,370 (carry figure to Line 16).
(Line 3 + Line 4)

PART 3 - MATRIX VALUE FOR THE GRAVITY-BASED COMPONENT

Potential for Harm: Moderate

Extent of Deviation: Major

6. Matrix Value (MV): $750 (from document page 16 or Appendix A)

7. Per-tank MV: $750
(Line 2 x Line 6)

(if violation is per facility, the amount on Line 7 will be the same as the amount on Line 6)

PART 4 - VIOLATOR-SPECIFIC ADJUSTMENTS TO MATRIX VALUE

	Percentage Change (+ or -)	x Matrix Value	= Dollar Adjustment (+ or -)	Justification for Adjustment:
8. Degree of cooperation/ noncooperation	+40%	$750	+$300	Owner unwilling to negotiate terms of compliance.
9. Degree of willfulness or negligence:	+25%	$750	+$188	Owner was aware of requirement and able to comply.
10. History of noncompliance:	+20%	$750	+$150	Previous violation
11. Unique factors:	0	$750	0	N/A
12. Adjusted Matrix Value (Line 7 + Lines 8-11)			$1388	

UST PENALTY COMPUTATION WORKSHEET

PART 5 - GRAVITY-BASED COMPONENT

Level of Environmental Sensitivity Low

Justification: Potential impact of a release on the environment and drinking-water supplies would be minimal. Clay soil would limit migration of product.

13. ESM (from document Page 21) 1

14. DNM (from document Page 21) 3

GRAVITY-BASED COMPONENT = Adjusted Matrix Value x Environmental Sensitivity Multiplier x Days of Noncompliance Multiplier

GBC = $1388 x 1 x 3 = $4,164

15. Gravity-Based Component: $4164
(Line 12 x Line 13 x Line 14)

PART 6 - INITIAL PENALTY TARGET FIGURE

16. Economic Benefit Component $22,370
(from Line 5)

17. Gravity-Based Component $4164
(from Line 15)

18. Initial Penalty Target Figure $26,534
(Line 16 + Line 17)

SIGNATURE ____________________ DATE ____________

EXAMPLE 3

BACKGROUND

Inspection Date: N/A

Facility Name and Description: Kelly's Kwik Stop is a convenience store that recently had its three USTs taken out of operation. Prior to their removal, the USTs were operated by the owner of the convenience store, Karen Kelly, and owned by Darby Distributors, an oil jobber. The taxable income of Darby Distributors was $400,000 in 1989.

Violations: On May 20, 1989, Ms. Kelly reported the presence of petroleum vapors outside of her convenience store. The Agency investigated the site and confirmed the presence of a petroleum release. Ms. Kelly reported that Darby Distributors had removed the 3 USTs located at her place of business on March 17, 1989; she was not aware of the requirement to notify the Agency prior to permanent closure or of the requirement to conduct a site assessment. Ms. Kelly also could not say whether Darby Distributors had fulfilled these requirements. Upon a review of the Agency's records, it was determined that Darby Distributors had failed to notify the Agency of the closure, thereby constituting a violation of 40 CFR section 280.71. The distributor was also unable to produce records demonstrating compliance with the closure site assessment requirements, constituting a violation of 40 CFR section 280.74. The distributor also failed to assess the site for the presence of a release before permanent closure, in violation of 40 CFR section 280.72(a).

Owner/Operator Response: When the Agency contacted Darby Distributors, they indicated that they would initiate corrective action only if they, and not Ms. Kelly, were actually responsible for the release. The Agency informed them that as the owner of the USTs formerly in operation at Kelly's Kwik Stop they as well as Ms. Kelly are responsible for addressing any release from those USTs. The Agency also informed Darby Distributors that administrative orders were being prepared to compel them to clean up the release and pay penalties for violations of the closure requirements (the Agency was dealing separately with Ms. Kelly). At that time, the company requested to enter into negotiations with the Agency in order to establish a corrective action schedule and determine the amount of the penalties to be assessed.

Previous Actions at Facility: There were no previous incidents of violation at the facility.

Current Status at Site: Kelly's Kwik Stop is located in a rural part of the county. There are, however, two private drinking-water wells within a mile of the facility and several others within 4 miles of the facility. The facility is located one-half mile from a river that is used for recreational purposes as well as by various wildlife as a source of water. The geology in the area of the site is silt.

PENALTY CALCULATION DATA

Violation: 40 CFR section 280.71(a)

Days of Violation: 94 days, from the latest required date of compliance (February 17, 1989) to the actual date of compliance (May 20, 1989), where actual compliance is assumed to be coincident with Ms. Kelly's report to the Agency.

Avoided expenditures: Deemed negligible.

Delayed expenditures: None.

Interest rate: 18.1% (the equity discount rate used in the BEN model for 1989).

Tax rate: 34% (the weighted average rate for a company with taxable income greater than $340,000).

PENALTY CALCULATION DATA

Violation: 40 CFR section 280.72(a)

Days of Violation: 64 days, from the latest required date of compliance (March 17, 1989) to the actual date of compliance (May 20, 1989), where actual compliance is assumed to be coincident with Ms. Kelly's report to the Agency.

Avoided expenditures: $8,500 x 3 USTs = $25,500 (where the average cost for a site assessment at closure is $8,500 per UST).

Delayed expenditures: None.

Interest rate: 18.1% (the equity discount rate used in the BEN model for 1989).

Tax rate: 34% (the weighted average rate for a company with taxable income greater than $340,000).

PENALTY CALCULATION DATA

Violation: 40 CFR section 280.74

Days of Violation: 64 days, from the latest required date of compliance (March 17, 1989) to the actual date of compliance (May 20, 1989), where actual compliance is assumed to be coincident with Ms. Kelly's report to the Agency.

Avoided expenditures: None.

Delayed expenditures: Deemed negligible.

Interest rate: 18.1% (the equity discount rate used in the BEN model for 1989).

Tax rate: 34% (the weighted average rate for a company with taxable income greater than $340,000).

[NOTE: The numbers used to determine avoided and delayed expenditures were chosen for convenience only. They do not necessarily represent true costs in any State or Region in the country.]

UST PENALTY COMPUTATION WORKSHEET

Assessments for each violation should be determined on separate worksheets and totaled. (If more space is needed, attach separate sheet.)

PART 1 - BACKGROUND

Company name Darby Distributors

Regulation violated 40 CFR section 280.71 (a) - Failure to notify 30 days prior to tank closure.

Previous violations None

Date of requirement 2/15/89

Date of inspection N/A

Date of compliance 5/20/89

Explanation (if appropriate):

1. Days of noncompliance 94
2. Number of tanks 3

PART 2 - ECONOMIC BENEFIT COMPONENT

Avoided Expenditures 0 — Basis: Costs for notification negligible.

Delayed Expenditures N/A — Basis:

Weighted Tax Rate N/A — Source:

Interest Rate N/A — Source:

$$\text{AVOIDED COSTS} = \left[\text{Avoided Expenditures} + \frac{\text{Avoided Expenditures} \times \text{Interest} \times \text{Number of Days}}{365 \text{ Days}}\right] \times (1 - \text{Weighted Tax Rate})$$

3. Calculated Avoided Cost: $0

UST PENALTY COMPUTATION WORKSHEET

DELAYED COSTS = $\frac{\text{Delayed Expenditures} \times \text{Interest} \times \text{Number of Days}}{\text{365 Days}}$

4. Calculated Delayed Cost: $0

5. Economic Benefit Component: $0 (carry figure to Line 16).
(Line 3 + Line 4)

PART 3 - MATRIX VALUE FOR THE GRAVITY-BASED COMPONENT

Potential for Harm: Major

Extent of Deviation: Major

6. Matrix Value (MV): $1500 (from document page 16 or Appendix A)

7. Per-tank MV: $1500 (if violation is per facility, the amount on Line 7 will be the same as the amount on Line 6)
(Line 2 x Line 6)

PART 4 - VIOLATOR-SPECIFIC ADJUSTMENTS TO MATRIX VALUE

	Percentage Change (+ or -)	x Matrix Value	= Dollar Adjustment (+ or -)	Justification for Adjustment:
8. Degree of cooperation/ noncooperation	+ 10%	$1500	+ $150	Owner requested negotiations only after being warned of impending administrative order
9. Degree of willfulness or negligence:	+ 40%	$1500	+ $600	Owner appeared to take advantage of operators ignorance of requirements
10. History of noncompliance:	0	$1500	0	N/A
11. Unique factors:	0	$1500	0	N/A
12. Adjusted Matrix Value (Line 7 + Lines 8-11)			$2250	

UST PENALTY COMPUTATION WORKSHEET

PART 5 - GRAVITY-BASED COMPONENT

Level of
Environmental Sensitivity High

Justification: Release could impact several drinking-water wells and a river used by humans for recreation and by wildlife as a source of drinking water.

13. ESM (from document Page 21) 2

14. DNM (from document Page 21) 1.5

GRAVITY-BASED COMPONENT = Adjusted Matrix Value x Environmental Sensitivity Multiplier x Days of Noncompliance Multiplier

GBC = \$2250 x 2 x 1.5 = \$6750

15. Gravity-Based Component: \$6750
(Line 12 x Line 13 x Line 14)

PART 6 - INITIAL PENALTY TARGET FIGURE

16. Economic Benefit Component 0
(from Line 5)

17. Gravity-Based Component \$6750
(from Line 15)

18. Initial Penalty Target Figure \$6750
(Line 16 + Line 17)

SIGNATURE________________________ DATE______________

UST PENALTY COMPUTATION WORKSHEET

Assessments for each violation should be determined on separate worksheets and totaled. (If more space is needed, attach separate sheet.)

PART 1 - BACKGROUND

Company name Darby Distributors

Regulation violated 40 CFR section 280.72(a)- Failure to assess site at tank closure.

Previous violations None

Date of requirement 3/17/89 Date of inspection N/A

Date of compliance 5/20/89 Explanation (if appropriate):

1. Days of noncompliance 64
2. Number of tanks 3

PART 2 - ECONOMIC BENEFIT COMPONENT

Avoided Expenditures $25,500 Basis: $8500 per UST site assessment

Delayed Expenditures N/A Basis:

Weighted Tax Rate 0.34 (34%) Source: MTR for income > $335,000

Interest Rate 0.181 (18.1%) Source: BEN model (equity discount rate)

$$\text{AVOIDED COSTS} = \left[\text{Avoided Expenditures} + \frac{\text{Avoided Expenditures} \times \text{Interest} \times \text{Number of Days}}{365 \text{ Days}}\right] \times (1 - \text{Weighted Tax Rate})$$

$$AC = \left[\$25,500 + \frac{\$25,500 \times .181 \times 64}{365}\right] \times (1 - .34) = \$17,364$$

3. Calculated Avoided Cost: $17,364

UST PENALTY COMPUTATION WORKSHEET

DELAYED COSTS = $\frac{\text{Delayed Expenditures} \times \text{Interest} \times \text{Number of Days}}{\text{365 Days}}$

4. Calculated Delayed Cost: 0

5. Economic Benefit Component: $17,364 (carry figure to Line 16).
(Line 3 + Line 4)

PART 3 - MATRIX VALUE FOR THE GRAVITY-BASED COMPONENT

Potential for Harm: Major

Extent of Deviation Major

6. Matrix Value (MV): $1500 (from document page 16 or Appendix A)

7. Per-tank MV: $6000 (if violation is per facility, the amount on Line 7 will be the same as the amount on Line 6)
(Line 2 x Line 6)

PART 4 - VIOLATOR-SPECIFIC ADJUSTMENTS TO MATRIX VALUE

	Percentage Change (+ or -)	x Matrix Value	= Dollar Adjustment (+ or -)	Justification for Adjustment:
8. Degree of cooperation/ noncooperation	+10%	$6000	+$600	Owner requested negotiations only after being warned of impending administrative orders
9. Degree of willfulness or negligence:	+40%	$6000	+$2400	Owner appeared to take advantage of operators ignorance of requirements.
10.* History of noncompliance:	0	$6000	0	N/A
11. Unique factors:		$6000	0	N/A
12. Adjusted Matrix Value (Line 7 + Lines 8-11)			$9000	

1500
3000
1500

UST PENALTY COMPUTATION WORKSHEET

PART 5 - GRAVITY-BASED COMPONENT

Level of
Environmental Sensitivity High

Justification: Release could impact several drinking-water wells and a river used by humans for recreation and by wildlife as a source of drinking water.

13. ESM (from document Page 21) 2

14. DNM (from document Page 21) 1

GRAVITY-BASED COMPONENT = Adjusted Matrix Value x Environmental Sensitivity Multiplier x Days of Noncompliance Multiplier

GBC = $9000 x 2 x 1 = $18,000

15. Gravity-Based Component: $18,000
(Line 12 x Line 13 x Line 14)

PART 6 - INITIAL PENALTY TARGET FIGURE

16. Economic Benefit Component $17,364
(from Line 5)

17. Gravity-Based Component $18,000
(from Line 15)

18. Initial Penalty Target Figure $35,364
(Line 16 + Line 17)

SIGNATURE____________________ DATE____________

UST PENALTY COMPUTATION WORKSHEET

Assessments for each violation should be determined on separate worksheets and totaled. (If more space is needed, attach separate sheet.)

PART 1 - BACKGROUND

Company name: Darby Distributors

Regulation violated: 40 CFR section 280.74 - Failure to maintain records capable of demonstrating compliance with tank closure requirements.

Previous violations: None

Date of requirement: 3/17/89

Date of inspection: N/A

Date of compliance: 5/20/89

Explanation (if appropriate):

1. Days of noncompliance: 64
2. Number of tanks: 3

PART 2 - ECONOMIC BENEFIT COMPONENT

Avoided Expenditures: N/A

Basis:

Delayed Expenditures: 0

Basis: Cost of record keeping negligible.

Weighted Tax Rate: N/A

Source:

Interest Rate: N/A

Source:

AVOIDED COSTS = [**Avoided Expenditures** + (**Avoided Expenditures** x **Interest** x **Number of Days**) / **365 Days**] x **(1 - Weighted Tax Rate)**

3. Calculated Avoided Cost: $0

UST PENALTY COMPUTATION WORKSHEET

DELAYED COSTS = $\frac{\text{Delayed Expenditures} \times \text{Interest} \times \text{Number of Days}}{\text{365 Days}}$

4. Calculated Delayed Cost: $0

5. Economic Benefit Component: $0 (carry figure to Line 16).
(Line 3 + Line 4)

PART 3 - MATRIX VALUE FOR THE GRAVITY-BASED COMPONENT

Potential for Harm: Major   Extent of Deviation: Major

6. Matrix Value (MV): $1500 (from document page 16 or Appendix A)

7. Per-tank MV: $1500 (if violation is per facility, the amount on Line 7 will be the same as the amount on Line 6)
(Line 2 x Line 6)

PART 4 - VIOLATOR-SPECIFIC ADJUSTMENTS TO MATRIX VALUE

	Percentage Change (+ or -)	x Matrix Value	= Dollar Adjustment (+ or -)	Justification for Adjustment:
8. Degree of cooperation/ noncooperation	+10%	$1500	+$150	Owner requested negotiations only after being warned of impending administrative orders
9. Degree of willfulness or negligence:	+40%	$1500	+$600	Owner appeared to take advantage of operator's ignorance of requirements.
10. History of noncompliance:	0	$1500	0	N/A
11. Unique factors:	0	$1500	0	N/A
12. Adjusted Matrix Value (Line 7 + Lines 8-11)			$2250	

UST PENALTY COMPUTATION WORKSHEET

PART 5 - GRAVITY-BASED COMPONENT

Level of
Environmental Sensitivity High

Justification: Release could impact several drinking-water wells and a river used by humans for recreation and by wild life as a source of drinking water.

13. ESM (from document Page 21) 2

14. DNM (from document Page 21) 1

GRAVITY-BASED COMPONENT = Adjusted Matrix Value x Environmental Sensitivity Multiplier x Days of Noncompliance Multiplier

GBC = $2250 × 2 × 1 = $4500

15. Gravity-Based Component: $4500
(Line 12 x Line 13 x Line 14)

PART 6 - INITIAL PENALTY TARGET FIGURE

16. Economic Benefit Component $0
(from Line 5)

17. Gravity-Based Component $4500
(from Line 15)

18. Initial Penalty Target Figure $4500
(Line 16 + Line 17)

Total Initial Penalty Target for Darby Distributors:
= Violation #1 + Violation #2 + Violation #3
= $6750 + $35,364 + $4500 = $46,614

SIGNATURE__________ DATE__________

EXAMPLE 4

BACKGROUND

Inspection Date: December 15, 1991

Facility Name and Description: Jerry's Gas and Grocery is a medium-sized facility in a commercial section of town. The facility has 4 USTs, 3 of which were installed in 1968 and one in 1989. It was estimated that the company's taxable income was $70,000 in 1990.

Violations: On October 16, 1991, the Agency discovered that Jerry's Gas and Grocery had a release. At the time of the release, an adequate method of release detection was not in use at the facility, constituting a violation of 40 CFR section 280.40(c) for the 3 tanks installed in 1968. The Agency sent written notification (after informing the owner of the release by telephone) of the release to the facility and requested, among other things, that the facility report evidence of financial responsibility within 30 days. While conducting a file review on December 15, the compliance staff observed that the facility had failed to report this evidence, in violation of 40 CFR section 280.106(a)(1). A site inspection conducted on this date indicated that an adequate method of release detection was still not in use.

Owner/Operator Response: When notified of these violations, the owner submitted evidence that he had acquired a letter of credit from a bank to meet the FR requirement and began to conduct inventory control and daily monitoring immediately, and arranged for tank tightness tests. The owner, however, had failed to initiate corrective actions (beyond the initial abatement measures) for lack of funds. The owner's failure to report his financial assurance mechanism within the required time period, therefore, delayed the contacting of the bank and the collection of funds with which to initiate corrective action.

Previous Actions at Facility: In 1989, the facility was assessed penalties for failure to notify the Agency of the new UST installation.

Current Status at Site: Because an adequate method of release detection was not in operation, the release went undetected for a matter of months. The geology in the area of the facility is fractured shale. The facility is located in a commercial area. There are no drinking water wells or sensitive wildlife receptors within a 5-mile radius of the site.

PENALTY CALCULATION DATA

Violation: 40 CFR section 280.40(c)

Days of violation: 358 days, from the latest required date of compliance (December 22, 1990) to the actual date of compliance (December 15, 1991).

Avoided expenditures: $2455 total = $895 labor for 358 days, at $2.50 per day (estimated cost for labor needed to conduct daily inventory control based on 1/2 hour labor at $5.00 per hour) + $1560 for tightness testing for 3 tanks (where the average cost for tank tightness testing is $520 per tank).

Delayed expenditures: None.

Interest rate: 18.1% (the equity discount rate used in the BEN model for 1991).

Tax rate: 18% (the weighted average rate for a company with taxable income of $70,000).

PENALTY CALCULATION DATA

Violation: 40 CFR section 280.106(a)(1)

Days of Violation: 30 days from the latest required date of compliance (November 15, 1991) to the actual date of compliance (December 15, 1991).

Avoided expenditures: $8219 = Amount of interest avoided on $1,000,000 letter of credit because of failure to provide the Agency with evidence of financial responsibility (based on 30 days of interest at 10%, the rate charged by Jerry's bank for letter of credit drawdown).

Delayed expenditures: None.

Interest rate: 18.1% (the equity discount rate used in the BEN model for 1990 and 1991).

Tax rate: 18% (the weighted average rate for a company with taxable income of $70,000).

[NOTE: The numbers used to determine avoided and delayed expenditures were chosen for convenience only. They do not necessarily represent true costs in any State or Region in the country.]

UST PENALTY COMPUTATION WORKSHEET

Assessments for each violation should be determined on separate worksheets and totaled. (If more space is needed, attach separate sheet.)

PART 1 - BACKGROUND

Company name Jerry's Gas & Grocery

Regulation violated 40 CFR section 280.40 (a)(1) - Failure to have release detection by compliance date (12/22/90).

Previous violations Notification (1989) - penalties assessed for failure to notify of new UST installation.

Date of requirement 12/22/90

Date of inspection 12/15/91

Date of compliance 12/15/91

Explanation (if appropriate):

1. Days of noncompliance 358
2. Number of tanks 4 (or 3)* — *(only 3 tanks require release detection).

PART 2 - ECONOMIC BENEFIT COMPONENT

Avoided Expenditures $2455 — Basis: $2.50 per day for monitoring; $520 per UST tightness test } × 3

Delayed Expenditures N/A — Basis: N/A

Weighted Tax Rate 0.18 (18%) — Source: MTR for income of $70,000

Interest Rate 0.181 (18.1%) — Source: BEN model (equity discount rate)

$$\text{AVOIDED COSTS} = \left[\text{Avoided Expenditures} + \frac{\text{Avoided Expenditures} \times \text{Interest} \times \text{Number of Days}}{365 \text{ Days}}\right] \times (1 - \text{Weighted Tax Rate})$$

$$AC = \left[\$2455 + \frac{\$2455 \times .181 \times 358}{365}\right] \times (1 - .18) = \$2370$$

3. Calculated Avoided Cost: $2370

UST PENALTY COMPUTATION WORKSHEET

DELAYED COSTS = Delayed Expenditures x Interest x Number of Days / 365 Days

4. Calculated Delayed Cost: 0

5. Economic Benefit Component: $2370 (carry figure to Line 16).
(Line 3 + Line 4)

PART 3 - MATRIX VALUE FOR THE GRAVITY-BASED COMPONENT

Potential for Harm: Major

Extent of Deviation Major

6. Matrix Value (MV): $1500 (from document page 16 or Appendix A)

7. Per-tank MV: $4500 (if violation is per facility, the amount on Line 7 will be the same as the amount on Line 6)
(Line 2 x Line 6)

PART 4 - VIOLATOR-SPECIFIC ADJUSTMENTS TO MATRIX VALUE

	Percentage Change (+ or -)	x Matrix Value	= Dollar Adjustment (+ or -)	Justification for Adjustment:
8. Degree of cooperation/ noncooperation	0	$4500	0	Complied as required following notification.
9. Degree of willfulness or negligence:	0	$4500	0	N/A
10. •History of noncompliance:	+30%	$4500	+$1350	Previous violation involving penalties
11. Unique factors:	0	$4500	0	
12. Adjusted Matrix Value (Line 7 + Lines 8-11)			$5850	

UST PENALTY COMPUTATION WORKSHEET

PART 5 - GRAVITY-BASED COMPONENT

Level of Environmental Sensitivity *Moderate*

Justification: *Release is not likely to have impact on ground or surface water. Potential impact on the environment is minimal, although potential human receptors are present. Fractured shale would complicate remediation.*

13. ESM (from document Page 21) *1.5*

14. DNM (from document Page 21) *2.5*

GRAVITY-BASED COMPONENT = Adjusted Matrix Value x Environmental Sensitivity Multiplier x Days of Noncompliance Multiplier

GBC = $5850 x 1.5 x 2.5 = $21,938

15. Gravity-Based Component: *$21,938*
(Line 12 x Line 13 x Line 14)

PART 6 - INITIAL PENALTY TARGET FIGURE

16. Economic Benefit Component *$2370*
(from Line 5)

17. Gravity-Based Component *$21,938*
(from Line 15)

18. Initial Penalty Target Figure *$24,308*
(Line 16 + Line 17)

SIGNATURE ____________________ DATE ____________

UST PENALTY COMPUTATION WORKSHEET

Assessments for each violation should be determined on separate worksheets and totaled. (If more space is needed, attach separate sheet.)

PART 1 - BACKGROUND

Company name: Jerry's Gas & Grocery

Regulation violated: 40 CFR section 280.106(a)(1) - Failure to report evidence of financial assurance within 30 days of discovering a release.

Previous violations: Notification (1989) - penalties assessed for failure to notify of new UST installation.

Date of requirement: 11/15/91 Date of inspection: 12/15/91

Date of compliance: 12/15/91 Explanation (if appropriate):

1. Days of noncompliance: 30

2. Number of tanks: 4

PART 2 - ECONOMIC BENEFIT COMPONENT

Avoided Expenditures: $8219 Basis: Avoided interest that would have been paid on $1,000,000 letter of credit for 30 days

Delayed Expenditures: 0 Basis: Negligible

Weighted Tax Rate: 0.18 (18%) Source: MTR for income of $70,000

Interest Rate: 0.181 (18.1%) Source: BEN model (equity discount rate)

$$\text{AVOIDED COSTS} = \left[\text{Avoided Expenditures} + \frac{\text{Avoided Expenditures} \times \text{Interest} \times \text{Number of Days}}{365 \text{ Days}}\right] \times (1 - \text{Weighted Tax Rate})$$

$$AC = \left[\$8219 + \frac{\$8219 \times .181 \times 30}{365}\right] \times (1 - .18) = \$6840$$

3. Calculated Avoided Cost: $6840

UST PENALTY COMPUTATION WORKSHEET

DELAYED COSTS = $\frac{\text{Delayed Expenditures} \times \text{Interest} \times \text{Number of Days}}{\text{365 Days}}$

4. Calculated Delayed Cost: 0

5. Economic Benefit Component: $6840 (carry figure to Line 16).
(Line 3 + Line 4)

PART 3 - MATRIX VALUE FOR THE GRAVITY-BASED COMPONENT

Potential for Harm: Moderate

Extent of Deviation Major

6. Matrix Value (MV): $750 (from document page 16 or Appendix A)

7. Per-tank MV: $750 (if violation is per facility, the amount on Line 7 will be the same as the amount on Line 6)
(Line 2 x Line 6)

PART 4 - VIOLATOR-SPECIFIC ADJUSTMENTS TO MATRIX VALUE

	Percentage Change (+ or -)	x Matrix Value	= Dollar Adjustment (+ or -)	Justification for Adjustment:
8. Degree of cooperation/ noncooperation	0	$750	0	Complied as required following notification
9. Degree of willfulness or negligence:	0	$750	0	N/A
10. History of noncompliance:	+30%	$750	+$225	Previous violation involving penalties
11. Unique factors:	0	$750	0	
12. Adjusted Matrix Value (Line 7 + Lines 8-11)			$975	

UST PENALTY COMPUTATION WORKSHEET

PART 5 - GRAVITY-BASED COMPONENT

Level of
Environmental Sensitivity Moderate

Justification: Release is not likely to have impact on ground or surface water. Potential impact on the environment is minimal, although potential human receptors are present. Fractured shale would complicate remediation.

13. ESM (from document Page 21) 1.5

14. DNM (from document Page 21) 1.0

GRAVITY-BASED COMPONENT = Adjusted Matrix Value x Environmental Sensitivity Multiplier x Days of Noncompliance Multiplier

GBC = \$975 x 1.5 x 1 = \$1462

15. Gravity-Based Component: \$1462
(Line 12 x Line 13 x Line 14)

PART 6 - INITIAL PENALTY TARGET FIGURE

16. Economic Benefit Component \$6840
(from Line 5)

17. Gravity-Based Component \$1462
(from Line 15)

18. Initial Penalty Target Figure \$8302
(Line 16 + Line 17)

Total Initial Penalty Target for Jerry's Gas & Grocery
= Violation #1 + Violation #2
= \$24,308 + \$8302
= \$32,610

SIGNATURE______________________ DATE______________

EXAMPLE 5

BACKGROUND

Inspection Date: January 8, 1990

Facility Name and Description: The Mammoth Oil facility located at 345 Pine Street has 5 USTs and is owned and operated by Mammoth Oil Company, a national petroleum marketer with taxable income over $335,000.

Violations: Upon inspection of the facility, the Agency discovered that 2 new bare steel USTs were installed on November 15, 1989 without cathodic protection. This omission constituted a violation of 40 CFR section 280.20(a)(2)(ii). The tanks failed to meet the performance standards specified in section 280.20(a)(2)(ii), or any of the codes or standards outlined by the regulations as acceptable for compliance.

Owner/Operator Response: When notified of the violation, the company's attorneys asked to enter into negotiations to determine the schedule and terms of compliance, as well as any penalties that might be assessed. The result of the negotiations was a consent order in which the owner agreed to install properly designed cathodic protection (in accordance with the National Association of Corrosion Engineers Standard RP-02-85) and pay the penalty by March 1, 1990.

Previous Actions at Facility: The facility was issued a notice of violation in 1987 for failure to notify the Agency of a new UST installation. In 1988, the company was issued two administrative orders, one compelling remediation of a release and the other assessing penalties for failure to report the release to the Agency.

Current Status at Site: At the time of the inspection, the facility was conducting a method of release detection in accordance with the requirements. The Agency determined that it was unlikely that there was a release at the present time. The geology in the area of the facility is gravel. The facility is located in an urban residential area. There are no drinking water wells or sensitive wildlife receptors within a 3-mile radius of the area.

PENALTY CALCULATION DATA

Violation: 40 CFR section 280.20(a)(2)(ii)

Days of violation: 105 days, from the required date of compliance (November 15, 1989) to the actual date of compliance (March 1, 1990).

Avoided expenditures: None.

Delayed expenditures: $3,050 x 2 USTs = $6,100 (where the average cost for installation of a cathodic protection system is $3,050 per UST).

Interest rate: 18.1% (the equity discount rate used in the BEN model for 1990).

Tax rate: 34% (the weighted average rate for a company with taxable income of $335,000).

[NOTE: The numbers used to determine avoided and delayed expenditures were chosen for convenience only. They do not necessarily represent true costs in any State or Region in the country.]

UST PENALTY COMPUTATION WORKSHEET

Assessments for each violation should be determined on separate worksheets and totaled. (If more space is needed, attach separate sheet.)

PART 1 - BACKGROUND

Company name Mammoth Oil Company

Regulation violated 40 CFR section 280.20(a)(2) - Failure to meet performance standards for cathodic protection

Previous violations Release notification (1987) - two administrative orders issued (one to compel cleanup & one to assess penalties)

Date of requirement 11/15/89

Date of inspection 1/8/90

Date of compliance 3/1/90

Explanation (if appropriate):

1. Days of noncompliance 105
2. Number of tanks 2

PART 2 - ECONOMIC BENEFIT COMPONENT

Avoided Expenditures N/A

Basis:

Delayed Expenditures $6100

Basis: Cost for cathodic protection

Weighted Tax Rate 0.34 (34%)

Source: MTR for income > $335,000

Interest Rate 0.181 (18.1%)

Source: BEN model (equity discount rate)

$$\textbf{AVOIDED COSTS} = \left[\text{Avoided Expenditures} + \frac{\text{Avoided Expenditures} \times \text{Interest} \times \text{Number of Days}}{365\ \text{Days}}\right] \times (1 - \text{Weighted Tax Rate})$$

3. Calculated Avoided Cost: 0

UST PENALTY COMPUTATION WORKSHEET

DELAYED COSTS = $\frac{\text{Delayed Expenditures} \times \text{Interest} \times \text{Number of Days}}{\text{365 Days}}$

$$DC = \frac{\$6100 \times .181 \times 105}{365} = \$318$$

4. Calculated Delayed Cost: $318

5. Economic Benefit Component: $318 (carry figure to Line 16).
(Line 3 + Line 4)

PART 3 - MATRIX VALUE FOR THE GRAVITY-BASED COMPONENT

Potential for Harm: Moderate

Extent of Deviation Moderate

6. Matrix Value (MV): $500 (from document page 16 or Appendix A)

7. Per-tank MV: $1000 (if violation is per facility, the amount on Line 7 will be the same as the amount on Line 6)
(Line 2 x Line 6)

PART 4 - VIOLATOR-SPECIFIC ADJUSTMENTS TO MATRIX VALUE

	Percentage Change (+ or -)	x Matrix Value	= Dollar Adjustment (+ or -)	Justification for Adjustment:
8. Degree of cooperation/ noncooperation	0	$1000	0	Company agreed to enter into negotiations and pay penalty
9. Degree of willfulness or negligence:	+50%	$1000	+$500	As national marketers, company would have been aware of the requirements
10. History of noncompliance:	+50%	$1000	+$500	Previous violation with two administrative orders.
11. Unique factors:	0	$1000	0	N/A
12. Adjusted Matrix Value (Line 7 + Lines 8-11)			$2000	

UST PENALTY COMPUTATION WORKSHEET

PART 5 - GRAVITY-BASED COMPONENT

Level of
Environmental Sensitivity Moderate

Justification: Facility is located in residential area with no nearby drinking-water wells or wildlife receptors. However, gravel would permit migration of released product.

13. ESM (from document Page 21) 1.5

14. DNM (from document Page 21) 1.5

GRAVITY-BASED COMPONENT = Adjusted Matrix Value x Environmental Sensitivity Multiplier x Days of Noncompliance Multiplier

GBC = $2000 x 1.5 x 1.5 = $4500

15. Gravity-Based Component: $4500
(Line 12 x Line 13 x Line 14)

PART 6 - INITIAL PENALTY TARGET FIGURE

16. Economic Benefit Component $ 318
(from Line 5)

17. Gravity-Based Component $4500
(from Line 15)

18. Initial Penalty Target Figure $4818
(Line 16 + Line 17)

SIGNATURE_______________ DATE_______________

United States
Environmental Protection
Agency

Office of
Solid Waste and
Emergency Response

EPA

DIRECTIVE NUMBER: 9610.16

TITLE: Guidance for Federal Field Citation Enforcement

APPROVAL DATE: October 6, 1993

EFFECTIVE DATE: October 6, 1993

ORIGINATING OFFICE: Office of Underground Storage Tanks (OUST)

☒ **FINAL**

☐ **DRAFT**

STATUS:

REFERENCE (other documents):

OSWER Directive 9610.11 "UST/LUST Enforcement Procedures Guidance Manual"

OSWER Directive 9610.12 "U.S. EPA Penalty Guidance for Violations of UST Regulations"

OSWER OSWER OSWER
E DIRECTIVE DIRECTIVE D

United States Environmental Protection Agency
Washington, DC 20460

EPA OSWER Directive Initiation Request

1. Directive Number: 9610.16

2. Originator Information

Name of Contact Person	Mail Code	Office	Telephone Code
Jerry Parker	5402W	OUST	703-308-8884

3. Title

Guidance for Federal Field Citation Enforcement

4. Summary of Directive (include brief statement of purpose)

Provides guidance to U.S. EPA Regional Offices on setting up and implementing a federal field citation program. Field citations represent a new enforcement tool to be used in certain situations where owners/operators of underground storage tanks violate UST regulations

5. Keywords

underground storage tanks, enforcement, citation, penalty, settlement offer

6a. Does This Directive Supersede Previous Directive(s)? [ ] No [X] Yes What directive (number, title) 9610.14

b. Does It Supplement Previous Directive(s)? [ ] No [X] Yes What directive (number, title) 9610.12

U.S. EPA Penalty Guidance for Violations of UST Regulations

7. Draft Level

[ ] A – Signed by AA/DAA [X] B – Signed by Office Director [ ] C – For Review & Comment [ ] D – In Development

8. Document to be distributed to States by Headquarters? [ ] Yes [x] No

This Request Meets OSWER Directives System Format Standards.

9. Signature of Lead Office Directives Coordinator	Date
Shushona Clark (signature) Shushona Clark, OUST Directives Coordinator	10/6/93
10. Name and Title of Approving Official	**Date**
David Ziegele, Director, OUST (signature)	10/6/93

EPA Form 1315-17 (Rev. 5-87) Previous editions are obsolete.

OSWER OSWER OSWER O

VE DIRECTIVE DIRECTIVE DIRECTIVE

GUIDANCE FOR

FEDERAL FIELD CITATION

ENFORCEMENT

U.S. ENVIRONMENTAL PROTECTION AGENCY

OFFICE OF UNDERGROUND STORAGE TANKS

October 1993

GUIDANCE FOR FEDERAL FIELD CITATION ENFORCEMENT

I. Federal Enforcement

Overview

The Office of Underground Storage Tanks' (OUST) program implementation approach is to build UST programs at the state level since states will be primarily responsible for the enforcement of UST regulations. Regions perform compliance inspections at UST sites or take enforcement actions, generally in conjunction with or in the place of a state when the state lacks enforcement resources, and on Indian Lands or at federal facilities.

One enforcement option is the use of field citations, "traffic ticket"-styled citations issued on-site by inspectors, generally carrying a penalty. Field citations are currently being used by a number of environmental programs on the federal, state, and local levels, including UST programs. In the experience of many state and local UST enforcement programs, field citations are extremely useful in addressing many prevalent, clear-cut violations that are relatively easy to correct. Addressing these violations using established enforcement methods, such as formal administrative proceedings under 40 CFR Part 22, requires a greater commitment of staff time and resources, which may be difficult to obtain or which must compete with time and resources that staff directs toward releases or violations that are not appropriately addressed by the field citation program. When a citation program is properly designed, violators issued citations for clear-cut violations have greater incentive to correct problems and pay penalties than to contest. Thus, in appropriate circumstances, field citation enforcement is less resource-intensive than traditional methods of UST enforcement. Resources are saved as citations are issued on the spot, and preparation of formal legal documents and procedures, such as administrative appeals, are minimized.

While field citations were developed to expedite the enforcement process, they also encourage owners and operators to come into compliance in an effective and resource-efficient manner. By removing the incentive to expend their time and resources litigating the large penalties typical of more formal enforcement actions, owners and operators who receive field citations should see a clear advantage in focusing their energy and economic resources on achieving compliance. Thus, field citations are a critical component of OUST's efforts to achieve high rates of compliance among regulated entities with minimal expenditure of public and private resources.

This guidance is intended to help the regions establish and operate field citation programs, but should not be construed to mandate the use of field citations in place of other existing enforcement mechanisms. When an inspection is conducted, there are in essence three potential outcomes: (1) compliance (no enforcement action taken), (2) non-compliance followed by issuance of warnings or field citations, or (3) non-compliance followed by more traditional enforcement. (Of course, in theory there is a fourth outcome, where there is non-compliance and no action is taken. However, this outcome is neither common nor desirable.) Field citations therefore are not separate from more formal enforcement mechanisms; they are complementary aspects of the enforcement program. (See Figure 1)

Field citations and more formal enforcement are procedurally distinct, but they share an identical objective: increasing the rate of regulated community compliance. Each should be chosen when the appropriate conditions for its use exist. For example, field citations should be chosen when the violation is clear-cut, easily verifiable, easily correctable, and on the list of citable violations. Generally, field citations should only be issued to first-time violators. More formal enforcement must be pursued for all other violations. If a field citation is ignored, or the owner or operator otherwise refuses to settle, more formal enforcement should be pursued.

UST program staff and legal counsel from several regions participated in a workgroup effort to develop procedures for federal enforcement using field citations. This guidance document is a result of that effort. It attempts to serve the workgroup participants' interest in using field citations in a variety of circumstances and addresses concerns that an enforcement program be fairly and uniformly applied across regions. Some key components of the field citation program are identical from region to region. However, the flexibility provided in this guidance and the relationship between field citations and existing enforcement capabilities should provide considerable room for accommodating local needs. On this score, it is important to emphasize that field citation enforcement will not supplant existing enforcement options. Discretion to exercise existing options for warnings and other enforcement tools remains unchanged by the introduction of field citations, which should blend into regional enforcement choices. Also, regions will continue to select which violations or facilities to target, within the parameters of this guidance, based on local needs and subject to previously issued enforcement guidance. Finally, the availability of federal field citations should not diminish the regions' efforts to assist states and localities in building strong UST enforcement programs.

In February 1993, OUST, the Office of Enforcement, and the Office of General Counsel held a conference to review the federal field citation program, assess its effectiveness, identify potential improvements, and provide valuable "how-to" information to those regions in the process of developing field citation programs. After presentation of data from the regions that had implemented field citation programs, reports, and general discussion, the conference addressed a number of issues, including implementation of regional programs, regional targeting and tracking procedures, standard operating procedures, use of field citations at federal facilities and on Indian lands, and legal issues associated with field citations.

Figure 1:

THE ROLE OF FIELD CITATIONS IN FEDERAL UST ENFORCEMENT [IDEALIZED AND SIMPLIFIED PROCESS]

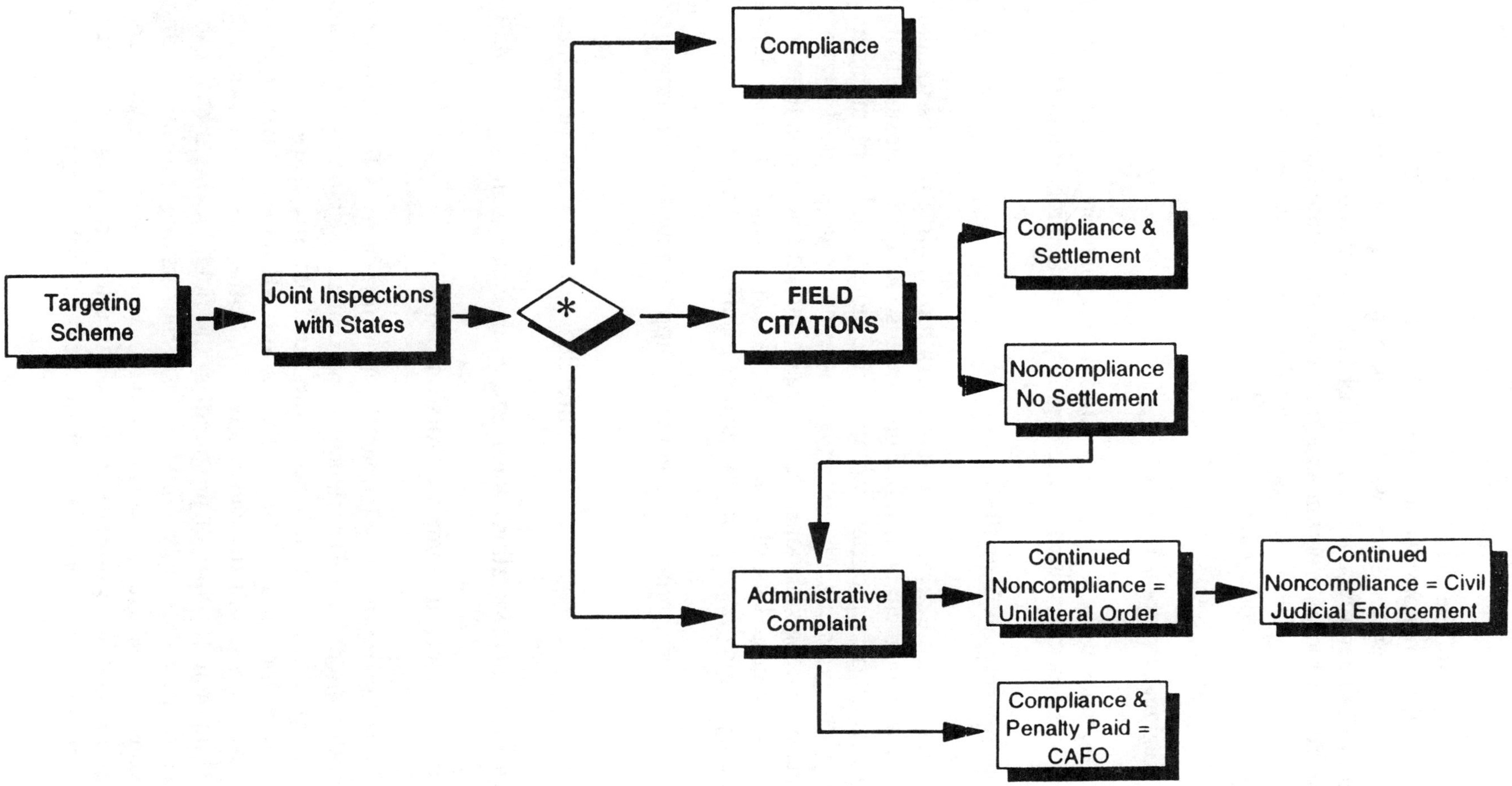

Refer to guidance for appropriate violations to be addressed by field citations.

One goal of the conference was to publicize the considerable success of field citations in those regions where they had been used, and thus encourage all regions to initiate development of their own field citation programs. Similarly, this guidance document has been revised as a result of input from participants at the conference in order to reflect recent program experience and facilitate development of regional field citation programs. The Office of General Counsel and the Office of Enforcement have endorsed the use of field citations and have reviewed and concurred with all changes to the document.

II. Regional Program Elements

Guidance for regions is presented in the following sections. The guidance should be considered in the context of the region's overall enforcement strategy and priorities.

Selecting Appropriate Violations

This guidance provides a framework for allowing regions to address certain violations with field citations. The guidance is intended to ensure that each region develops its list of appropriate violations judiciously and implements its program reasonably by providing a list of violations appropriate for field citations and guidelines for selection among violations. Each region should select violations to be cited only from those on the attached list of violations for which a penalty amount is indicated. Consistency among regions will be further assured by training.

The following criteria are generally appropriate for selecting the violations to be cited:

- Select violations which are clear-cut and easily verifiable.
- Select violations which are easily correctable.
- Select first-time violators only.

Determining which violations are appropriate for a field citation program requires considerable discretion. Experience shows that field citation programs work most effectively in achieving compliance if the violations are clear-cut and the inspectors exercise little discretion in citing the violations. Established field citation programs have found that easily identifiable violations (e.g., "either they have it or they don't") require the least amount of inspector judgment in the field, making it easier to provide clear guidance to inspectors and facilitate consistency among inspectors. On the other hand, the regions may believe that certain violations, while clear-cut, are very serious in terms of environmental harm threatened and require a more formal enforcement response. The list of violations appropriate for the field citation program, which accompanies this guidance, relieves the regions of some of the burden of making these decisions. However, it is the responsibility of each region to designate which of these violations will be appropriate candidates for its field citation program given specific regional needs and resources, and list the violations in the region's Standard Operating Procedures or use the attached list verbatim (see section below on Regional Standard Operating Procedures).

In selecting a preferred approach, a region may choose to target a certain prevalent or high priority violation or violations, e.g., the leak detection requirements. This may be a good strategy for a region to use if a state program lacks enforcement authority or regulations in a certain program area and the region needs to fill a key gap in coverage or send an important message to violators. However, if a region enforces in the place of the state, the region may find it advantageous to include all appropriate violations in the field citation enforcement program, as long as they meet the above-referenced criteria.

Guidance for When to Use Citations

This guidance establishes procedures for issuing citations, and describes some appropriate circumstances for inspectors to issue citations. Since the inspector is the one who must implement the program in the field, the regions must clearly establish the extent of discretion allowed to inspectors in determining whether to issue field citations within the general parameters set forth here. Field citations provide an additional enforcement tool, and inspectors must be instructed in how to respond when they find violations which are addressed appropriately with field citations.

The proper use of field citations must be measured against the backdrop of the regions' existing authority to issue warnings or pursue other existing enforcement measures for all violations of UST requirements. Although the primary objective of any enforcement program is to achieve compliance, formal enforcement mechanisms, such as those found in 40 CFR Part 22, normally will be more appropriate in particular circumstances. These circumstances include, among others, instances involving repeat violations or where payment of a more significant penalty may be more effective in achieving EPA's enforcement goals.

This guidance is intended to provide a framework for the inspector's discretionary use of the field citation enforcement option. Therefore, the guidance is phrased in terms of the action an inspector would take in the typical case, but leaves room for exception if the circumstances in the inspector's judgment so warrant.

The following discussion outlines the three basic enforcement options available to address violations of UST requirements:

Warnings

- Although warnings can be useful as a first step in the enforcement process, regional inspectors generally should consider issuing citations in all cases where violations which the region has determined are appropriately addressed with a field citation are discovered. Field citations are designed to uniformly address certain violations and promote a quick resolution of the violation with the assessment of a small penalty. Therefore, when a region inspects a facility, inspectors should consider issuing a field citation rather than a warning for a violation or violations which the region has determined may be an appropriate candidate for its field citation program. In order to keep the field citation program distinct from existing enforcement programs, if a warning is issued, it <u>may not</u> be in the form of a field citation without a penalty.

Citations

There are several situations in which inspectors will typically issue citations:

- Inspectors may issue citations for as many violations as are identified at a site. However, if the number of violations found at a site exceeds "x" (a number set by each region), the inspector should forego issuing a field citation and use more formal, existing enforcement methods instead.

Once a region has selected its list of violations appropriate for the field citation program and trained inspectors in procedures for issuing field citations, inspectors may routinely issue field citations for all appropriate violations found at a facility. Each region will have the discretion to place an upper limit on the number of violations that may be cited at one site. The threshold should be set below the point beyond which the number of violations, regardless of the nature of those violations, suggests that a facility was seriously out of compliance and requires a more formal enforcement response. At this point, a more formal enforcement response is likely to be more effective than the use of field citations. As a general matter, a suggested threshold is between three and ten violations.

This guidance does not suggest any predetermined cap on the cumulative dollar amount of penalties that may be set out in a field citation. However, there is a natural cap to the extent that each region will be foregoing issuing field citations if the number of citable violations at a site exceeds a number fixed by the region (see preceding section). The region may want to consider the practical effect that the total cumulative penalty amount proposed in a field citation may have on settlement incentives. This amount should be below the point above which a violator no longer has an incentive to correct the violations and pay the penalty instead of resisting compliance.

- During joint inspections, regional inspectors should usually not cite for violations that are cited by the state inspector where state penalties are at least equivalent.

As states are the primary enforcers in the program, regions usually will take enforcement actions only in the circumstances noted in the first paragraph of this guidance document. Therefore, it is likely that during joint inspections regional inspectors will defer to the state program's regulations or authorities and not cite for violations that state inspectors cite. Generally, this will be the case where state penalties are at least equivalent with federal penalties. On the other hand, there may be cases where a field citation would serve an important federal enforcement objective, for example, sending a signal to the regulated community about an enforcement initiative. In these cases, a field citation or other federal enforcement measure might reinforce the state's message.

- Inspectors will usually issue citations to first-time violators only. If upon follow-up inspection a cited violation has not been corrected, the inspector should generally use Part 22 procedures, or, if a later inspection uncovers a different violation, the inspector should not use a field citation.

Limiting the use of field citations to first-time violators makes sense if it appears to the inspector that the citation and penalty will convince a violator to bring a facility into compliance and to keep it in compliance. The inspector should be guided by the goal of the field citation program, which is to achieve rapid and resource-efficient compliance, rather than to penalize owners and operators for regulatory violations. When conducting inspections, it is critical that the inspector fully conduct the inspection and thoroughly complete the inspection report. If a field citation is not issued because the number of violations is above the threshold for field citations, or the field citation settlement form is not returned, the Agency may choose to pursue standard enforcement based on the inspection report. Therefore, while field citations may expedite the correction and penalty phases of enforcement, the quality and effort applied to the underlying inspection should not be abbreviated.

Standard Enforcement

If an inspector discovers not only violations that are appropriate for the field citation program, but other violations as well, the inspector should address all of the violations at the site using more formal, existing enforcement methods or refer the case to the state for appropriate action. As used in this guidance, more formal enforcement typically refers to the procedures for issuing administrative complaints/compliance orders (including those assessing civil penalties) and conducting the administrative enforcement process governed by 40 CFR Part 22, the Consolidated Rules of Practice Governing the Administrative Assessment of Civil Penalties and the Revocation or Suspension of

Permits ("Part 22" or "CROP"). CROP outlines the major steps in the administrative adjudication process and presents the various authorities and duties of Agency officials in the process. More formal enforcement methods also include issuance of corrective action orders pursuant to 40 CFR Part 24 or civil judicial enforcement of the UST requirements.

In selecting those violations which are appropriate for field citations, the regions will, in effect, also be identifying violations which, because of their potential for environmental harm or other characteristics (i.e., not clear-cut), should be addressed using the more formal, existing enforcement mechanisms. The more formal enforcement methods may also be the appropriate response in some circumstances where field citations would otherwise be appropriate (for example, if the total number of individual violations which are appropriate for a field citation surpasses the threshold for multiple violations or are repeated). Another case where a clear-cut violation might be addressed by more formal enforcement is the case of a clear-cut but not easily correctable violation. In these cases, a field citation may not serve the goal of encouraging compliance and might appear to treat the violator mildly compared to penalties applicable under the penalty policy. In general, the regions will need to assess how to maximize resources while bringing as many facilities into compliance as possible.

This guidance is phrased so that inspectors will know what action to take in the typical case. When in the exercise of their enforcement discretion they determine that deviation from this guidance will result in more effective compliance or a more efficient use of enforcement resources, inspectors are not bound to follow this guidance. This approach is consistent with the guidance found in other EPA penalty policies and procedures.

Guidance for Penalty Amounts

In order to ensure that penalty amounts applied by different regions for the same violations are consistent, standard penalty amounts have been set by this guidance. These penalties are in the amounts of either $50, $150, or $300. Consistency among regions is important to achieve fairness in the treatment of the regulated community in selecting regional penalty amounts. In the case of multiple violations, penalties should be totaled. In general, field citation programs set penalty amounts according to the severity of each violation or category of violations. Penalties should be assessed per facility rather than per tank.

The size of the penalties attached to violations is important to the success of a field citation program. Penalties that are relatively high (e.g., greater than $500 per violation) may discourage owners and operators from agreeing to settle. On the other hand, penalties need to be high enough to convince owners and operators that EPA is serious about obtaining compliance and deterring others when issuing field citations. In general, the field citation program should operate optimally when the penalties are geared primarily to achieving compliance rather than to penalizing violators.

Form of the Citation

While each region will have considerable discretion in tailoring its field citation program within the boundaries set forth in this guidance, the regions must use the approved field citation form or obtain approval for any region-specific citation form from OUST in writing, after first having obtained approval of Regional Counsel. OUST will obtain concurrence for any proposed change from both the Office of General Counsel and the Office of Enforcement before authorizing such a change.[1] This approach will ensure that the field citations used are legally supportable and designed to accommodate the program elements described in this guidance. In addition, use of a standard citation form will promote uniformity across regions in the issuance of field citations.

The field citation form developed by OUST is entitled "Expedited Enforcement Compliance Order and Settlement Agreement". There are two versions of the field citation form. The first version was designed for use in states that have not yet received state program approval; it instructs inspectors to cite for violations of the federal regulations. The second version of the form was designed for use in states that have received state program approval; it instructs inspectors to cite for violations of the state regulations. Otherwise, the two versions of the citation and the instructions for their use are identical. Because inspectors in approved states still rely upon federal authorities for enforcement, the references to federal authorities which appear on the second version still apply. A copy of each citation form and the instructions for its use are attached at the end of this guidance.

Regions should specify in their SOP and on the citation form the location where the citation form and penalty payment should be sent. Citation forms and penalties are often sent to different addresses. Generally, penalties should be sent to the same location where enforcement monies are normally sent within the region. The inspector should clearly explain this distinction to the violator if the locations to which payment and the citation form should be sent are not the same.

Procedures For Issuance of Citations, Follow-up, and Extensions

The field citation represents the issuance of an order pursuant to RCRA § 9006 to address violations listed in RCRA § 9006(d), coupled with a short-form settlement agreement. Each region, as it determines is appropriate, must delegate to individual inspectors the authority necessary to issue the citation form. The violator is given an

---

[1] OUST approved an alternate citation form which provides for issuance of a Notice of Violation prior to issuance of a compliance order and settlement agreement.

opportunity to resolve the enforcement action expeditiously by correcting the violation and settling for a lesser penalty amount than might be assessed according to the penalty policy if formal administrative proceedings were initiated. The lower penalty amount reflects the time and expense saved by the Agency over that normally incurred in utilizing more formal enforcement methods. If the violator does not accept the settlement agreement within the time provided in the field citation, the compliance order is automatically withdrawn. The Agency's policy is then to pursue other enforcement options for the violations cited.

Thus, the violator has only two options: accept the field citation or risk more formal enforcement proceedings. If a violator refuses to accept the terms of the field citation or if it is determined that a violator has not fully complied with the terms of a signed settlement agreement, follow-up enforcement should be initiated by EPA. Such follow-up enforcement should be more stringent than the field citation settlement terms in order to achieve compliance and ensure the integrity of the field citation program. Follow-up enforcement is particularly crucial in environmentally sensitive areas, as these areas are most likely to suffer severe adverse impacts from prolonged violations that increase the likelihood of a release.

While it is essential that the regions take steps to conduct follow-up inspections in those cases where the violator does not settle and come into compliance, some program of follow-up inspections is also important for monitoring those violators who do accept the settlement offer, pay the fine and certify that they have achieved compliance. The regions should institute follow-up inspections at a subset of those facilities that have reported a return to compliance as a result of a field citation, as appropriate, to assure that owners and operators are taking the compliance actions that they claim to be taking. In addition, these inspections can be used to determine whether the compliance actions are completed within the 30-day period. Without proper follow-up, the region cannot be sure the field citation program is achieving its goal of assuring compliance through cost-effective means.

There might be circumstances where a 30-day extension of the 30-day period provided to pay the fine and correct the violations would be appropriate. The region should condition the grant of a 30-day extension on the following: (1) the owner or operator files a formal request for the extension, (2) the owner or operator demonstrates that there are factors beyond the control of the owner or operator that necessitate an extension, and (3) the region believes that compliance will be achieved within the period of the extension. The circumstances justifying the extension will typically involve unusual difficulties in obtaining parts or securing and scheduling contractors to install equipment within the initial 30-day period, or delays in securing loan approval to finance repairs. Merely neglecting to seek expert assistance or equipment in a timely fashion should not in itself justify an extension. The region should document the reasoning for granting any extension in the file.

In certain circumstances, the region might also consider extending the 30-day extension for cases in which a *force majeure* event occurs. "*Force majeure*," for the purpose of this guidance, is defined as any event arising from causes beyond the control of the owner or operator or of any entity controlled by the owner or operator (including, but not limited to, the owner or operator's contractors and subcontractors) that delays or prevents the performance of any obligation under the field citation, despite the owner or operator's best efforts to fulfill the obligation. The owner or operator's "best efforts to fulfill the obligation" include using best efforts to anticipate any potential *force majeure* event, and best efforts to address the effects of any potential *force majeure* event (1) as it is occurring and (2) following the *force majeure* event, such that the delay is minimized to the greatest extent possible.

Examples of *force majeure* events include extreme weather conditions that render scheduled excavation of tanks or piping impossible, or an act of God, such as flooding or an earthquake that disrupts normal commerce. Events not constituting *force majeure* include, but are not limited to, financial inability to perform any actions required by the field citation and unanticipated or increased costs or expenses associated with the implementation of the field citation. The owner or operator should, nevertheless, provide written justification for the second extension and the region should document its reason for granting the extension for the file.

Regional Standard Operating Procedures

Each region should develop a set of Standard Operating Procedures (SOP) for the use of field citations. The SOP generally should include information on what training is required for inspectors; what procedures are required for issuing the citation and conducting follow-up activities; how to handle requests for extensions; and what steps to follow when the terms of the field citation are not met. Training requirements may include a list of required and advisable courses to be completed by each inspector prior to certification. When providing guidelines for determining violations and penalties, the SOP should include a list of citable violations and their associated penalties, or the regions must use the attached list verbatim. If a region is using field citations in approved states, the region must include in its SOP a list of state violations and penalties that corresponds to the attached list of federal violations and penalty amounts eligible for field citations. Procedures for issuing the citation and follow-up should explain what information the inspector needs to convey to the owner or operator, and set forth the steps to be followed from issuance of the citation through settlement and case closure. The SOP should also indicate the criteria to consider when determining if a 30-day extension will be granted for paying the fine and/or coming into compliance, and separate criteria for determining whether an additional extension for a period to be determined by the region under unusual circumstances is justified. Finally, the SOP should indicate what steps, if any, should be completed before initiation of standard enforcement procedures when the terms of the field citation are not met.

A number of additional topics may be included in the regional SOP, such as regional caps on the maximum number of citable violations and penalty amount; procedures for setting up and maintaining a file system; policies for interaction with other government agencies; procedures for targeting, notifying, and entering facilities; tips on document identification and handling; and requirements for background information. It is up to the region to decide what will be included in the SOP; the goal should be to include sufficient information and guidance to allow effective and legally defensible implementation of the program in accordance with the requirements outlined in this guidance.

Issuance of Field Citations at Hazardous Substance Tanks

Issuance of federal field citations is an appropriate response to violations involving hazardous substance UST systems, as long as the violations also could be addressed with field citations at a petroleum UST site. Field citations may be issued for such violations if certain conditions are met. Specifically, the facility where the UST system is located should have no more than twelve tanks, and total hazardous substance UST capacity at the facility should be less than 40,000 gallons. If the site exceeds either of these recommended thresholds, more formal enforcement methods should be selected. The two thresholds reflect the greater danger to human health and the environment potentially posed by violations that increase the likelihood of releases from hazardous substance tanks.

Issuance of Field Citations on Indian Lands

If a region wishes to issue field citations against tribally owned or operated facilities, it should coordinate with the Senior Legal Advisor of the Office of Federal Activities, within the Office of Enforcement, before taking action. The Senior Advisor coordinates policy and management issues and legal issues in consultation with the Office of General Counsel, and will be responsible for coordinating Headquarters and regional review of the proposed enforcement action. (For further information, refer to the October 21, 1992 Office of Enforcement Memorandum from Thomas L. McCall, Jr. to Deputy Regional Administrators, Enforcement Counsels, and Regional Counsels.)

Issuance of Field Citations at Federal Facilities

[Reserved.]

Hearing Requirements

Subtitle I of RCRA provides for an opportunity for a hearing where an order is issued -- the hearing process is outlined in Part 22. As described above, the field citation has been designed as a compliance order and short-form settlement agreement.

The field citation compliance order is not an adjudicatory proceeding under 40 CFR Part 22. The violator has no right to a hearing under Part 22, since those procedures have not been invoked through issuance of a field citation. Violators who accept the terms of the settlement offer will have expressly waived their rights to a public hearing under § 9006 of RCRA. If the violator does not accept the settlement offer, the compliance order is withdrawn.

A region initiating administrative actions against a violator should follow the Part 22 procedures if a violator forgoes the settlement offered through the field citation process. The federal procedures guidance (OSWER Directive 9610.11 "UST/LUST Enforcement Procedures Guidance Manual") describes appropriate procedures in detail. Judicial enforcement may also be appropriate in certain instances, in which case the region should follow appropriate referral procedures for judicial actions.

Outreach

Regions are encouraged to publicize their field citation programs while they are still in the developmental stage, as well as when they actually begin issuing citations. It is important to inform the members of the regulated community as well as the public that the programs are being developed so they know what to expect, and to publicize violations addressed using citations, in order to demonstrate that the program is being actively implemented. A variety of methods can be employed to publicize the program, including public meetings, cooperation with trade associations, press releases, radio or television announcements, and written outreach materials.

Developing State Field Citation Programs

OUST's ultimate goal is to see state and local governments develop the authorities and capabilities needed to implement their own UST enforcement programs. As a way to help realize that goal, OUST is working with the regions to issue federal field citations to demonstrate the success of this expedited enforcement approach to the states. Federal field citations are therefore an interim step, to be used primarily until state enforcement programs are fully developed. Federal field citations represent a low-cost method of maintaining a field presence in states without their own field presence, and of improving state capabilities by training state inspectors through joint inspections. All regions should develop the capability to issue federal field citations and continuously work with states to streamline state enforcement processes, which may include developing state and local field citation programs.

Helping states to develop their own field citation programs raises a number of questions. For example, the regions will need to determine the mechanics of issuing federal field citations in states that have received state program approval. When conducting enforcement activities in approved states, federal officials rely upon federal statutory authority, but enforce the state regulatory requirements. Therefore, federal inspectors in approved states will need to use the model citation OUST has developed for use in approved states. Regions also will need to amend their SOPs to include all citable violations, identified in terms of each approved state's regulatory requirements.

Purposes and Use of This Guidance

This guidance is intended to provide overall direction for establishing regional field citation programs. As such, the role of the guidance is to enunciate the general principles that should underlie an appropriately designed field citation program; further details not contained in this guidance will be developed and transmitted to program staff through subsequent training or guidance.

This guidance and any internal procedures adopted for its implementation, are intended solely as guidance for employees of the U.S. Environmental Protection Agency. They do not constitute rulemaking or final Agency action by the Agency and may not be relied on to create a right or a benefit, substantive or procedural, enforceable at law or in equity, by any person. The Agency may take action at variance with this memorandum or its internal implementing procedures.

SELECTED VIOLATIONS OF FEDERAL UNDERGROUND STORAGE TANK REGULATIONS

Regulatory Citation	Violation	Violations Appropriate for Field Citations (Penalty Amount)
	SUBPART B -- UST SYSTEMS: DESIGN, CONSTRUCTION, INSTALLATION, AND NOTIFICATION	
	§280.20 Performance standards for new UST systems	
§280.20(a)(1)	Installation of an improperly constructed fiberglass-reinforced plastic tank	$300
§280.20(a)(2)	Installation of an improperly designed and constructed metal tank that fails to meet corrosion protection standards	$300
§280.20(a)(2)(i)	Installation of a metal tank with unsuitable dielectric coating	$150
§280.20(a)(2)(ii)	Installation of an improperly designed cathodic protection system for a metal tank	$300
§280.20(a)(2)(iii)	Improper installation of cathodic protection system for a metal tank	$150
§280.20(a)(2)(iv)	Improper operation and maintenance of tank cathodic protection system	$150
§280.20(a)(3)	Installation of an improperly constructed steel-fiberglass-reinforced-plastic tank	$300
§280.20(b)(1)	Installation of improperly constructed fiberglass-reinforced plastic piping	$300
§280.20(b)(2)	Failure to provide any cathodic protection for metal piping	$300
§280.20(b)(2)(i)	Installation of piping with unsuitable dielectric coating	$150
§280.20(b)(2)(ii)	Installation of improperly designed cathodic protection for metal piping	$300
§280.20(b)(2)(iii)	Improper installation of cathodic protection system for piping	$150
§280.20(b)(2)(iv)	Improper operation and maintenance of cathodic protection system for metal piping	$150
§280.20(c)	Failure to use a spill prevention system and an overfill prevention system	$300
§280.20(c)(1)(i)	Installation of inadequate spill prevention equipment in a new tank	$150
§280.20(c)(1)(ii)	Installation of inadequate overfill prevention equipment in a new tank	$150
§280.20(d)	Failure to install tank in accordance with accepted codes and standards	$150
§280.20(d)	Failure to install piping in accordance with accepted codes and standards	$150
§280.20(e)	Failure to provide any certification of UST installation	$150

SELECTED VIOLATIONS OF FEDERAL UNDERGROUND STORAGE TANK REGULATIONS (continued)

Regulatory Citation	Violation	Violations Appropriate for Field Citations (Penalty Amount)
	SUBPART B -- UST SYSTEMS: DESIGN, CONSTRUCTION, INSTALLATION, AND NOTIFICATION (continued)	
	280.21 Upgrading of existing UST systems	
§280.21 (a)	Failure to perform replacement, upgrade, or closure for existing substandard tank systems	$300
§280.21 (b)	Failure to meet all tank upgrade standards	$300
§280.21 (b) (1) (i)	Improper installation of interior lining for tank upgrade requirements	$150
§280.21 (b) (1) (ii)	Failure to meet interior lining inspection requirements for tank upgrade	$150
§280.21 (b) (2) (i)	Failure to ensure that tank is structurally sound before installing cathodic protection	$150
§280.21 (b) (2) (ii)	Failure to conduct monthly release detection monitoring for upgraded tank under 10 years of age	$300
§280.21 (b) (2) (iii)	Failure to meet tightness test requirements for a tank upgraded with cathodic protection	$150
§280.21 (b) (2) (iv)	Failure to meet requirements for testing for corrosion holes for a tank upgraded with cathodic protection	$150
§280.21 (c)	Failure to install any cathodic protection for metal piping upgrade requirements	$300
§280.21 (c)	Failure to meet piping tightness test requirements for metal piping after upgrade with cathodic protection	$150
§280.21 (d)	Failure to provide spill or overfill prevention system for an existing tank	$300
	280.22 Notification requirements	
§280.22(a) or	Failure to notify state or local agency within 30 days of bringing an UST system into use	$300
§280.22(b)	Failure to notify designated state or local agency of existing tank	$300
§280.22(c)	Failure to submit a separate notification form identifying all known tanks for each site where tanks are located	$150

SELECTED VIOLATIONS OF FEDERAL UNDERGROUND STORAGE TANK REGULATIONS (continued)

Regulatory Citation	Violation	Violations Appropriate for Field Citations (Penalty Amount)
	SUBPART B -- UST SYSTEMS: DESIGN, CONSTRUCTION, INSTALLATION, AND NOTIFICATION (continued)	
§280.22(e)	**Failure to certify on notification form UST system requirements of proper installation, cathodic protection, financial responsibility, and release detection**	**$150**
§280.22(f)	**Failure to provide installer certification of compliance with installation requirements on notification form**	**$150**
§280.22(g)	Failure to inform tank purchaser of notification requirements	
	SUBPART C -- GENERAL OPERATING REQUIREMENTS	
	280.30 Spill and overfill control	
§280.30(a)	**Failure to take necessary precautions to prevent overfill/spillage during the transfer of product**	**$300**
§280.30(b)	Failure to report a spill/overfill	
§280.30(b)	Failure to investigate and clean up a spill/overfill	
	280.31 Operation and maintenance of corrosion protection	
§280.31 (a)	**Failure to operate and maintain corrosion protection system continuously**	**$150**
§280.31 (b)	**Failure to ensure proper operation of cathodic protection system**	**$150**
§280.31 (c)	**Failure to inspect impressed current systems every 60 days**	**$150**
§280.31 (d)	**Failure to maintain records of cathodic protection inspections**	**$50**
	280.32 Compatibility	
§280.32	**Failure to ensure that UST system is made of or lined with materials compatible with substance stored**	**$150**

SELECTED VIOLATIONS OF FEDERAL UNDERGROUND STORAGE TANK REGULATIONS (continued)

Regulatory Citation	Violation	Violations Appropriate for Field Citations (Penalty Amount)
	SUBPART C -- GENERAL OPERATING REQUIREMENTS (continued)	
	280.33 Repairs allowed	
§280.33(a)	Failure to repair UST system in accordance with accepted codes and standards	$150
§280.33(b)	Failure to repair fiberglass-reinforced UST in accordance with accepted codes and standards	$150
§280.33(c)	Failure to replace metal piping that has released product	$150
§280.33(c)	Failure to repair fiberglass-reinforced piping in accordance with manufacturers specifications	$150
§280.33(d)	Failure to ensure that repaired tank systems are tightness tested within 30 days of completion of repair	$300
§280.33(e)	Failure to test cathodic protection system within 6 months of repair of an UST system	$150
§280.33(f)	Failure to maintain records of each repair to an UST system	$50
	280.34 Reporting and recordkeeping	
(For violations of reporting and recordkeeping, see appropriate regulatory section (e.g., reporting of releases will be under Subpart D)).		
§280.34(a)(1) or	Failure to submit notification for UST system	$300
§280.34(a)(1)	Failure to submit certification of a new installation with notification form	$300
§280.34(b)(1)	Failure to maintain analysis of site corrosion potential if corrosion protection equipment is not used	$50
§280.34(b)(2)	Failure to maintain corrosion protection equipment operation documentation	$50
§280.34(b)(3)	Failure to maintain documentation of UST system repairs	$50
§280.34(b)(4)	Failure to maintain documentation of compliance with release detection requirements	$50
§280.34(c)(1) or	Failure to maintain records at UST site and immediately available for inspection	$50
§280.34(c)(2)	Failure to maintain records at a readily available alternative site	$50

SELECTED VIOLATIONS OF FEDERAL UNDERGROUND STORAGE TANK REGULATIONS (continued)

Regulatory Citation	Violation	Violations Appropriate for Field Citations (Penalty Amount)
SUBPART D -- RELEASE DETECTION		
	280.40 General requirements for all UST systems (Applies only to petroleum tanks)	
§280.40(a)	Failure to provide adequate release detection method	$300
§280.40(b)	Failure to notify implementing agency when release detection indicates release	
§280.40(c)	Failure to provide any release detection method by phase-in date	$150
§280.40(d)	Failure to close any UST system that cannot meet release detection requirements	$300
	280.41 Requirements for petroleum UST systems	
§280.41(a) or	Failure to monitor tanks at least every 30 days, if appropriate	$300
§280.41(a)(1) or	Failure to conduct tank tightness testing every 5 years, if appropriate	$300
§280.41(a)(2)	Failure to conduct annual tank tightness testing, if appropriate	$300
§280.41(b)(1)(i)	Failure to equip pressurized piping with automatic line leak detector	$300
§280.41(b)(1)(ii)	Failure to have annual line tightness test or perform monthly monitoring on pressurized piping	$300
§280.41(b)(2)	Failure to conduct line tightness test or use monthly monitoring on suction piping	$300
	280.42 Requirements for hazardous substance UST systems	
§280.42(a)	Failure to provide release detection for an existing hazardous substance tank system	$300
§280.42(b)	Failure to provide adequate release detection for a new hazardous substance UST system	$300
§280.42(b)(1)	Failure to provide adequate secondary containment of tank for a hazardous substance UST	$300
§280.42(b)(2)	Failure to provide adequate double-walled tank/adequate lining for a hazardous substance UST	$300
§280.42(b)(3)	Failure to provide adequate external liners for a hazardous substance UST	$300
§280.42(b)(4)	Failure to provide adequate secondary containment of piping for a hazardous substance UST	$300

SELECTED VIOLATIONS OF FEDERAL UNDERGROUND STORAGE TANK REGULATIONS (continued)

Regulatory Citation	Violation	Violations Appropriate for Field Citations (Penalty Amount)
	SUBPART D -- RELEASE DETECTION (continued)	
	280.43 Methods of release detection for tanks	
§280.43(a)	Inadequate operation or maintenance of inventory control	$300
§280.43(a)(1)-(6)		$50 each
§280.43(b)	Inadequate operation or maintenance of manual tank gauging	$300
§280.43(b)(1)-(4)*		$50 each
§280.43(c)	Inadequate operation or maintenance of tank tightness testing	$150
§280.43(d)	Inadequate operation or maintenance of automatic tank gauging	$300
§280.43(d)(1)-(2)		$150 each
§280.43(e)	Inadequate operation or maintenance of vapor monitoring	$300
§280.43(e)(1)-(7)#		$150 each
§280.43(f)	Inadequate operation or maintenance of ground-water monitoring	$300
§280.43(f)(1)-(8)@		$150 each
§280.43(g)	Inadequate operation or maintenance of interstitial monitoring	$300
	280.44 Methods of release detection for piping	
§280.44	Failure to provide any release detection for underground piping	$300
§280.44(a) or	Failure to provide adequate line leak detector system for underground piping	$150
§280.44(b)	Failure to provide adequate line tightness testing system for underground piping system	$150
§280.44(c)	Inadequate use of applicable tank release detection methods	$150

*If citing more than 3 subsections, cite instead §280.43(b) or §280.41(a)
#If citing more than 1 subsection, cite instead §280.43(e)
@If citing more than 1 subsection, cite instead §280.43(f)

SELECTED VIOLATIONS OF FEDERAL UNDERGROUND STORAGE TANK REGULATIONS (continued)

Regulatory Citation	Violation	Violations Appropriate for Field Citations (Penalty Amount)
	SUBPART D -- RELEASE DETECTION (continued)	
	280.45 Release detection recordkeeping	
§280.45	**Failure to maintain records of release detection monitoring**	**$150**
§280.45(a)	**Failure to document all release detection performance claims for 5 years after installation**	**$50**
§280.45(b)	**Failure to maintain results of sampling, testing or monitoring for release detection for at least 1 year or failure to retain results of tightness testing until next test is conducted**	**$50**
§280.45(c)	**Failure to document calibration, maintenance, and repair of release detection**	**$50**
	SUBPART E -- RELEASE REPORTING, INVESTIGATION, AND CONFIRMATION	
	280.50 Reporting of suspected release	
§280.50(a)-(c)	Failure to report a suspected release within 24 hours to the implementing agency	
	280.52 Release investigation and confirmation steps	
§280.52(a)-(b)	Failure to investigate and confirm a release (if appropriate) using accepted procedures	
	280.53 Reporting and cleanup of spills and overfills	
§280.53(a)	Failure to report a spill/overfill (if appropriate) to implementing agency within 24 hours (or other specified time period)	
§280.53(b)	Failure to contain and immediately clean up a spill/overfill of less than 25 gallons	
§280.53(b)	Failure to contain and immediately clean up a hazardous substance spill/overfill	

SELECTED VIOLATIONS OF FEDERAL UNDERGROUND STORAGE TANK REGULATIONS (continued)

Regulatory Citation	Violation	Violations Appropriate for Field Citations (Penalty Amount)
	SUBPART F -- RELEASE RESPONSE AND CORRECTIVE ACTION	
§280.61	Failure to take initial response actions within specified time period after a release is confirmed	
§280.62	Failure to perform initial abatement measures and submit report within 20 days (or other specified time) of release confirmation	
§280.63	Failure to submit report on initial site characterization within 45 days (or other specified time) of release confirmation	
§280.64	Failure to perform free product removal and submit report within 45 days (or other specified time) of release confirmation	
	SUBPART G -- OUT-OF-SERVICE UST SYSTEMS AND CLOSURE	
	280.70 Temporary closure	
§280.70(a)	**Failure to continue operation and maintenance of corrosion protection system in a temporarily closed tank system**	**$150**
§280.70(a)	**Failure to continue operation and maintenance of release detection in a temporarily closed tank system**	**$300**
§280.70(b)	**Failure to comply with temporary closure requirements for a tank system for 3 or more months**	**$300**
§280.70(b)(1)-(2)		**$150 each**
§280.70(c)	**Failure to permanently close or upgrade a temporarily closed tank system after 12 months**	**$300**
	280.71 Permanent closure and changes-in-service	
§280.71(a)	**Failure to notify implementing agency of a closure or change-in-service**	**$300**
§280.71(b)	**Failure to remove all liquids and sludges for tank closure**	**$300**
§280.71(b)	**Failure to remove closed tank from the ground or fill tank with an inert solid for tank closure**	**$300**
§280.71(c)	Failure to empty and clean tank system and conduct a site assessment prior to a change-in-service	

SELECTED VIOLATIONS OF FEDERAL UNDERGROUND STORAGE TANK REGULATIONS (continued)

Regulatory Citation	Violation	Violations Appropriate for Field Citations (Penalty Amount)
	SUBPART G -- OUT-OF-SERVICE UST SYSTEMS AND CLOSURE (continued)	
	280.72 Assessing the site at closure or change-in-service	
§280.72(a)	Failure to measure (if required) for the presence of a release before a permanent closure	
§280.72(b)	If contaminated soil, contaminated ground water, or free product is discovered, failure to begin corrective action	
	280.74 Closure records	
§280.74	**Failure to maintain closure records for at least 3 years**	**$300**
§280.74	**Failure to maintain change-in-service records for at least 3 years**	**$300**
	SUBPART H -- FINANCIAL RESPONSIBILITY	
§280.93(a)	**Failure to comply with financial responsibility requirements by the required phase-in time**	**$150**
§280.93(a)(1)-(2)	**Failure to meet the requirement for per-occurrence coverage of insurance.**	**$150**
§280.93(b)(1)-(2)	**Failure to meet the requirement for annual aggregate coverage of insurance.**	**$150**
§280.93(f)	**Failure to review and adjust financial assurance after acquiring new or additional USTs**	**$150**
§280.94	**Use of an unapproved mechanism or combination of mechanisms to demonstrate financial responsibility**	**$150**
§280.95	Use of falsified financial documents to pass financial test of self-insurance	
§280.106(a)(1)	**Failure to report evidence of financial responsibility to the implementing agency within 30 days of detecting a known or suspected release**	**$150**
§280.106(a)(2)	**Failure to report evidence of financial responsibility to the implementing agency if the provider becomes incapable of providing financial assurance and the owner or operator is unable to obtain alternate coverage within 30 days.**	**$150**
§280.106(b)	**Failure to report evidence of financial responsibility to the implementing agency when new tanks are installed**	**$150**
§280.107	**Failure to maintain copies of the financial assurance mechanism(s) used to comply with financial responsibility rule and certification that the mechanism is in compliance with the requirements of the rule at the UST site or place of business**	**$150**

UNITED STATES ENVIRONMENTAL PROTECTION AGENCY
REGION ____, MAIN STREET, USA

EXPEDITED ENFORCEMENT
COMPLIANCE ORDER AND SETTLEMENT AGREEMENT

PART I: COMPLIANCE ORDER

COMPLIANCE ORDER NO. ____________________

On ____________________ Time ____________
(Date of Violation) (a.m. or p.m.)

At ______________________________________
(Name of Facility)

____
(Address of Facility)

Facility Identification Number____________________

Name of Owner, Operator or
On-site Representative ____________________
(Circle one)

____
(Address of Owner, Operator, or On-Site Representative)

An authorized representative of the United States Environmental Protection Agency (EPA) inspected this facility to determine compliance with underground storage tank regulations promulgated under Subtitle I of the Resource Conservation and Recovery Act of 1976 (42 U.S.C. § 6912 et seq.). During this inspection, the following violations of underground storage tank regulations were found, with corresponding penalty amounts:

40 CFR ______________ Penalty $ __________

Nature of Violation: ____________________

40 CFR ______________ Penalty $ __________

Nature of Violation: ____________________

40 CFR ______________ Penalty $ __________

Nature of Violation: ____________________

Penalty Total $ __________

The owner or operator of the above facility is hereby ordered to correct the violations and pay the penalties described above.

This Compliance Order is not an adjudicatory proceeding under 40 CFR Part 22, the Consolidated Rules of Practice Governing the Administrative Assessment of Civil Penalties and the Revocation or Suspension of Permits, but is issued solely with reference to the Settlement Agreement in Part II of this form. If the Settlement Agreement in Part II is not returned in correct form by the owner or operator within 30 days of the date of signature below by the Authorized Representative of EPA, this Compliance Order is hereby withdrawn, without prejudice to EPA's ability to file additional enforcement actions for the above or any other violations.

I have personally observed the above violations and find the owner or operator in violation of the above-referenced UST regulations.

______________________________ Date: ________
(Signature of Authorized Representative of EPA)

I hereby acknowledge receipt of this Compliance Order and Settlement Agreement.

______________________________ Date: ________
(Signature of Owner, Operator or On-site Representative)

PART II: SETTLEMENT AGREEMENT

The United States Environmental Protection Agency (EPA) offers this Settlement Agreement under its expedited enforcement procedures in order to settle the violations found in the Compliance Order in Part I of this form subject to the following terms and conditions:

The Owner or Operator signing below certifies, under civil and criminal penalties for making a false submission to the United States Government, that Owner or Operator has corrected the violation(s) and enclosed a certified check for $__________ in payment of the full penalty amount, as described in the Compliance Order.

Upon EPA final approval of this Settlement Agreement, EPA will take no further action against the Owner or Operator for the violations described in the Compliance Order. EPA does not waive any enforcement action by EPA, the State, or Indian tribe or authorized tribal organization where the facility is located or any local agencies for any other past, present or future violations of the underground storage tank requirements or any other violations under any other statute not described in the Compliance Order.

Also, upon EPA final approval of this Settlement Agreement, the Owner or Operator waives the opportunity for a public hearing pursuant to Section 9006 of the Resource Conservation and Recovery Act.

This Settlement Agreement is binding on the EPA and the Owner or Operator signing below. The Owner or Operator signing below waives any objections to EPA's jurisdiction with respect to the Compliance Order and this Settlement Agreement, and consents to EPA's final approval of this Settlement Agreement without further notice. This Settlement Agreement is effective upon EPA's final approval below. Upon final approval, EPA shall mail a copy of the approved Settlement Agreement to the Owner or Operator signing below.

Final approval of the Settlement Agreement is in the sole discretion of the Regional Administrator, Region ___, EPA, or authorized delegate.

SIGNATURE BY OWNER OR OPERATOR:

Name (print) ______________________________

Title (print) ______________________________

Signature ____________________ Date: ________

FINAL APPROVAL BY EPA:

Name (print) ______________________________

Title (print) ______________________________

Signature ____________________ Date: ________

UNITED STATES ENVIRONMENTAL PROTECTION AGENCY
REGION ____, MAIN STREET, USA

EXPEDITED ENFORCEMENT
COMPLIANCE ORDER AND SETTLEMENT AGREEMENT

INSTRUCTIONS

The United States Environmental Protection Agency has authority under Section 9006 of the Resource Conservation and Recovery Act to issue compliance orders and pursue civil penalties for violations of underground storage tank regulations. However, the EPA encourages the expedited settlement of easily verifiable violations of underground storage tank requirements, such as the violations cited in the Expedited Enforcement Compliance Order and Settlement Agreement for which these instructions are provided, by agreeing to these settlement terms that include expedited correction of the violation and payment of penalties.

You may resolve the cited violations quickly by signing and returning the Settlement Agreement and paying the penalty amount within 30 days of the issuance of the Compliance Order. You must correct the violations within 30 days of the issuance of the Compliance Order. EPA at its discretion, may grant one 30 day extension for the period to come into compliance where the owner or operator satisfactorily demonstrates that it is technically infeasible or impracticable to achieve compliance within 30 days. The Settlement Agreement is binding on EPA and the Owner or Operator upon EPA final approval. Upon EPA final approval of the Settlement Agreement, a copy of which will be returned to you, EPA will take no further action against you for these violations. EPA will not accept or approve any Settlement Agreement returned more than 30 days after the date of the Compliance Order unless an extension has been granted by EPA. This Compliance Order is not an adjudicatory proceeding under 40 CFR Part 22, the Consolidated Rules of Practice Governing the Administrative Assessment of Civil Penalties and the Revocation or Suspension of Permits, but is issued solely with reference to the Settlement Agreement in Part II of this form.

If you do not return the Settlement Agreement with payment of the penalty amount 30 days after issuance, unless an extension has been granted by EPA, the Compliance Order will be automatically withdrawn, without prejudice to EPA's ability to file additional enforcement actions for the above or any other violations. Failure to return the Settlement Agreement within the approved time does not relieve you of the responsibility to comply fully with the regulations, including correcting the violations that have been specifically identified by the inspector. If EPA pursues administrative enforcement measures in order to correct the violation(s) or to seek penalties, you will receive instructions describing your rights under the Consolidated Rules of Practice Governing the Administrative Assessment of Civil Penalties and the Revocation or Suspension of Permits (40 CFR Part 22).

You are required under the Settlement Agreement to certify that you have corrected the violations found in the Compliance Order and paid the penalty amount. The payment for the penalty amount must be in the form of a certified check payable to the "Treasurer of the United States of America," with the number of the Compliance Order written on the check.

The Settlement Agreement and copy of the check shall be sent to:

Underground Storage Tank Program
U.S. Environmental Protection Agency
Region __
Main Street
USA

Payment of the penalty amount shall be sent to:

U.S. Environmental Protection Agency
Region __
P.O. Box __
Main Street
USA

By the terms of the Settlement Agreement, and upon EPA's final approval of the Settlement Agreement, you waive the opportunity for a public hearing pursuant to Section 9006 of the Resource Conservation and Recovery Act. EPA will treat any response to the citation, other than acceptance of the Settlement Offer, as an indication that the recipient is not interested in pursuing this expedited settlement procedure.

Final approval of the Settlement Agreement is at the sole discretion of the Regional Administrator, Region ___, EPA, or authorized delegate.

If you have any questions, you may contact the EPA Regional Office of Underground Storage Tanks at ______________________.

UNITED STATES ENVIRONMENTAL PROTECTION AGENCY
REGION ____, MAIN STREET, USA

EXPEDITED ENFORCEMENT
COMPLIANCE ORDER AND SETTLEMENT AGREEMENT

INSTRUCTIONS

The United States Environmental Protection Agency has authority under Section 9006 of the Resource Conservation and Recovery Act to issue compliance orders and pursue civil penalties for violations of underground storage tank regulations. However, the EPA encourages the expedited settlement of easily verifiable violations of underground storage tank requirements, such as the violations cited in the Expedited Enforcement Compliance Order and Settlement Agreement for which these instructions are provided, by agreeing to these settlement terms that include expedited correction of the violation and payment of penalties.

You may resolve the cited violations quickly by signing and returning the Settlement Agreement and paying the penalty amount within 30 days of the issuance of the Compliance Order. You must correct the violations within 30 days of the issuance of the Compliance Order. EPA, at its discretion, may grant one 30 day extension for the period to come into compliance where the owner or operator satisfactorily demonstrates that it is technically infeasible or impracticable to achieve compliance within 30 days. The Settlement Agreement is binding on EPA and the Owner or Operator upon EPA final approval. Upon EPA final approval of the Settlement Agreement, a copy of which will be returned to you, EPA will take no further action against you for these violations. EPA will not accept or approve any Settlement Agreement returned more than 30 days after the date of the Compliance Order unless an extension has been granted by EPA. This Compliance Order is not an adjudicatory proceeding under 40 CFR Part 22, the Consolidated Rules of Practice Governing the Administrative Assessment of Civil Penalties and the Revocation or Suspension of Permits, but is issued solely with reference to the Settlement Agreement in Part II of this form.

If you do not return the Settlement Agreement with payment of the penalty amount 30 days after issuance, unless an extension has been granted by EPA, the Compliance Order will be automatically withdrawn, without prejudice to EPA's ability to file additional enforcement actions for the above or any other violations. Failure to return the Settlement Agreement within the approved time does not relieve you of the responsibility to comply fully with the regulations, including correcting the violations that have been specifically identified by the inspector. If EPA pursues administrative enforcement measures in order to correct the violation(s) or to seek penalties, you will receive instructions describing your rights under the Consolidated Rules of Practice Governing the Administrative Assessment of Civil Penalties and the Revocation or Suspension of Permits (40 CFR Part 22).

You are required under the Settlement Agreement to certify that you have corrected the violations found in the Compliance Order and paid the penalty amount. The payment for the penalty amount must be in the form of a certified check payable to the " Treasurer of the United States of America," with the number of the Compliance Order written on the check.

The Settlement Agreement and copy of the check shall be sent to:

Underground Storage Tank Program
U.S. Environmental Protection Agency
Region __
Main Street
USA

Payment of the penalty amount shall be sent to:

U.S. Environmental Protection Agency
Region __
P.O. Box __
Main Street
USA

By the terms of the Settlement Agreement, and upon EPA's final approval of the Settlement Agreement, you waive the opportunity for a public hearing pursuant to Section 9006 of the Resource Conservation and Recovery Act. EPA will treat any response to the citation, other than acceptance of the Settlement Offer, as an indication that the recipient is not interested in pursuing this expedited settlement procedure.

Final approval of the Settlement Agreement is at the sole discretion of the Regional Administrator, Region ____, EPA, or authorized delegate.

If you have any questions, you may contact the EPA Regional Office of Underground Storage Tanks at ________________________.

UNITED STATES ENVIRONMENTAL PROTECTION AGENCY
REGION ____, MAIN STREET, USA

EXPEDITED ENFORCEMENT
COMPLIANCE ORDER AND SETTLEMENT AGREEMENT

APPROVED FOR STATES

PART I: COMPLIANCE ORDER

COMPLIANCE ORDER NO. ____________________

On ____________________ **Time** ____________
(Date of Violation) **(a.m. or p.m.)**

At ____
(Name of Facility)

____
(Address of Facility)

Facility Identification Number ____________________

Name of Owner, Operator or On-site Representative ____________________
(Circle one)

____
(Address of Owner, Operator, or On-Site Representative)

An authorized representative of the United States Environmental Protection Agency (EPA) inspected this facility to determine compliance with underground storage tank regulations promulgated under Subtitle I of the Resource Conservation and Recovery Act of 1976 (42 U.S.C. § 6912 et seq.). During this inspection, the following violations of underground storage tank regulations were found, with corresponding penalty amounts. Because the state program has received federal approval, violations are cited in terms of the State regulations.

(State Reg. No.) ____________________ **Penalty $** ________

Nature of Violation: ____________________

(State Reg. No.) ____________________ **Penalty $** ________

Nature of Violation: ____________________

(State Reg. No.) ____________________ **Penalty $** ________

Nature of Violation: ____________________

Penalty Total $ ________

The owner or operator of the above facility is hereby ordered to correct the violations and pay the penalties described above.

This Compliance Order is not an adjudicatory proceeding under 40 CFR Part 22, the Consolidated Rules of Practice Governing the Administrative Assessment of Civil Penalties and the Revocation or Suspension of Permits, but is issued solely with reference to the Settlement Agreement in Part II of this form. If the Settlement Agreement in Part II is not returned in correct form by the owner or operator within 30 days of the date of signature below by the Authorized Representative of EPA, this Compliance Order is hereby withdrawn, without prejudice to EPA's ability to file additional enforcement actions for the above or any other violations.

I have personally observed the above violations and find the owner or operator in violation of the above-referenced UST regulations.

______________________________ **Date:** ________
(Signature of Authorized Representative of EPA)

I hereby acknowledge receipt of this Compliance Order and Settlement Agreement.

______________________________ **Date:** ________
(Signature of Owner, Operator or On-site Representative)

PART II: SETTLEMENT AGREEMENT

The United States Environmental Protection Agency (EPA) offers this Settlement Agreement under its expedited enforcement procedures in order to settle the violations found in the Compliance Order in Part I of this form subject to the following terms and conditions:

The Owner or Operator signing below certifies, under civil and criminal penalties for making a false submission to the United States Government, that Owner or Operator has corrected the violation(s) and enclosed a certified check for ____________ in payment of the full penalty amount as described in the Compliance Order.

Upon EPA final approval of this Settlement Agreement, EPA will take no further action against the Owner or Operator for the violations described in the Compliance Order. EPA does not waive any enforcement action by EPA, the State, or Indian tribe or authorized tribal organization where the facility is located or any local agencies for any other past, present or future violations of the underground storage tank requirements or for violations under any other statute not described in the Compliance Order.

Also, upon EPA's final approval of this Settlement Agreement, the Owner or Operator waives the opportunity for a public hearing pursuant to section 9006 of the Resource Conservation and Recovery Act.

This Settlement Agreement is binding on the EPA and the Owner or Operator signing below. The Owner or Operator signing below waives any objections to EPA's jurisdiction with respect to the Compliance Order and this Settlement Agreement, and consents to EPA's final approval of this Settlement Agreement without further notice. This Settlement Agreement is effective upon EPA's final approval below. Upon final approval, EPA shall mail a copy of the approved Settlement Agreement to the Owner or Operator signing below.

Final approval of the Settlement Agreement is in the sole discretion of the Regional Administrator, Region ___, EPA, or authorized delegate.

SIGNATURE BY OWNER OR OPERATOR:

Name (print) ____________________

Title (print) ____________________

Signature ____________________ Date: ________

FINAL APPROVAL BY EPA:

Name (print) ____________________

Title (print) ____________________

Signature ____________________ Date: ________

SONREEL Monographs

Qty	Title	SONREEL price	Nonmember price	Amount
	On oil & gas issues			
	Drafting Indemnity Provisions in Oil and Gas Contracts (5350046)	$29.95	$39.95	$
	Legal Research Checklists for International Petroleum Operations (5350042)	$29.95	$39.95	$
	Federal Oil & Gas Lease Appeals in the Department of the Interior (5350036)	$29.95	$39.95	$
	Legal Aspects of the Purchase & Sale of Oil & Gas Properties (5350034)	$29.95	$39.95	$
	International Oil & Gas Joint Ventures (5350033)	$29.95	$39.95	$
	Improved 1989 Joint Operating Agreement (5350030)	$29.95	$39.95	$
	Drafting Oil & Gas Exploration Agreements	$19.95	$29.95	$
	PACKAGE: 1989 Joint Operating Agreement *and* Drafting Oil & Gas Exploration Agreements (5350031)	$39.95	$49.95	$
	Drafting Natural Gas Contracts After Order 436 (5350021)	$19.95	$29.95	$
	Federal Oil & Gas Leasing Reform Act of 1987 (5350018)	$29.95	$39.95	$
	Analysis of Rules & Policies-OCS Operations (5350027)	$19.95	$29.95	$
	Offshore Bidding Agreements (5350010)	$19.95	$29.95	$
	Protecting Oil & Gas Lien and Security Agreements (5350014)	$19.95	$29.95	$
	1982 Joint Operating Agreement (5350008)	$19.95	$29.95	$
	Standard Form Farmout Agreements (5350007)	$19.95	$29.95	$
	PACKAGE: 1982 Joint Operating Agreement *and* Farmout Agreements (5350009)	$29.95	$39.95	$
	On land patents			
	Railroad Land Grants (5350011)	$19.95	$29.95	$
	Environmental issues			
	The Application of RCRA and CERCLA to Wastes Produced by Oil & Gas Operations (5350039)	$29.95	$39.95	$
	Identification of RCRA-Related Substances (5350016)	$19.95	$29.95	$
	Checklists for Preparing NEPA Documents (5350013)	$19.95	$29.95	$

The Year In Review. Information on pertinent legislation, administrative, and regulatory actions in such areas as oil; natural gas; electric power; alternate energy sources; water, marine and forest resources; air, water, and environmental quality; and solid and hazardous wastes.

	1993 Year in Review (5350041)	$29.95	$39.95	$
	1992 Year in Review (5350038)	$29.95	$39.95	$

SUBTOTAL	$
TAX*	$
HANDLING**	$
TOTAL	$

* IL residents add 8.75%; DC residents add 5.75%; MD residents add 5%.

** For orders of $10-24.99 add $3.95; orders of $25-49.99 add $4.95; orders of $50 and more add $5.95.